Surface-Active Processes in Materials

Related titles published by The American Ceramic Society:

Finishing of Advanced Ceramics and Glasses (Ceramic Transactions Volume 102)
Edited by Robert Sabia, Victor A. Greenhut, and Carlo G. Pantano
©1999, ISBN 1-57498-082-3

Ceramic Material Systems with Composite Structures (Ceramic Transactions Volume 99)
Edited by Nubuo Takeda, Laurel M. Sheppard, and Jun-ichi Kon
©1998, ISBN: 1-57498-065-3

Sol-Gel Synthesis and Processing (Ceramic Transactions Volume 95)
Edited by Sridhar Komarneni, Sumio Sakka, Pradeep P. Phulé, and Richard M. Laine
©1998, ISBN 1-57498-063-7

Electrochemistry of Glass and Ceramics (Ceramic Transactions Volume 92)
Edited by S.K. Sundaram, D.F. Bickford, and E.J. Hornyak, Jr.
©1999, ISBN: 1-57498-062-9

For information on ordering titles published by The American Ceramic Society, or to request a publications catalog, please contact our Customer Service Department at 614-794-5890 (phone), 614-794-5892 (fax),<customersrvc@acers.org> (e-mail), or write to Customer Service Department, 735 Ceramic Place, Westerville, OH 43081, USA.

Visit our on-line book catalog at <www.ceramics.org>.

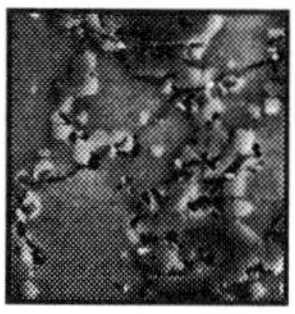

Ceramic Transactions
Volume 101

Surface-Active Processes in Materials

Edited by

David E. Clark
University of Florida

Diane C. Folz
University of Florida

Joseph H. Simmons
University of Florida

Engineering Ceramics Division
The American Ceramic Society

Published by
The American Ceramic Society
735 Ceramic Place
Westerville, Ohio 43081

Proceedings of the Larry Hench Symposium on Surface-Active Processes in Materials, held at the 23rd Annual Cocoa Beach Conference, Cocoa Beach, Florida, January 25–29, 1999.

COVER PHOTO: "SEM micrograph of fracture surface featues for specimen tested at 700°C/250 MPa in air," is courtesy of C.E. Olson, T.D. McGee, W.D. Hoefle, and R. Kudej, and appears as figure 1 in their paper "Osteoceramic Bone Replacement and Grafting," which begins on page 187.

Library of Congress Cataloging-in-Publication Data
A CIP record for this book is available from the Library of Congress.

For information on ordering titles published by The American Ceramic Society, or to request a publications catalog, please call 614-794-5890.

Printed in the United States of America.

4 3 2 1–03 02 01 00

ISSN 1042-1122
ISBN 1-57498-079-3

Dedication

Larry L. Hench

This book is dedicated to Larry L. Hench, a scientist, engineer, and educator who has devoted his life to the field of materials. Presently a Professor in the Department of Materials at Imperial College in London, Larry initiated the research program and curriculum in ceramic engineering at the University of Florida where he served as a faculty member in the Department of Materials Science and Engineering from 1964 to 1995.

The chemical and structural makeup of materials' surfaces plays a major role in their environmental stability, reactivity, and performance. No one has done more to understand the mechanisms that control the surface reactivity in glasses than Larry. Many important technologies with direct impact on the quality of our lives were influenced by Larry's work (e.g. bioactive materials, durable glasses for long-term storage of nuclear wastes and micro/macro porous materials for filters, sensors, and optics). The objective of this book is to provide a perspective of this field with papers that represent state-of-the-science studies and offer a glimpse of the discoveries that the future holds in store.

Contents

Biomaterials and Bioreactivity

Porous Materials

Coatings and Composites

Preface

This book is the result of a symposium titled "Surface-Active Processes in Materials," to honor Professor Larry Hench for his significant contributions to this field. The symposium was held in Cocoa Beach, Florida, in January 1999 as part of the 23rd Annual Cocoa Beach Conference and Exposition: An International Conference on Engineering Ceramics and Structures, sponsored by the Engineering Ceramics Division of The American Ceramic Society. Over the years, Larry has been active in several divisions and classes of ACerS. We are grateful for the financial and organizational support provided by the Engineering Ceramics Division, Glass and Optical Materials Division, Nuclear and Environmental Technology Division, National Institute of Ceramic Engineers, and the Florida Section of The American Ceramic Society. It's appropriate that this symposium to honor Larry was held at the Engineering Ceramics Division's Cocoa Beach Conference since Larry was one of its founders nearly 25 years ago. Several past and present presidents of ACerS and past and present chairs of divisions, sections, and classes attended the symposium and banquet and gave short presentations on Larry's contributions.

The keynote paper, "Advances in Surface-Active Processes in Materials," was presented by W. David Kingery. Professor Kingery provided an entertaining and historical perspective of Larry's contributions to the field. A highlight of the symposium was the banquet, where Alice Holt, Larry's long-time secretary, was presented with flowers and the organizing committee presented a plaque to Larry containing the names and graduation dates of the 88 M.S. and Ph.D. students that he supervised during his 30+ years at the University of Florida. USBiomaterials Corp. president David Greenspan provided $1000 toward the cost of the banquet.

The symposium was attended by over 100 participants from the United States, United Kingdom, Japan, Brazil, Germany, Italy, Switzerland, Sweden, and Thailand. There were 26 oral presentations and 14 posters of which 28 are published in these proceedings. Papers covered a wide range of topics, including fracture, microwave processing, characterization, sol-gel processing, composites, thin films and coatings, corrosion, nuclear waste glasses, biomaterials and bioreactivity, glass crystallization and dielectrics, with emphasis on the role of the surfaces and interfaces. The papers in this book are peer-reviewed. Each paper underwent at least two reviews: one review by a member of technical community and one by the book editors. All papers were technically edited prior to publication.

David E. Clark
Diane C. Folz
Joseph H. Simmons

Organizing Committee

David E. Clark, University of Florida, Gainesville, Florida
Diane C. Folz, University of Florida, Gainesville, Florida
Catherine J. Simmons, University of Florida, Gainesville, Florida
Joseph H. Simmons, University of Florida, Gainesville, Florida
Stephen W. Freiman, National Institute of Standards and Technology, Gaithersburg, Maryland
David C. Greenspan, USBiomaterials Corporation, Alachua, Florida
Martin Wilson,

Session Chairs

D.E. Clark, University of Florida
E.E. Douek, Guy's Hospital
S. Freiman, National Institute of Standards and Technology
D. Greenspan, USBiomaterials Corporation
C. Pantano, Pennsylvania State University
J.J. Mecholsky, Jr., University of Florida
J.H. Simmons, University of Florida
G.G. Wicks, Westinghouse Savannah River Company

GLASS SURFACES FOR FUN AND FAME

Professor Larry L. Hench
Professor of Ceramic Materials
Director, Centre for Tissue Regeneration and Repair
Imperial College of Science, Technology and Medicine
Prince Consort Road
London
England, SW7 2BP

THE EARLY DAYS

My interest in ceramics was born in 1957 when members of the Ohio State University student chapter of the American Ceramic Society took me to the basement of Lord Hall and opened the door to a large gas-fired kiln firing a batch of whitewares. I can still visualise the glowing beauty of that fiery furnace.

The white-hot 2400° heat not only transformed the clay of the pots into a ceramic but also transformed my life. "This is for me", I said, and the next day I transferred from Engineering Physics into the Ceramic Engineering Department. We had lots of labs at Ohio State and I especially enjoyed firing glazes and enamels and melting and casting colored glasses.

I became intrigued about the mechanisms of high temperature processing in the summer of my junior year, 1958, when I had the good fortune to be working at the GE Evandale plant in Cincinnati on the AEC Atomic Jet Engine program. My minor role was to operate a CVD reaction chamber that was supposed to produce a uniform coating of Al_2O_3 from $AlCl_3$ vapor on UO_2 tubular fuel rods. I was curious as to why the process sometimes produced beautiful transparent dense coatings, but in the same run opaque, porous areas occurred. I had no answers, just questions. No one else did either, as I discovered later, and the program was eventually cancelled.

My senior research project at OSU in 1961 continued this quest to find answers to thermal processing-environmental questions. I built an apparatus to study the effect of electric fields on the sintering of NaCl, pressed into bars, and heated in a gradient furnace with and without an electric field applied via a rheostat. The large ionic conductivity of NaCl at high temperature and high humidity led to some dramatic results, and an unexpected "hands on" introduction to dielectric breakdown, as well as evaporation-condensation and surface diffusion-controlled sintering.

Following graduation in 1961, my AEC Q clearance from GE landed me another great summer position at Lawrence Livermore Lab (LLL) to work on the sintering of BeO for the AEC atomic rocket program, which also never got off the ground. But, at LLL I learned the practical side of working with toxic materials under the supervision of Ray Cooperstein and benefited greatly from the stimulation of talking ceramic science with Birch Holt and Ralph Condit. Professor Dave Kingery's classic text "Introduction to Ceramics" had just been published and I pestered Birch and Ralph almost daily about topics in the book which I read while the BeO samples were sintering. This was a great learning experience and led to my first paper given at the 1962 American Ceramic Society Annual Meeting.

I returned to OSU in the Fall to do a PhD, under the supervision of Prof Ralston Russell, Past President and winner of almost all the American Ceramic Society Honors. My topic "Sintering and Solid State Reactions of MgO and Cr_2O_3" was chosen to attempt to answer an intriguing sintering question raised by one of the many papers by Ivan Cutler at University of Utah, a great ceramic educator. He, with Nelson, showed that adding M_2O_3 oxides to MgO generally enhanced sintering, due to creation of extrinsic defects; Cr_2O_3 was an exception. During my thesis research, I learned the importance of surface-environment interactions: $2\ Cr_2O_3 + 3/2(O_2)\ \tau 2\ CrO_3$. The high vapor pressure of CrO_3 led to diffusion through the pores of the MgO- Cr_2O_3 compact, coated the MgO grains, and led to rapid interfacial reactions to form Mg Cr_2O_4 spinel which stopped grain boundary diffusion and solid state sintering, a satisfying answer to the question.[1]

Good fortune had also come to me with funding by an Owens-Illinois PhD Fellowship. I used the freedom offered by the fellowship and my interest in glass-ceramics to initiate a highly speculative project, an attempt to study the mechanisms of nucleation and crystallization of glasses by use of dielectric spectroscopy. My hypothesis was that structural relaxations must occur in the glass network during rearrangement to form stable nuclei and the net motion of the more mobile modifier ions should alter local dipole moments and thereby induce a permanent shift in the dielectric loss spectra of the glasses. I chose to study the Li_2O-SiO_2 system; the emphasis was on 33 mole % Li_2O, 67% mole% SiO_2 (33L), due to its importance in the new field of glass-ceramics and the high electric field strength of the Li^+ ion. A glass of the 33L composition could crystallize to the stable Li_2O-$2SiO_2$ (L2S) disilicate phase without a shift in composition.

GLASS-CERAMICS

The dielectric relaxation apparatus was assembled at OSU and sample holders designed, using input from Von Hippel's books, but little publishable data was obtained. Thus, continuing this area of research became my top priority after accepting a position as Assistant Professor at the University of Florida upon graduation with my PhD in June 1964. This turned out to be a very fortunate decision because the dielectric aspects interested Don Kinser and the mechanical properties of the crystallized 33L glass-ceramics interested Steve Freiman, both new graduate students in the Materials Department. I raised the money; Steve and Don made the glasses, assembled the equipment and collected the data and together with Bob Gould and John Hren and their students, we even built new laboratories in an off-campus swamp. The fun of research has seldom been greater than in those early days.

We learned from the dielectric spectroscopy that a metastable lithum-metasilicate (LS) phase was thermodynamically more stable as a transition product during nucleation and crystallization.[(2)] Large dielectric losses occurred due to Maxwell-Wagner-Sillars interfacial polarization events as the Li^+ ions underwent rearrangement. Hot stage TEM, Gunier-DeWulff X-ray diffraction and X-ray small angle scattering data were obtained to confirm the reaction mechanisms, as summarized in references 2 - 6. Collaborations with John Hren and Bob Gould were invaluable. The kinetics analysis was extended to include quantitative stereology, aided by Bob DeHoff[4]. Integrating these different approaches was a lot of fun, especially so when it was possible to use the results to produce a series of L2S glass-ceramics with controlled volume fractions and sizes of crystal phases and use them to establish microstructure-dielectric and conductivity relationships[6] (Don Kinser's Ph.D) and microstructure-mechanical property relationships (Steve Freiman's PhD)[7]. A similar approach was used to understand the structure-property relationships in the polaronic conducting V_2O_5-P_2O_5 amorphous semi-conductor system (Herb Schaake's'PhD)[8-10].

GLASS SURFACE – ENVIRONMENT INTERACTIONS

However, our efforts in the late 1960's to use Mozzi and Warren's new method of X-ray Pair Function Distribution Analysis (XRPFD) to quantify glass nucleation mechanisms at a molecular level were failing in spite of Bob Gould's creative design of the X-ray system and Bert Warren's expert advice as a Visiting Professor. We had selected high purity amorphous silica for the study, sponsored by Art Diness's ONR program. The problem we eventually realized was water-enhanced heterogeneous surface nucleation, as just identified by Frank Wagstaff at the GE labs in Schenectady. C.V. Gokulanatham did the XRPFD measurements and completed his PhD with a study of fast neutron radiation induced structural damage in a-SiO_2.[11]

David A. Sanders had recently joined the team as a new PhD student and made the key discovery that launched our subsequent program in glass surface-environment interactions.

David came in one morning in his first year of graduate school and asked permission to buy \$80 of lenses from the Edmund Scientific Catalogue. I asked, "Why?" He explained that a lecture on infrared (IR) spectroscopy in Bill Pearsons's Chemistry course suggested to him a new way to look at the effect on water on surface crystallization. To do the measurements required construction of an IR reflection stage suitable for glass samples. He was eager to build the stage and Prof. Pearson was willing to let him use a spectrometer for one weekend. I said "Go ahead, give it a try." The results were a series of six papers using infrared reflection spectroscopy[12-17] (IRRS) and David's PhD dissertation. This was research at its most fun. David designed a static corrosion test for our 33 L and 33N glass samples, as well as pure SiO_2 standards, and was able to quantify the surface spectral changes due to Stage 1 de-alkalization as described by Douglas and El Shamy a few years earlier. Dave was also able to monitor onset of Stage 2 attack, network dissolution, and determine the effects of initial glass composition, surface roughness and other variables on glass corrosion kinetics.

Confirmation of the IRRS-based surface compositional profiles came soon thereafter with the studies of David E. Clark,[18-20] Wayne Acree, Ed Ethridge, and Morris Dilmore using cryogenically cooled electron microprobe (EMP) analysis and SEM-EDS analysis. Faculty colleagues Derek Dove and Paul Holloway built an Auger electron spectrometer (AES) with cryogenic cooling and Ar ion beam milling. PhD student Carlo Pantano used the AES to quantify the surface compositional changes due to glass–environment[21] interactions. The AES compositional profiles combined with the IRRS spectral changes[22] and glass dissolution data made it possible to generalize glass surface environment interactions to consist of 5 types of surfaces.[23] (Fig.1)

Our group found this multi-instrumental approach to studying glass-environment interactions very productive, in part because the IRRS method was so rapid and inexpensive. We used it for all preliminary tests and then selected key samples from the kinetics runs for the quantitative analyses using AES, SEM-EDS, EMP etc. Ed Ethridge established the importance of surface area (SA) to solution volume (V), i.e.: SA/V on glass water reaction kinetics for his PhD[24]. Morris Dilmore together with David Clark quantified the mixed alkali effect on glass durability.[25] Morris Dilmore's PhD thesis established the important role of Al^{3+} ions on stabilizing the reaction layer that forms on silicate glasses and showed that a Type III A protective Al_2O_3 –SiO_2 layer could be formed by either release of Al^{3+} from the glass or addition of Al^{3+} to the solution.[26, 27]

These systematic studies were successful largely because they all used simple two and three component glasses and the same experimental protocols.

Application of our findings to the real world of commercial glasses was the next challenge. The Glass Container Research Corp. funded a project conducted by David E. Clark, Carlo Pantano and myself, which used the same methodology to analyze glass containers removed from the Anchor Hocking production line in Jacksonville, Florida. The bottles were driven to Gainesville at 70 mph, were broken and analyzed within 12-24 hours. The effects of water, moist atmosphere, surface treatments, static versus dynamic test environments, temperature, time, etc were all examined using the full range of analytical instruments. The results were too voluminous for journal submission so they were published in 1979 as a book, "Corrosion of Glass", by D.E. Clark, C.G. Pantano, and L.L. Hench, by Magazines for Industry, Inc.

NUCLEAR WASTE GLASSES

I received a telephone call about this time (1977) that had a big impact on my life and seven years of effort that was often not much fun and certainly did not lead to any fame. The call was from Dennis Ready, then at the Department of Energy. He said, "Larry, we're having a conference on glass environment interactions up here in Germantown in a couple of weeks. Can you come and talk about glass corrosion?" I looked out the window and saw the warm Florida sunshine and replied, "Dennis, don't you have a foot of snow up there? Why would I want to leave Florida in December to go to D.C. for a meeting?"

He answered, "Because there are 30 million gallons of high level radioactive waste sitting just north of you on the Savannah River in rusting steel tanks and its time for the country to do something permanent like put the waste in glass. We need you to help." I put on a wool coat and got on the plane.

For seven to eight years it seemed as if I was up to my neck in nuclear waste. Together with David Clark we worked with the Pacific Northwest Labs, the Nuclear Regulatory Commission, and especially the Savannah River Laboratory (SRL) in setting up and running corrosion tests for nuclear waste glasses. A real joy was visiting various DOE-related facilities with John Crandall and Bob Garvin from SRL and learning first hand about the "ole days" of the US nuclear weapons program. The Cold War took on a personal meaning when we visited northern Canadian labs in February with -30° F temperatures.

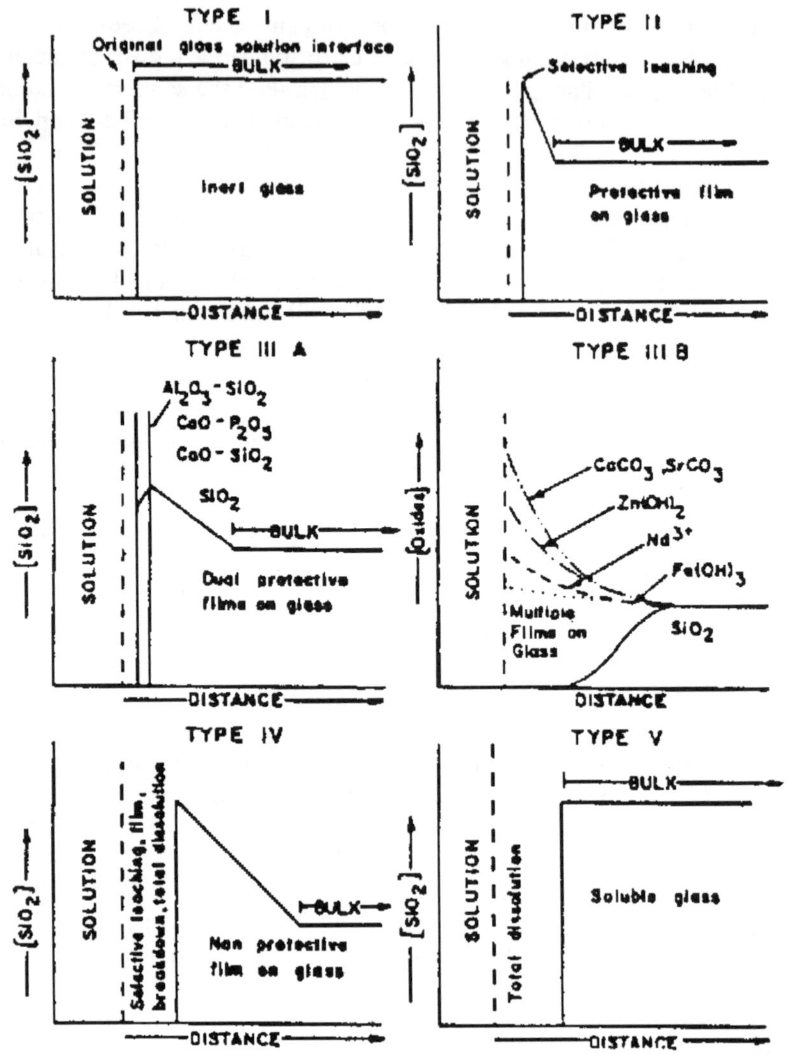

Figure 1: Five types of silicate glass surface reactions

The site visits revealed that DOE was funding research on more than 20 various approaches to solidification of radioactive waste. Alternatives varied from simply mixing high level waste with concrete to sophisticated ceramic processing that incorporated the radionuclide elements within molecularly tailored crystal phases. (Fig. 2).

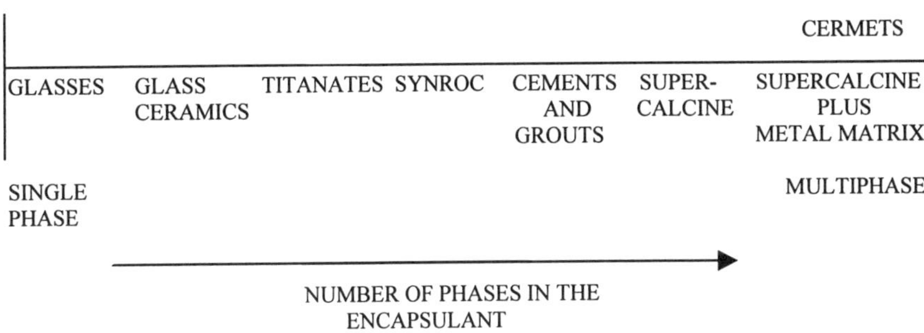

Figure 2. Materials spectrum of alternative nuclear waste encapsulants.

I was asked to testify before the United States House and Senate Sub-Committees debating the US nuclear waste policy bill. This was one of the least fun experiences of my career. Explaining the scientific difference between glasses and crystals to politicians who drink rare wines out of crystal glasses was a daunting challenge.

Seeing German police dragging Green Party demonstrators blocking the entrance to an International Symposium on Safe Disposal of High Level Radioactive Waste, held in Berlin, followed by a green paint bombing of the Symposium tour boat was also not so fun. The issues were primarily political and rational scientific debate was often not possible. In the US the perceived large amount of Department of Energy (DOE) research funds often produced an emotional defense of sparse scientific data. The Savannah River Plant (SRP) convened a Peer Review Panel for DOE to evaluate and quantify the alternative disposal systems. I was asked to chair the Panel and we met frequently for three years and recommended borosilicate glass for the the host matrix for SRP waste, due to processing simplicity, developed by George Wicks and his colleagues at SRL, and the high level of durability of glasses containing the very large number (250) nuclear waste elements.

Our work on nuclear waste glasses at the University of Florida led to a series of papers[28-32] and attracted the attention of Lars Werme, a manager at SKBF in Sweden, which led to a marvellous collaboration with him and Alex Lodding from Chalmers University in Gothenburg, Sweden that lasted for five years. We designed a test system for *in situ* burial of nuclear waste glasses deep within bore holes in STRIPA granite in Sweden. Heater rods in the center of the test assemblies made it possible to compare effects of deep leaching at ambient 5^0C with 90° C, the thermal period of storage of high level radioactive waste. We analysed samples that represented both defense waste and waste from commercial reprocessing of UO_2 nuclear fuel rods.[33-46]

Alex Lodding used secondary ion mass spectroscopy (SIMS) to obtain compositional profiles of the glass interfaces and leach rates.[37-39, 43, 44] At the University of Florida we used IRRS[41] and SEM-EDS and EMP data and Rutherford Back Scattering (RBS)[40] to quantify glass structural changes and calculate leach rates. Burial system variables were compared and made it possible to publish a risk assessment analysis that indicated long term disposal of high level radioactive waste has a low level of risk when it is incorporated as part of a borosilicate glass network.[42, 45, 46] (Fig.3)

BIOACTIVE GLASSES

The most significant event in my professional life occurred prior to any of these studies on glass-surface interactions and provided a great stimulus to them. The event was a trip in 1967 to the Army Materials Conference in Sagamore, N.Y. I shared the bus ride to the conference center with a US Army Colonel who had just returned from Vietnam. He described, in graphic detail, the damage done to young men in war and the army's inability to repair them due to the body's rejection of metals and plastics. When he learned that I was studying glasses and glass-ceramics he issued a challenge, "Why don't young guys like you work on ceramics to help people rather than do research on materials for weapons?" On my return to Gainesville I discussed this with Ray Splinter, a former student assistant, who was in Medical School. He agreed there was a need for better materials in surgery and arranged a meeting with two young orthopaedic surgeons, at the University of Florida Medical School, Dr. Ted Greenlee and Dr. Bill Allen.

We found little common knowledge to discuss the problem at that first meeting. I asked Ray what I needed to read to be able to understand why the body rejects foreign materials. He recommended: Ham's Histology, Gray's Anatomy, Stevenson's Biochemistry and Lehninger's Bioenergetics. Following six months of evening study I was much better prepared to discuss a research proposal with Bill and Ted. A proposal submitted to the US Army Medical R & D Command was funded for one year starting October 1, 1969 with myself as Principal Investigator and Drs Allen and Greenlee as Co-PI's. The hypothesis was simple. "Since the human body has a well established rejection process for metals and plastics and since bones contain calcium and phosphorus, perhaps the body may not reject glasses, ceramics or glass-ceramics that contain calcium and

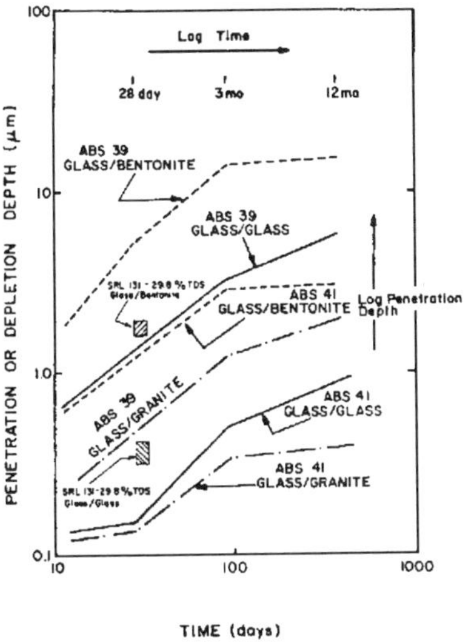

Log Time
28 day
3 mo
12 mo
ABS 39 GLASS/BENTONITE
ABS 39 GLASS/GLASS
ABS 41 GLASS/BENTONITE
Log Penetration Depth
ABS 39 GLASS/GRANITE
ABS 41 GLASS/GLASS
ABS 41 GLASS/GRANITE
PENETRATION OR DEPLETION DEPTH (μm)
TIME (days)

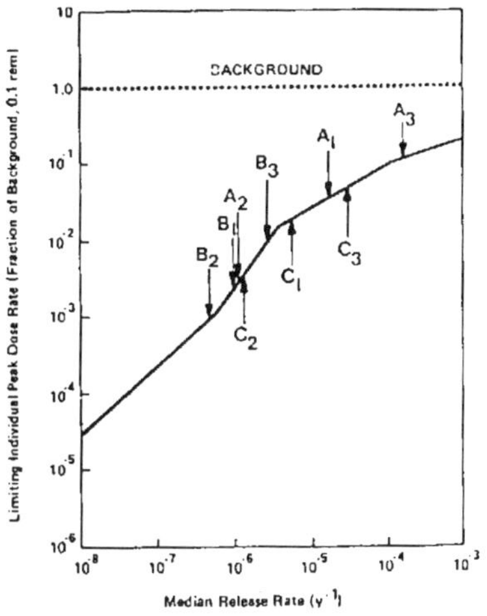

BACKGROUND
Limiting Individual Peak Dose Rate (Fraction of Background, 0.1 rem)
Median Release Rate (y⁻¹)

Figure 3.

phosphorus." Based on this hypothesis, I made three compositions of glass with of 45 weight percent of SiO_2 and 24.5 weight percent Na_2O. The amount of CaO and P_2O_5 was varied to bracket the relative ratios of Ca/P in bones and body fluids. One of the three compositions, termed 45S5 Bioglass®, contained 45% SiO_2, 24.5% Na_2O, 24.5% CaO and 6% P_2O_5. The 45S5 composition is in the center of the bioactive bone bonding boundary, shown in Fig.4.

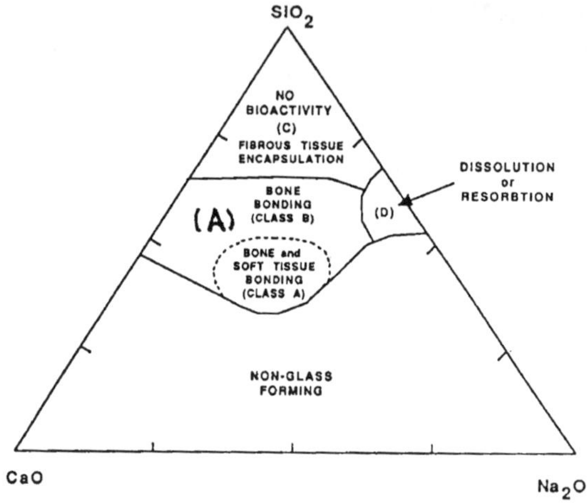

Figure 4. Compositional range for Class B bone bonding bioactive glasses and Class A bone & soft tissue bonding bioactive glasses (Region A). All compositions contain 6 weight % P_2O_5. Compositions in Region B resorb in the body and those in Region C have no bioactive bonding.

The three glass batches were cast into slabs, cut into small rectangles and given to Ted Greenlee to implant into the mid-femur of adult rats. Six weeks later Ted called me. He was very agitated. I said, assuming the worst, "Ted, calm down there are hundreds of other glasses we can try. These are only the first". He replied, "You can forget the others, these won't come out. I've pulled on them, I've pushed on them, I've hammered on them and they are stuck in the bone". Ted was right. The very first compositions of Bioglass® bonded to bone.[47-49] They bonded faster than any other composition of the hundreds tested world-wide since November 1969.[50-64]

The important question became "How does a man-made material bond to living tissue?" Ted Greenlee and Cassandra Beckham's TEM and SEM studies helped answer this question.[47-49] Details were revealed early in A. E. (Buddy) Clark's PhD research.[52, 55, 56] X ray diffraction analysis showed that the bioactive glass surface heterogeneously nucleated bone mineral from *in vitro* solutions containing Ca and P ions. Buddy applied the new IRRS spectroscopy techniques being developed by David Sanders and David Clark and showed the rapid exchange of Na+ ions from the glass and the growth of a SiO_2 -rich film which provided nucleation sites for the crystalline biological hydroxyapatite layer on the glass surface that bonded to the living bone.[56] The series of papers that followed, including additional TEM evidence from Homer Paschall's lab[50,51,56] at the Gainesville Veterans Administration Hospital, who replaced Ted Greenlee on the Army project, established much of the methodology for studying bioactive materials that is in use today.

Carlo Pantano, Fumio Ohuichi, pursuing PhD's using Auger Electron Spectroscopy and Mike Ogino, visiting scientist from Nikon in Japan, David E. Clark and Wayne Acree and Mary Harrell using EMP and SEM-EDS, all combined to establish the quantitative surface compositional profiles of the bioactive glasses following various times of implantation in bone.[52,55,58,59] The discovery of bioactive bonding became a science.

David Greenspan, Paul Buscemi, Gary Miller and Bill Lacefield began PhD studies developing clinical orthopaedic applications with continued funding by the US Army Medical R &D command and funds from a new bioengineering program at the National Science Foundation.[60-66]

In April 1978 another critical event happened; Dr.June Wilson from Imperial Chemical Industries Central Toxicology Lab, visited the University of Florida and our Bioglass® research program. She concluded from reviewing our data that collagen from soft connective tissues should bond to bioactive glasses as well as bone, if the interface was properly immobilized. She did the experiments, developed the methodology for proving soft tissue adherence to Bioglass®[67,68] and pioneered the clinical use of the concept in middle ear prostheses, together with Gerry Merwin, ENT surgeon.[69-76] The testing was done in a project funded by the NIH. FDA gave permission in 1985 for clinical sales of Bioglass® MEP's (Middle Ear Prostheses). June conducted the toxicology and biocompatibility testing required for the FDA submission.[67] For nearly twenty years she has co-ordinated histopathology and animal testing of Bioglass® products and helped USBiomaterials Corp. and the University of Florida to maintain a high standard of excellence in the field. We were married in 1980 and have co-directed Bioglass® research every since.

Throughout much of this period our academic studies have been translated into the clinical marketplace under the direction of Dr. David Greenspan, V.P., USBiomaterials Corp. Alachua, Florida. A proud achievement is that after 14 years of clinical use of Bioglass® MEP's there are no reported cases of extrusion through the ear drum, the most common mode of failure of metal and plastic devices for ossicular replacement.[76, 77]

Research leading to the first Bioglass® dental product, the ERMI® (Endosseous Ridge Maintenance Implant) began early in the 1970's in baboons, sponsored by the National Institute of Dental Research and directed by Professor Harold Stanley, of the Department of Oral Biology, University of Florida College of Dentistry.[78-82] Stan has been involved in the Bioglass® research program ever since, as have Dr. Buddy Clark and Dr. Cal King, periodontist, who also worked on the baboon study. The very high success rate of the ERMI ® implants in stabilizing the jaw bone of denture wearing patients[82] provided the incentive to develop the most important dental product, Perioglas®. The importance of Perioglas® is that it saves teeth. Large amounts of bone can be lost due to resorption around teeth that have periodontal (gum) disease. It was discovered by Drs. Sam Low, Buddy Clark, Alan Fentner, and Cal King of the College of Dentistry, June Wilson and myself in 1989 that a particulate array of Bioglass® has the important attribute of forming a cohesive mass when mixed with blood or when placed in a bleeding bone defect and leads to rapid repair of bone and periodontal tissues.[83-89] FDA approved clinical sale of Perioglas® in 1992 and it is now widely used in 26 countries. By the end of this century more than 1,000,000 teeth will have been saved from extraction by use of 45S5 Bioglass® periodontal products.

In 1997 limited orthopaedic clinical trials of Bioglass® particulate (Osteoglas®) began; uses include repair of non-unions, regeneration of bone in large defects and augmentation by bones in revision surgery. Use of Bioglass® as a segmental bone replacement in our US Army Medical R & D Command program did not result in a clinically useful product but the basic research understanding of the mechanisms of bonding of bone has culminated in numerous important clinical applications.[90--102]

My conclusion is that applications oriented research is important because it provides a direction. It is even more important to understand the physical principles and the mechanisms underlying your research because they provide the means to *change* direction towards a reachable goal.

Through the thirty years of Bioglass® research we have been fortunate to have available the experimental methodology developed in studying the environmental interaction of the binary Li_2O-SiO_2, Na_2O-SiO_2 and K_2O-SiO_2 glasses. The IRRS method is used to obtain quality assurance (QA) and ISO 9000 certification of bioactivity of commercial Bioglass ® batches. The QA test has a high level of reproducibility using an *in-vitro* simulated body fluid environment and thereby eliminates most of the need for animal testing, a very important ethical advance for the field.[103-106] Every test method developed in our basic research studies on glass-environment interactions has been applied to the field of bioactive glasses. When we began to develop Perioglas® it was necessary to have a reliable QA test for the Bioglass® particulate. Our previous studies on SA/V effects on rates of reaction of Type II simple binary and ternary glasses and then complex Type III B nuclear waste glasses (Fig.1) made it possible to establish our inexpensive *in vitro* test for Bioglass ® powders. Learning how to keep QA costs low and reproducibility and reliability high is one of the most significant commercial pay-offs of our research program.

SOL-GEL GLASSES

I bring this overview of our thirty years of research on glass surface-environment interactions to a close with another key personal encounter. It occurred in 1977 on the shores of a lake in New Hampshire at a Gordon Research Conference. Don Ulrich and I met for the first time at lunch and we continued our talk at the lake. He had just accepted a position at the Division of Chemical and Atmospheric Sciences of the Air Force Office of Scientific Research (AFOSR) and had the mandate to sponsor a long term research program that included environmental interactions with ceramic and glass materials. He asked what I thought were some interesting basic science questions in the field. I responded that we knew little about the effect of microstructures on the rate of heterogeneous environmental reactions at phase boundaries in ceramics and glass-ceramics. We knew even less about the combined effects of microstructure, applied stress and reactive environments on mechanical properties. He inquired as to how we could approach this problem and my response was that it would be possible to combine the results of the early Kinser and Freiman studies on nucleation and crystallisation of 33L glasses with the Sanders, Clark and Ethridge studies on glass corrosion kinetics to quantify microstructural effects on the heterogeneous corrosion and fracture mechanics of glasses and glass-ceramics. The notes of the lakeside discussion, written on a paper napkin, became a proposal and led to 20 years of AFOSR funding of our research on the theme of control of surface-environment interactions. The results included Walt McCracken's PhD on corrosion of 33L glass ceramics[107-108] and Ron Palmer's PhD on environmental sensitivity of 33L glass-ceramic fracture mechanics.

Our AFOSR funded research took a fateful and fruitful change in emphasis in 1981 when I talked with Don Ulrich, from a phone box on the Rennselaer Polytechnic Institute campus while attending a Society for Biomaterials meeting.

Don asked if we were doing any research on materials processing that was relevant to long term Air Force space missions. I told him of the exciting results that I had achieved on sabbatical at Prof. Jerzy Zarzycki's lab in Montpellier, France. I had learned, largely by accident, how to make 33N glasses by a sol-gel route with gel-glass conversion at only one hour at 450 to 500^{0}C in an ambient air atmosphere. The glasses were stable and beautifully transparent. I suggested to Don that low temperature sol-gel processing might be possible for a whole range of composite lightweight assembled structures for space (i.e., a new CLASS of materials). He said "Send me a proposal". I did, and the proposal to study sol-gel processing at a molecular level was funded and began the development of what has been commercialized as Gelsil® optical products by Geltech Inc., Orlando Florida. A large number of PhD and MSc students worked on the processing-properties-environmental studies of gel-silica with particular emphasis on net shape, net surface processing of optics. The history of our gel-silica program has recently been published* so I will not dwell on any details here. I must add that our success would have been impossible without Don Ulrich's vision of long range research from the government's perspective and the commercial technical leadership of Dr. Jean-Luc Nogues (Geltech, Inc.), Dr. David Greenspan (USBiomaterials Corp.) and the University of Florida's technical and laboratory management of Dr. Jon West, Guy LaTorre and Alice Holt (Advanced Materials Research Center) and Dr. June Wilson and Waverly Munn (Bioglass® Research Center). Our multi-disciplinary team was able to identify long range directions for research, which Don always approved, and then direct graduate students over a fifteen year period of time to reach most of the long range goals. A large portion of the portfolio of 24 patents obtained over my forty years of research resulted from this freedom to pursue long-range research goals with minimal intervention from the funding agencies. Hopefully, this freedom to be creative will continue for the generations of researchers to follow. It has certainly been a lot of fun for me.

*Sol-Gel Silica: "Properties, Processing and Technology Transfer", Larry L. Hench, Noyes Publications (1998).
Note: This book discusses the important contributions of the more than twenty graduate students that pursued degrees in sol-gel science at the University of Florida from 1980-1998.

REFERENCES

[1]L.L. Hench and R. Russell, Jr., "Effect of Cr_2O_3 Vaporization on the Sintering of MgO," *Trans. British Ceram. Soc.* **67** [9] 377-390 (1968).

[2]L.L. Hench, D.L. Kinser and S.W. Freiman, "The Early Stages of Crystallization of a Li_2O-$2SiO_2$ Glass," *Phys. Chem. of Glasses* **12**[2] 58-63 (1971).

[3]S.W. Freiman and L.L. Hench, "Kinetics of Crystallization of Lithia-Silica Glass," *J. Am. Ceram. Soc.* **51**[8] 382-387 (1968).

[4]S.W. Freiman and L.L. Hench, "Further Analysis of Glass Crystallization Kinetics," *J. Am. Ceram. Soc.* **52** 111 (1969).

[5]D.L. Kinser and L.L. Hench, "Hot-Stage Transmission Electron Microscopy of Crystallization of a Lithium Silicate Glass," *J. Materials Science* 5, 369 (1970).

[6]D.L. Kinser and L.L. Hench, "Effect of a Metastable Precipitate on the Electrical Properties of a Li_2O-SiO_2 Glass" *J. Am. Ceram. Soc.* **51**[8] 445-449 (1968).

[7]S.W. Freiman and L.L. Hench, "Effect of Crystallization on the Mechanical Properties of Li_2O-SiO_2 Glass-Ceramics," *J. Am. Ceram.Soc.* **55** [2] 86-90 (1972).

[8]H.F. Schaake and L.L. Hench, "Polaronic Effects in Potassium Vanadium Phosphate Glasses," *Int. J. of Non-Crystalline Solids* **2** 292-306 (1970).

[9]L.L. Hench, "Semiconducting Glass-Ceramics," *Int. J. of Non-Crystalline Solids* **1** 250-278 (1970).

[10]L.L. Hench and H.F. Schaake "Electrical Properties of Glass." pp 589-659 in *Introduction to Glass Science* Edited by D. Pye and H. Simpson, Plenum Press, New York, 1972.

[11]C.V. Gokularathnam, R.W. Gould and L.L. Hench, "X-ray Diffraction Analysis of Structural Changes in Vitreous Glass," pp. 61-70 in *Advances in Nucleation and Crystallization in Glass,* Am. Ceram. Soc., Columbus, Ohio, 1972.

[12]D.M. Sanders, W.H. Person and L.L. Hench, "New Methods for Studying Glass Corrosion Kinetics," *Appl. Spectroscopy,* **26** [5] 530-536 (1972).

[13]D.M. Sanders and L.L. Hench, "Mechanisms of Glass Corrosion," *J. Am. Ceram. Soc.* **56** [7] 373-377 (1973).

[14]D.M. Sanders and L.L. Hench, "Environmental Effects on Glass Corrosion Kinetics," *Bull. Am. Ceram. Soc.* **52** [9] 662-665 (1973).

[15]D.M. Sanders and L.L. Hench, "Surface Roughness and Glass Corrosion," *Bull. Am. Ceram. Soc.* **52**[9] 666-669 (1973).

[16]D.M. Sanders, W.B. Person and L.L. Hench, "Quantitative Analysis of Glass Structure with the Use of Infrared Reflection Spectra," *Appl. Spectroscopy* **28**[3] 247-255 (1974).

[17]L.L. Hench and D.M. Sanders, "Analysis of Glass Corrosion," *Glass Industry*, 13 (1974).

[18]D.E. Clark, L.L. Hench and W.A. Acree, "Electron Microprobe Analysis of Na_2O-CaO-SiO_2 Glass," *J. Am. Ceram. Soc.* **58** [11-12] 531-532 (1975).

[19]D.E. Clark, M.F. Dilmore, E.C. Ethridge and L.L. Hench, "Aqueous Corrosion of Soda-Silica and Soda-Lime-Silica Glass," *J. Am. Ceram. Soc.* **59** [1-2] 62-65 (1976).

[20] D.E. Clark, W.A. Acree, and L.L. Hench, "Electron Microprobe Analysis of Corroded Soda-Lime-Silica Glasses," *J. Am. Ceram. Soc.* **59** [9-10] 463-464 (1976).

[21]C.G. Pantano, Jr. and L.L. Hench, "Cleaning Borosilicate Glass for Biological Application," *J. Testing and Evaluation* **5** [1] 66-69 (1977).

[22]D.E. Clark, E.C. Ethridge, M.F. Dilmore and L.L. Hench, "Quantitative Analysis of Corroded Glass Using IRRS Frequency Shifts," *Glass Technol.* **18** [4] 121-124 (1977).

[23]L.L. Hench and D.E. Clark, "Physical Chemistry of Glass Surfaces," *J. Non-Crystalline Solids* 28[1] 83-105 (1978).

[24]D.E. Clark, E.C. Ethridge and L.L. Hench, "Effects of Glass Surface Area to Solution Volume Ratio on Glass Corrosion," *Physics and Chemistry of Glasses* **20** [2] 35-40 (1979).

[25]M.F. Dilmore, D.E. Clark and L.L. Hench, "Chemical Durability of a Na_2O-K_2O-CaO-SiO_2 Glass," *Bull. Am. Ceram. Soc.* **61** [9-10] 439-443 (1978).

[26]M.F. Dilmore, D.E. Clark and L.L. Hench, *Bull* "Aqueous Corrosion of Lithia-Alumina-Silicate Glasses," *Am. Ceram. Soc.* **57** 1040-1044 (1978).

[27]M.F. Dilmore, D.E. Clark and L.L. Hench, "Corrosion Behavior of Lithia Disilicate Glass in Aqueous Solutions of Aluminum Compounds," *Am. Ceram. Soc. Bull.* **58** [11] 1111-1114 (1979).

[28]L.L. Hench, D.E. Clark and E. Lue Yen-Bower, "Surface Leaching of Glasses and Glass-Ceramics," *Proceedings of the Conference on High Level Radioactive Solid Waste Forms*, L. A. Casey, ed., NUREG/CP-0005, published by U.S./NRC pp. 199-235 (1979).

[29]E. Lue Yen-Bower, D.E. Clark and L.L. Hench, "An Approach Towards Long Term Prediction of Stability of Nuclear Waste Forms," pp. 41-46 in *Ceramics in Nuclear Waste Management,* T.D.Chikalla and J.E. Mendel, eds., Tech. Info. Center, U. S. Dept. Energy, 1979.

[30]D.E. Clark, E. Lue Yen-Bower and L.L. Hench, "Corrosion Behavior of Zinc-Borosilicate Simulated Nuclear Waste Glass," pp. 256-262 in *Ceramics in Nuclear Waste Management,* T.D.Chikalla and J.E. Mendel, eds., Tech. Info. Center, U.S. Dept. Energy, 1979.

[31]D.E. Clark and L.L. Hench, "An Overview of the Physical Characterization of Leached Surfaces," *Nuclear and Chemical Waste Management* **2** 93-101 (1981).

[32]J. L. Nogues, L.L. Hench, and J. Zarzycki, "Comparative Study of Seven Glasses for Solidification of Nuclear Wastes," pp. 211-218 in *Scientific Basis for Radioactive Waste Management V,* W. Lutze, ed., Elsevier Sci. Pub. Co., New York, 1982.

[33]L. Werme, L.L. Hench and A. Lodding, "Effect of Overpack Materials on Glass Leaching in Geological Burial," pp. 135-144 in *Scientific Basis for Radioactive Waste Management V,* W. Lutze, ed., Elsevier Sci. Pub. Co., New York, 1982.

[34]L.L. Hench, L. Werme and A. Lodding, "Burial Effects on Nuclear Waste Glass," pp. 153-162 in *Scientific Basis for Radioactive Waste Management V,* W. Lutze, ed., Elsevier Sci. Pub. Co., New York, 1982.

[35]A.A. Hench and L.L. Hench, "Computer Analysis of Nuclear Waste Glass Composition Effects on Leaching," *J. Nuclear and Chemical Waste Management* **4** [4] 231-258 (1983).

[36]J.E. Mendel, R.P. Turcotte, T.D.Chikalla and L.L. Hench, "Evaluation of Conditioned High-Level Waste Forms," *Proceedings of the International Conf. on Radioactive Waste Management*, Seattle, WA, May 16-20, 1983.

[37]L.L. Hench, A. Lodding and L. Werme "Analysis of One Year in Situ Burial of Nuclear Waste Glasses in Stripa," pp. 310-323 in *Nuclear Waste Management, Advances in Ceramics Series,* Vol. 8, G. G. Wicks and W.A. Ross, eds., Am. Ceram. Soc., Columbus, Ohio, 1984.

[38]"Nuclear Waste Glass Interfaces After One Year Burial in Stripa, Part 1: Glass/Glass," L.L. Hench, A. Lodding, and L. Werme, *J. Nuclear Materials* **125** 273-279 (1984).

[39]A. Lodding, L.L. Hench, and L. Werme, "Nuclear Waste Glass Interfaces After One Year Burial in Stripa, Part 2: Glass/Bentonite," *J. Nuclear Materials* **125** 280-286 (1984).

[40]L.L. Hench, D.B. Spilman and A.D. Buonaquisti, "Rutherford Back Scattering Surface Analysis of Nuclear Waste Glasses After One Year Burial in STRIPA," *J. Nuclear and Chemical Waste Management* **5** 75-85 (1984).

[41]L.L. Hench, G. LaTorre, L. Storz, D.E. Clark and H. Yamanaka "Use of Fourier Transform IR Reflection Spectroscopy (FTIRRS) in Non-Destructive Quality Assurance of Glasses and Ceramics," pp. 97-107 in *Advances in Ceramics,* Vol. 11, K. M. Nair, ed., Am. Ceram. Soc., Columbus, Ohio, 1984.

[42]L.L. Hench, D.E. Clark, and J. Campbell, "High Level Waste Immobilization Forms," *Nuclear and Chemical Waste Management* **5** 149-173 (1984).

[43]L.L. Hench, A. Lodding, and L. Werme, "Nuclear Waste Glass Interfaces After One Year Burial in STRIPA, Part 3: Glass/Granite," *J. Nuclear Materials* **126** 226-233 (1984).

[44]L.L. Hench and M.J.R. Wilson, "Nuclear Waste Glass Interfaces After One Year Burial in STRIPA, Part 4: Comparative Surface Profiles," *J. Nuclear Materials* **136** 218-228 (1985).

[45]L.L. Hench, "International Collaboration in Nuclear Waste Solidification," *Nuclear Technology* **73** 188-198 (1986).

[46]L.L. Hench, D.E. Clark and A.B. Harker, "Nuclear Waste Solids," *Journal of Materials Science* **21** 1457-1478 (1986).

[47]L.L. Hench, R. J. Splinter, T.K. Greenlee and W.C. Allen, "Bonding Mechanisms at the Interface of Ceramic Prosthetic Materials," *J. Biomed. Mater. Res.* **November No 2 (part 1)** 117-141 (1971).

[48]C.A. Beckham, T.K. Greenlee,Jr., and A.R. Crebo, "Bone Formation at a Ceramic Implant Interface," *Calc.tiss.Res.* **8**, 165-171 (1971).

.[49]K. Greenlee, Jr., C.A. Beckham, A.R. Crebo and J.C. Malmorg, *J* "Glass Ceramic Bone Implants," T.*Biomed.Mater.Res.* **6**, 235-244 (1972).

[50]L.L. Hench and H.A. Paschall, "Direct Chemical Bond of Bioactive Glass-Ceramic Materials to Bone and Muscle," *J. Biomed. Mater. Res.* Symposium No 4, 25-42, (1973).

[51]L.L. Hench and H.A. Paschall, "Histochemical Responses at a Biomaterial's Interface," *J. Biomed.Mater.Res.* Symposium No 5 (Part 1) pp 49-54 (1974).

[52]G.G. Pantano, Jr., A.E. Clark, Jr., and L.L. Hench, "Multilayer Corrosion Films on Bioglass®," *J. American Ceramic Society,* **57** [9] 412-413 (1974).

[53]G. Piotrowski, L.L. Hench, W.C. Allen and G.J. Miller, *J* "Mechanical Studies of the Bone-Bioglass® Interfacial Bond,". *Biomed Mater. Res* **9** [6] 47-61 (1975).

[54]L.L. Hench, H.A. Paschall, W.C. Allen and G. Piotrowski, "Interfacial Behaviour of Ceramics Implants," *National Bureau of Standards Special Publication*, May 1975, 415 19-35.

[55]A.E. Clark, Jr., C. Pantano, Jr., and L.L. Hench, "Auger Spectroscopic Analysis of Bioglass® Corrosion Films," *J. American Ceramic Society,* **59** [1-2] 37-39 (1976).

[56]A.E. Clark, Jr., L.L. Hench and H.A. Paschall, "The Influence of Surface Chemistry on Implant Interface Histology: A Theoretical Basis for Implant Materials Selection," *J. Biomed. Mater. Res.,* **10** 161-174 (1976).

[57]J.J. Hren, P.F. Johnson, S.R. Bates and L.L. Hench, "STEM/EDX Analysis of a Bone-Bioglass® Interface," *EMSA*, 290-291 (1976).

[58]M. Ogino, F. Ohuchi and L.L. Hench, "Compositional Dependence of the Formation of Calcium Phosphate Films on Bioglass®," *J. Biomed. Mater. Res.,* **14** 55-64 (1980).

[59]C.Y. Kim, A.E. Clark, Jr., and L.L. Hench, "Early Stages of Calcium-Phosphate Layer Formation on Bioglasses," *J. Non-Crystalline Solids,* **113** 195-202 (1989).

[60]P. Griss, D.C. Greenspan, G. Heimke, B. Krempien, R. Buchinger, L.L. Hench and G. Jentschura,. "Evaluation of Bioglass® Coated Al_2O_3 Total Hip Prosthesis in Sheep," *J. Biomed. Mater. Res.,* **10** [4] 511-518 (1976).

[61]D.C. Greenspan and L.L. Hench, "Chemical and Mechanical Behavior of Bioglass®-Coated Alumina," *J.Biomed. Mater. Res.*Symposium No 7, 503-509 (1976).

[62]L.L. Hench, C.G. Pantano, Jr., P. Buscemi and D.C. Greenspan "Analysis of Bioglass® Fixation of Hip Prostheses," *J. Biomed. Mater. Res.* **11** 267-282 (1977).

[63]J.E. Ritter, Jr., D.C. Greenspan, R.A. Palmer and L.L. Hench, "Use of Fracture Mechanics Theory in Lifetime Predictions for Alumina and Bioglass®-Coated Alumina," *J. Biomed. Mater. Res.* **13** 251-263 (1979).

[64]L.L. Hench and A.E. Clark, Jr., "Adhesion to Bone," *in Biocompatibility of Orthopaedic Implants,* D. F. Williams Editor, CRC Press, Boca Raton, Florida, Vol. II, Chapter 6.

[65]W.R. Lacefield and L.L. Hench, "The Bonding of Bioglass® to a Cobalt-Chromium Surgical Implant Alloy," *Biomaterials*, **7** [2] 1986.

[66]L.L. Hench, "Stability of Ceramics in the Physiological Environment," in *Fundamental Aspects of Biocompatibility*, D. F. Williams, Editor, CRC Press, Boca Raton, Florida, 1.

[67]J. Wilson-Hench, F.J. Schoen, G.H. Pigott and L.L. Hench, "Toxicology and Biocompatibility of Bioglasses," *J. Biomed. Mater. Res.* **15** 805-817 (1981).

[68]L.L. Hench and J. Wilson-Hench, "Surface-Active Biomaterials," *Science* **226** 630-636 (1984).

[69]G.E. Merwin, J.S. Atkins, J. Wilson-Hench, L.L. Hench, "Comparison of Ossicular Replacement Materials in a Mouse Ear Model," *Otolaryngol. Head Neck Surg.* **90** 461-469 (1982).

[70]G.E. Merwin, J. Wilson-Hench and L.L. Hench "Current Status of the Development of Bioglass® Ossicular Replacement Implants," pp 220-229 in Biomaterials in Otology, J.J. Grote, Editor Martinus Nijhoff Publishers, The Hague – Boston- London (1983).

[71]L.L. Hench, J. Wilson-Hench and G. Merwin "Bioglass® Implants for Otology," in Biomaterials in Otology, J.J. Grote Editor, Martinus Nijhoff Publishers, The Hague – Boston – London.

[72]L.L. Hench and J. Wilson-Hench "Glasses in Medical Applications," in *Encyclopedia of Materials,* CRC Press, Boca Raton, FL, (1984).

[73]J. Wilson-Hench, G.E. Merwin, and L.L. Hench, "Machining Bioglass® Implants," *SAMPE Journal* **21** [3]: 6-8 (1985).

[74]G. Merwin, "Bioglass® Middle Ear Prosthesis: Preliminary Report," *Ann.Otol.Rhinol.Laryngol.* **95**: 78-82 (1986).

[75]G.E. Merwin, L.W. Rogers, J. Wilson-Hench, R. Martin, "Facial Bone Augmentation Using Bioglass® in Dogs," *Arch. Otolaryngol. Head Neck Surger.* **112** (1986).

[76]J. Wilson-Hench, E. Douek and K. Rust, "Bioglass® Middle Ear Devices; Ten Year Clinical Results," pp 234-245 in *Bioceramics, 8*, L.L. Hench, J. Wilson-Hench and D.C. Greenspan, Eds., Elsevier, (1995).

[77]L.L. Hench, "Bioceramics," *J. Am. Ceram. Soc.* **81**, [7] 1705-1728 (1998)

[78]H.R. Stanley, L.L. Hench, C.G. Bennett, Jr., S.J. Chellemi, C.J. King, III, R.G. Going, N.J. Ingersoll, E.C. Ethridge, K.L. Kreutziger, L. Leob, and A.E. Clark, Jr., "Dental Implantation of Natural Tooth form Bioglass® in Baboons – Long Term Results," *Internal J. Oral Implantology 2*:26-36 (1981).

[79]J. Wilson-Hench, H.R. Stanley and L.L. Hench "Dental Application of Bioglass®," in *Clinical Dentistry*, J.W. Clark, Editor, Harper & Row, Philadelphia, PA **4**[50] (1981).

[80]A.M. Weinstein, J.J. Klawitter, and S.D. Cook, "Implant-Bone Interface Characteristics of Bioglass® Dental Implants," *J. Biomed. Mater. Res.* **14** 23-29 (1980).

[81]J. Wilson-Hench, A.E. Clark, Jr., M. Hall and L.L. Hench, "Tissue Response to Bioglass® Endosseous Ridge Maintenance Implants," *J. Oral Implantology,* XIX [4] 295-302 (1993).

[82]H.R Stanley, M.B. Hall, A.E. Clark, C.J. King III, L.L. Hench, J.J. Berte, "Using 45S5 Bioglass® Cones as Endosseous Ridge Maintenance Implants to Prevent Alveolar Ridge Resorption: A 5-Year Evaluation," *Int. J. Oral Maxillofac. Implants,* **12** 95-105 (1997).

[83]J. Wilson-Hench, S. Low, A Fetner, and L.L. Hench "Bioactive Materials for Periodontal Treatment: A Comparative Study," pp. 223-228 in *Biomaterials and Clinical Applications*, A. Pizzoferrato, P.G. Marchetti, A Ravaglioli and A.J.C. Lee, Eds., Elsevier Science Publishers B. V. Amsterdam, (1987).

[84]J. Wilson-Hench and S.B. Low "Bioactive Ceramics for Periodontal Treatment: Comparative Studies in the Patus Monkey," *J. Applied Biomaterials*, **3** 123-129 (1992).

[85]H. Oonishi, S. Kushitani, E. Yasukawa, H. Iwaki, L.L. Hench, J. Wilson-Hench, E. Tsuji, and T. Sugihara, "Particulate Bioglass® Compared with Hydroxyapatite as a Bone Graft Substitute," *J. Clinical Orthopaedics and Related Research*, **334** 316-325 (1997).

[86]C.A. Shapoff, D.D. Alexander, A.E. Clark, Jr., "Clinical Use of a Bioactive Glass in the Treatment of Human Osseous Defects," *Compendium of Continuing Educaiton in Dentistry,* **18** [4] 352-362 (April 1997).

[87]J.S. Zamet, U.R. Darbar, G.S. Griffiths, J.S. Bulman, U Bragger, W. Burgin and H.N. Newman, "Particulate Bioglass® as a Grafting Material in the

Treatment of Periodontal Intrabony Defects," *J. Clin Periodontal* **24** 410-418 (1997).

[88]S.B. Low, C.J. King, J. Krieger, "An Evaluation of Bioactive Ceramic in the Treatment of Periodontal Osseous Defects," *Int. J. Periodont. Rest. Dent,* **17** 359-367 (1997).

[89]S.J. Froum, M.A. Weinberg and D. Tarnow, "Comparison of Bioactive Glass Synthetic Bone Graft Particles and Open Debridement in the Treatment of Human Periodontal Defects. A Clinical Study," *J. Periodontal* **69** 698-709 (1998).

[90]J. Wilson-Hench, L.T. Yu and B.S. Beale, "Bone Augmentation Using Bioglass® Particulates in Dogs: Pilot Study," *Bioceramics 5,* Yamamuro, Kokubo and Nakamura, Eds., Kyoto Kokunshi Kankakai, Kyoto (1992).

[91]B.Oguntebi, A.E. Clark and J. Wilson-Hench, "Pulp Capping with Bioglass® and Autologous Demineralised Dentin in Miniature Swine," *J. Dent. Res.***I** [72-92] 484-489 (1993).

[92]L.L. Hench and J.K. West, "Biological Applications of Bioactive Glasses," *Life Chemistry Reports*, **13** 187-241 (1996).

[93]L.L. Hench, "Sol-Gel Materials for Bioceramic Applications," *Current Opinion in Solid State & Materials Science*, **2** 604-610 (1997).

[94]J. Wilson-Hench, G. Lowery, S. Courtney, "Spinal Fusion using Titanium Spacers with Bioglass® and Autogenous Bone: A Comparative Study in Sheep" pp. 65-68 in *Bioceramics 10: Proceedings of the 10th Interntional Symposium on Ceramics in Medicine*, L. Sedel and C. Rey. Eds., Paris, Elsevier Science Ltd (1997).

[95]L.J. Litkowski, G.D Hack, H.B. Sheffer and D.C. Greenspan "Occlusion of Dentin Tubules by 45S5 Bioglass®," pp. 411-414 in *Bioceramics 10: Proceedings of the 10th International Symposium on Ceramics in Medicine* L. Sedel and C. Rey Eds., Paris, Elsevier Science Ltd, (1997).

[96]I. Thompson, C. Chan, P. Robinson, P. Revell, J. Wilson-Hench and L.L. Hench "Bioglass®-Dextran Composite Tested In Rabbit Femur," pp. 379-392 in *Bioceramics 11: Proceedings of the 11th International Symposium on Ceramics in Medicine*, R. LeGeros and J. LeGeros, Eds., World Scientific Pub. Co. Pte. Ltd., (1998).

[97]A.M. Gatti, L.L. Hench, E. Monari, J. Thompson, D. Tanza, "New Putty Composite Materials for Bone Defects in Dentistry," pp 349-352 in *Bioceramics 11: Proceedings of the 11th International Symposium on Ceramics in Medicine*, R. LeGeros and J. LeGeros, Eds., World Scientific Pub. Co. Pte. Ltd., (1998).

[98]S.E. Efflandt, P. Magne, W.H. Douglas, L.F. Francis, "Bonding of Bioactive Glasses to Human Dentin," pp 571-574 in *Bioceramics 11. Proceedings of the 11th International Symposium on Ceramics in Medicine*, R. LeGeros and J. LeGeros, Eds., World Scientific Pub. Co. Pte. Ltd., (1998).

[99]M.B. Hall et al "Placement of Dental Implants Into Alveolar Ridges Reconstructed With Particulate Bioglass® An Animal Study," AAOMS, 114-115 (1995).

[100]J. Wilson-Hench, A.E. Clark, E. Douek, J. Kreiger, W.K. Smith, J.S. Zamet, "Clinical Applications of Bioactive Glass Implants," pp 415-422 in *Bioceramics 7: Proceedings of the 7th International Symposium on Ceramics in Medicine*, O.H. Andersson Editor, Butterworth-Heinemann, London, (1994).

[101]J. Wilson-Hench, A.E. Clark, R.D. Walker and M. Ramer "Development and Use of an Injectable Form of Bioglass®," in *Bioceramics 4: Proceedings of the 4th International Symposium on Ceramics in Medicine*, W. Bonfield, G.W. Hastings and K.E. Tanner, Eds., Butterworth-Heinemann Ltd., pp 99-103 (1991).

[102]R. Dixon-Walker, J. Wilson-Hench and A.E. Clark, "Injectable Bioglass® As A Potential Substitute for Injectable Polytetrafluoroethylene," *J. of Urology,* **148**, 645-647 (1992).

[103]G. LaTorre and L.L. Hench, "Analysis of Bioactive Glass Interfacial Reactions Using Fourier Transform Infrared Reflection Spectroscopy," pp. 177-195 *in Handbook on Characterization Techniques for the Solid-Solution Interface,* J.H. Adair, J.A. Casey, and S. Venigalla, Eds., American Ceramic Soc., Westerville, OH, (1993).

[104]I. Rehman, L.L. Hench, W. Bonfield and R. Smith, "Analysis of Surface Layers on Bioactive Glasses," *Biomaterials,* **15** [10] 865-870 (1994).

[105]K.D. Lobel and L.L. Hench, "*In vitro* Adsorption and Activity of Enzymes on Reaction Layers of Bioactive Glass Substrates," *J. Biomed Mater Res,* **39** 575-579 (1998).

[106]L.D. Warren, A.E. Clark, Jr., and L.L. Hench, "Investigation of Bioglass® Powders: Quality Assurance Test Procedures and Test Criteria," *J.Biomed. Mater. Res. Applied Biomaterials*, **23** 285-703 (1989).

[107]W.J. McCracken, D.E. Clark and L.L. Hench, "Aqueous Durability of Lithium Disilicate Glass-Ceramics," A*m. Ceram. Soc. Bull.,* **61** [11] 1218-1223 (1982).

[108]W.J. McCracken, W.B. Person and L.L. Hench, "Polarized Infrared Reflection Spectroscopy of Single Crystal Lithia-Silicats and Quartz," *J of Materials Science* **20** 218-228 (1985).

Fracture and Dielectric Behavior

FRACTURE AND THE STRUCTURE OF SILICA

Stephen W. Freiman
Ceramics Division, NIST
Gaithersburg, MD 20899-8520

ABSTRACT

This paper traces the historical development of our understanding of the phenomenon of environmentally enhanced crack growth in glasses. Key experimental findings as well as models of the crack growth process are pointed out. A second theme of the paper is the demonstration of both the direct and indirect contributions of Dr. Larry Hench and his coworkers to the science of glass fracture.

INTRODUCTION

Not only because of its historic and widespread use in applications where its strength and brittle nature are significant factors, but also because it is homogeneous in composition and structure, and isotropic in its properties, glass is an ideal medium in which to apply concepts of brittle fracture. Vitreous silica is the simplest form of oxide glasses and so represents a model for other glass compositions. While there can be numerous aspects to brittle fracture, this paper will be focused primarily on the phenomenon termed "delayed failure,"* i.e., time-dependent fracture under constant load. We now know that this phenomenon is due to the combined effects of environment and stress, causing cracks in the glass (invariably surface damage) to grow slowly with time. From a practical point of view delayed failure significantly affects the design and use of glass structures for which survival over long periods of time is critical. What has been learned about delayed failure in glass is now being applied to other brittle materials as well.

The purpose of this paper is twofold, to show how the fracture process is intimately related to the atomic and molecular structure of glasses as well as to pay tribute to the many contributions that Larry Hench and his coworkers have made

*One must always recognize that the water dissolved in these other liquids will also cause cracks to grow in proportion to its partial pressure.

toward understanding glass structure and the fracture process.

HISTORICAL PERSPECTIVE

The year 1847 is the earliest reference that I have found to an investigation of glass strength (1). In this paper, Chevandier and Wertheim report on studies of the elastic modulus and strength of a series of glass compositions. Later, Fairbairn and Tate (1859)(2) used glass as a medium in which to study the collapse of structures. They formed glass globes and cylinders, and measured their resistance to bursting forces and external hydrostatic pressure.

In 1897 Brodmann (3) investigated the strength of glass in pure tension, in bending, and in torsion. Recognizing that failure of these specimens began at the surface, Brodmann etched the glass in an HF solution and retested. In all three modes of loading, etched strengths were 50 to 100% higher than their unetched counterparts, leading him to the conclusion that etching removed a damaged layer on the glass surface, and providing the first evidence that glass fails from surface defects.

From the point of view of this paper, a very significant finding was made in 1899 when Grenet (4) observed a testing rate dependence to the strength of glass. Grenet fractured glass plates in three-point bending, stressing them by hanging a bucket from their center-line and gradually filling the bucket with water at varying rates. He found that decreasing the loading rate from .25 kg/min to .25 kg/hr produced nearly a 50% decrease in strength. Grenet also observed that when specimens were left under a constant load, failure could occur after periods as long as 5 days.

Over the next twenty years no direct references to studies of glass strength were found. However, in 1920, A. A. Griffith published a paper (5) that revolutionized the way that we think about brittle fracture. Griffith, whose primary job was the design of aircraft structures, approached the study of glass as a model material from which he could learn why most glass structures are weak compared to, for example, fine glass fibers. Griffith demonstrated unequivocally that glass fails from small surface flaws, and that the density and severity of such flaws determine its strength.

Griffith's contribution toward the understanding of delayed failure was in quantifying the relationship between fracture strength and the "surface energy" of the fracture surfaces. He stated that a crack begins to grow when the energy supplied by the external stress equals that required to form two new surfaces. However, Griffith did not realize that environment, i.e. moisture, would reduce the stress necessary to form these surfaces. We now know that Griffith's estimate of the surface energy of glass, that he obtained by extrapolating from values for molten glass, is incorrect. We also know that the kinetics of the process, not just the thermodynamics is important. Nonetheless, the expression that he derived for the relationship between surface energy and strength leads directly to fracture mechanics.

There were large numbers of studies conducted on delayed failure over the next number of years. Although, I must skip most of these, four in particular caught my attention. Orowan in 1944 (6) made a significant contribution to the analysis of delayed failure by recognizing that the surface energy proposed by Griffith should be a function of the environment in which it is measured. He used data on the cleavage of mica to show that strengths measured in vacuum are 3.5 to 4 times greater than those tested in moist air. Further, he suggested that the rate of strength decrease was due to cracks deepening, with the rate of growth governed by the diffusion rate of water into the crack.

In 1947, Gurney (7) presented thermodynamic concepts to explain moisture enhanced fracture. He suggested that the rate of crack growth is controlled by the rate at which thermal motions overcome energy barriers. Even more significantly, Gurney suggested that, "Due to concentration of strain energy, the material at the end of the crack has a much higher free energy than normal unstressed glass, and is therefore much more chemically active. Atmospheric attack will result in the formation of a complex of glass and atmospheric constituents. The crack will extend continually if the strength of this complex, during or after its formation, is less than the load imposed on it."

A further important development is represented in the 1953 paper of Stuart and Anderson (8) who modeled delayed failure using chemical rate theory. They suggested that there would be forward and reverse components to bond rupture and stated that tensile stresses would be expected to modify the energy barrier to promote bond rupture rather than bond re-formation.

In 1962 Hillig and Charles (9) published work expanding on the concept that corrosive processes at a crack tip are responsible for its slow growth under stress. The key point in their model is that the contour of the flaw should change under the combined effects of stress and a suitable chemical environment. "Stress corrosion" is accompanied by flaw sharpening, leading to an increase in the stress concentration at a flaw tip. In steady state, the sharpening is balanced by rounding due to corrosion.

A major drawback to experimental studies on delayed failure is that they depended on changes in glass strength; there was no method by which crack growth could be directly observed. The development of the science of fracture mechanics solved this problem by allowing us to separate the variability of the flaw population in a material from the resistance to growth of a single flaw, and thereby to monitor the growth of a flaw over time in great detail.

The work of Irwin and Kies (1948) (10) and later that of Irwin (1953) (11) in developing the science of fracture mechanics were key to opening up a new realm of possibilities in brittle fracture research. In their earliest work Irwin and Kies (1948) developed the concept of strain-energy-release-rate, G, where G ≡ 2γ, the surface energy from Griffith's expression (5). Irwin later introduced the concept of stress intensity factor which describes the manner in which a crack intensifies the magnitude of a far-field stress at its tip.

IDENTIFICATION OF CRACK GROWTH MECHANISMS

Wiederhorn's work on crack growth in soda-lime glass, published in 1967 (12) represents a landmark in our understanding of environmentally enhanced fracture. Wiederhorn used a double cantilever beam specimen to which he applied a constant force and observed cracks growing through an optical microscope. The ability to directly observe a growing crack is a critical factor in understanding the mechanism of crack growth. (It is usually assumed that Wiederhorn used a fracture mechanics approach to this study; in fact he normalized his forces on the specimen to a specified crack length which in effect yields a fracture mechanics parameter. Only in later papers did he replot his data in terms of stress intensity factor rather than applied force.)

The key data in Wiederhorn's publication is reproduced in Figure 1. This data represents a number of "firsts": It is the first direct proof that crack growth rates are a function of applied stress. It is the first indication that the crack growth process is a very complex one, in this case showing three distinct stages. It is the first piece of evidence that relative humidity, i.e. chemical potential of water, plays a key role in the process. Wiederhorn was able to mathematically relate crack growth rates to chemical kinetics[**].

Wiederhorn also explained the so called "plateau" regime (II) in terms of the diffusion rate of water to the crack tip. Wiederhorn developed mathematical expressions to show that a zone formed at the crack tip such that when the crack moved at a velocity greater than that for a water molecule to diffuse through this zone, then this diffusion rate rather than the rate of chemical reaction is the limiting step in the process. Wiederhorn showed that the steep portion of the curve (Region III) was, in fact, independent of the presence of water. Later work by Wiederhorn et. al. (13) showed that crack growth in this region is controlled by the dielectric properties of the environment.

[**]Note that Wiederhorn's data is plotted exponentially as would be the case for chemical rate theory. More commonly, a power law dependence on K is used because of the ease of mathematical manipulation but there is no fundamental basis for such practice.

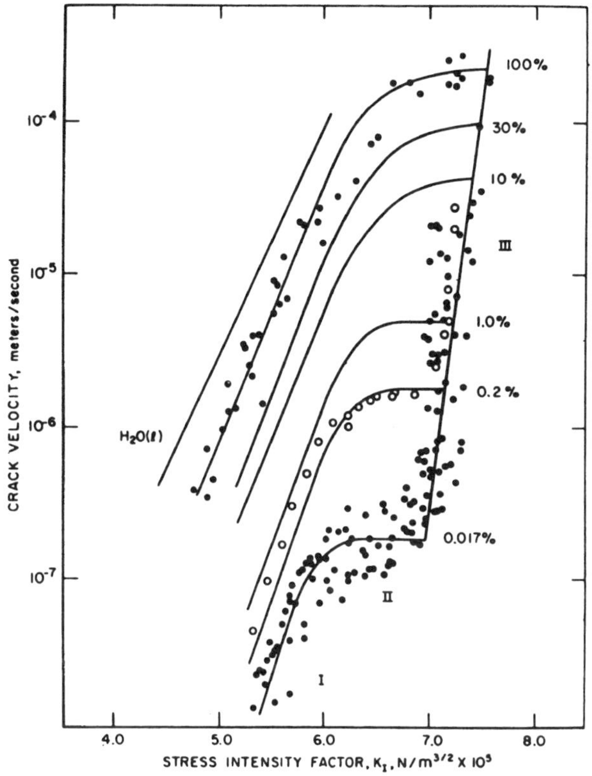

Figure 1. Crack growth curves for soda-lime-silica glass in N_2 gas in various relative humidities. (after Wiederhorn (12))

Freiman (14) later showed that cracks growing in organic liquids such as alcohols or alkanes also depended on the partial pressure of water in the liquid. Stemming from this work was the important concept that no liquid could be considered to be completely inert, i.e., it is the partial pressure rather than the absolute quantity of dissolved water which governs the rate of crack growth. This observation also suggests that one must be very cautious in ascribing crack-growth-enhancing properties to a liquid.

It was at the same time that Wiederhorn was conducting his crack growth studies (1965-66) that Larry Hench began considering the mechanical properties of glasses through his work with me on the fracture behavior of lithia-silica glass ceramics. As Larry Hench's graduate student, I was encouraged by him to investigate the effect of crystallization on the mechanical properties of Li_2O-SiO_2 glasses. The

Li_2O-SiO_2 system would play an important part in Hench's later research as well as my own work on fracture. However, in the years immediately following, Hench's work moved away from mechanical properties to focus on the surface properties of glasses, particularly corrosion. In collaboration with a number of his students, Hench developed a comprehensive picture of corrosion processes in glasses. Of particular importance was his work with David Sanders, which helped develop IR spectroscopy as a tool to follow the corrosion process (15). Such studies lead to later interest in the use of this tool to understand the molecular reactions leading to the development of silica from solutions, i.e. the sol-gel process, and its inverse, the fracture of silica in the presence of water.

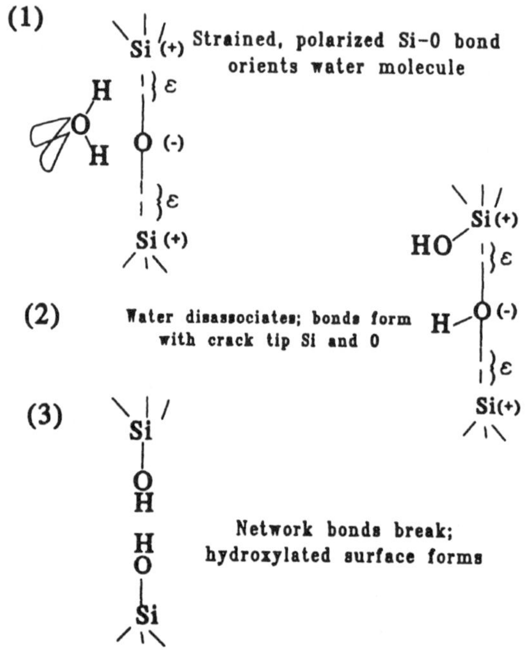

Figure 2. Michalske-Freiman model for environmentally enhanced fracture of silica due to bond stretching (16).

The work of Wiederhorn and others confirmed that water was indeed the medium that in the presence of stress accelerated the rupture of Si-O bonds. A further question was why is water so effective compared to other chemical environments, and what is the nature of the fracture mechanism.

In 1983, Michalske and Freiman (16) published a paper which described a molecular mechanism by which water (and nonaqueous environments) can react with a strained crack tip bond. Michalske and Freiman's model is based on the schematic shown in Figure 2, which depicts a strained crack tip bond in a silicate glass. Of particular importance is the fact that a water molecule contains one set of atomic orbitals which does not participate in the oxygen-hydrogen bonding. They described the bond rupture process in three steps:

1] A water molecule attaches to a bridging Si-O-Si bond at the crack tip. The water molecule is aligned by hydrogen bonding with the $O_{(bridging)}$ and interaction of the lone-pair orbitals from $O_{(water)}$ with Si.

2] A reaction occurs in which both proton transfer to the $O_{(br)}$ and electron transfer from the $O_{(w)}$ to the Si takes place. It is here that the original bridging bond between $O_{(br)}$ and Si is destroyed.

3] Rupture of the hydrogen bond between $O_{(w)}$ and transferred hydrogen occurs to yield Si-O-H groups on each fracture surface.

This model has some interesting implications. First, there is no requirement for prior dissociation of the water molecule, nor must any reaction products be removed. Second, the model suggests that environments other than water should enhance crack growth if the species possess structural and bonding features similar to water, i.e., proton donor sites at one end and lone-pair orbitals at the other, as well as being of comparable size to the Si-O bond.

By performing controlled crack growth experiments on vitreous silica, Michalske and Freiman demonstrated that molecules such as ammonia, hydrazine, and formamide, which possess the requisite structure noted above are effective in enhancing crack growth rates. Their test for effectiveness was the absence of a Region II plateau which would have indicated that dissolved water rather than the liquid itself was causing crack growth. Using this criterion, Michalske and Freiman observed that other molecules such as carbon monoxide and acetonitrile that do not have lone-pair orbitals opposite proton donor sites were not crack-growth-enhancing[***].

Michalske and Bunker (17) later expanded on the above model and suggested that chemicals that dissociatively adsorb on strained surface bonds will enhance crack growth. They suggested that there are bond sites in strained silica which can serve as active kink sites for fracture. Michalske and Bunker also suggested that bond angle strains rather than the direct tensile strains on the Si-O bond were the primary reaction sites in the silica structure(Figure 3). At this time we cannot say which of these two mechanisms best describes the environmentally enhanced fracture process.

Later work by Freiman, White, and Fuller on soda-lime glass(18) and White,

---

[***]One must always recognize that the water dissolved in these other liquids will also cause cracks to grow in proportion to its partial pressure.

Greenspan, and Freiman on binary alkali silicate glasses (19) showed that the addition of alkali ions to silica did not change the fundamental fracture mechanism. However, crack growth curves in all of these glasses were much more complex in shape than those in silica.

However, neither the experimental data nor the models developed to that point answered the question as to why tensile stress was able to increase the normally very slow rate of reaction between silica and water. Experimentally, it is hard to devise an experiment which can employ the high levels of strain that we know must exist at the tip of a crack and still avoid fracture. Because the stressed volume at the crack tip is quite small, identifying a probe which can separate what is happening at a crack tip from the rest of the silica structure is quite difficult. It is therefore natural to turn to molecular modeling in an attempt to understand the reaction mechanism.

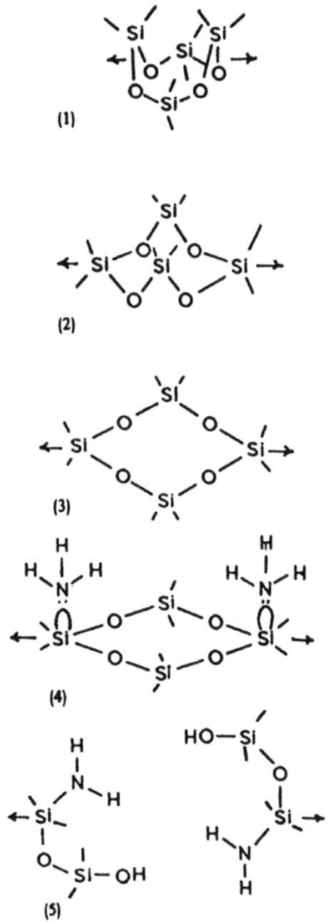

Figure 3. Michalske-Bunker mode for environmentally enhanced fracture of silica due to bond angle pinching (17).

MOLECULAR MODELING STUDIES

At NIST, we made the decision to use abinitio molecular orbital calculations (20) on a silica fragment (Figure 3) as a means of describing the environmentally enhanced reaction. While this choice was made in part by the availability of computer power (employing a smaller silica fragment simplifies the modeling problem), we also felt that rupture of the Si-O bond is the fundamental process governing crack growth.

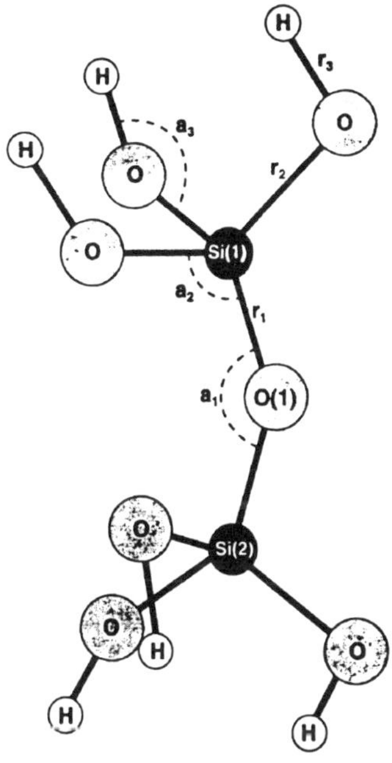

Figure 4. Fragment of the silica structure used as a model for the abinitio calculations performed by Wong-Ng et al. (21).

We approached the problem in a series, first considering what happens to the reactivity of the Si-O bond under strain without any external environment, then adding effects of environment in a later modeling study. In this first study (Wong-Ng et al. (21)), pyrosilicic acid ($H_2Si_2O_7$) was employed as the model (Figure 4). Strains up to 20% were simulated in terms of bond stretching and angle distortion. The results of the simulation indicated that, regardless of how the external strain was applied, the gross electron population on the bridged oxygen increased and that on the silicon decreased with increasing strain (Figure 5). Also of interest was the observation that using the Mulliken electron partition scheme (22), it was found that the bonding electrons associated with the bridging Si and O atoms decreased as the strain was increased, suggesting that these electrons may be available to interact with an environmental molecule.

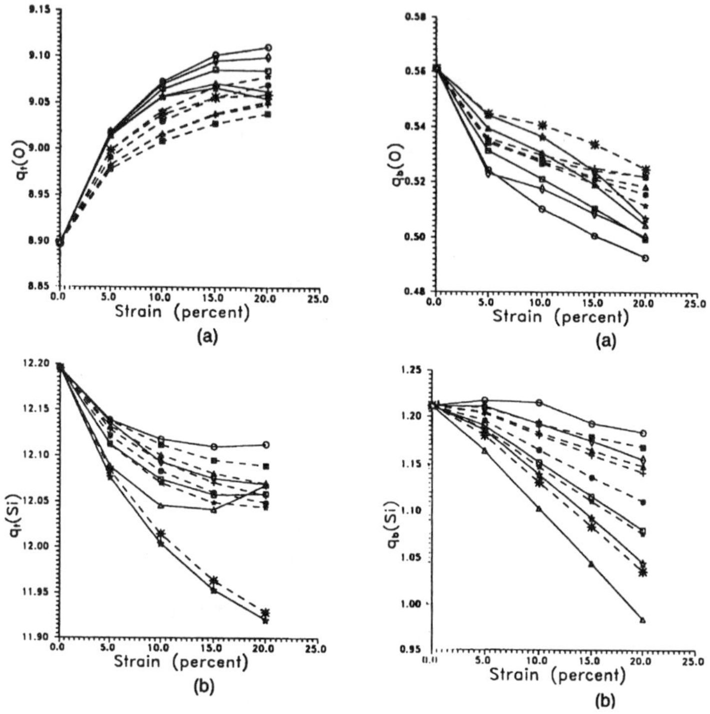

Figure 5. Calculated total number (q_t), and number of bonding electrons (q_b) on the oxygen atom (a) and the silicon atom (b) as the Si-O fragment shown in Figure 4 is pulled in tension . The various curves represent different strain configurations (after Wong-Ng et al. (21))

In the second paper in this series (Lindsay et. al. (23)), the effects of a Si-O bond strain were investigated, a la the model of Michalske and Freiman, in the presence of five different environments, H_2O, NH_3, $HCONH_2$, N_2, and Ar. The primary conclusion stemming from this modeling study was that oxygen atoms adjacent to the reaction site engage in anti-bonding interactions with certain of the environmental molecules or atoms. These anti-bonding interactions were much smaller in magnitude under strain, suggesting that a major effect of strain is to reduce stearic hindrance to the approach of environmental molecules to the reaction site.

In a third paper, White and Wong-Ng (24), used abinitio calculations to determine the reaction coordinate for the water - silica reaction under stress. They showed that the activation barrier to crack growth was reduced under stress because the initial state energy was raised.

At about the same time that these modeling studies were being undertaken at NIST, West and Hench (25-27) were using a semi-empirical method to model the fracture of silica rings. While a drawback of semi-emperical techniques is their reliance on experimental data, they can be used to model much larger groups of atoms. Part I of this series describes the reaction pathway for fracture of polysiloxane rings in the absence of any external environment. West and Hench calculate the transition state barriers to ring fracture for a series of ring geometries, and suggest that a crack tip in silica will move along the lowest energy pathways. They calculate that fracture by ring contraction in the absence of water has an activation barrier of 77kcal mol^{-1} for four-fold rings compared to 96 kcal mol^{-1} for three-fold and 103 kcal mol^{-1} for five-fold rings. Based upon this activation barrier data, West and Hench submit that fracture of four-fold rings should be expected in Region III of the crack growth curve.

The second paper in this series examines the effects of water in influencing the path of a crack through a silica ring structure (26). Again using activation energies as a guide, West and Hench conclude that in the presence of water, three-fold rings will be the primary site at which bond rupture will occur; i.e. cracks will seek out three-fold ring structures to follow as they grow. In a very recent paper (27) West and Hench describe the effects of a number of different environments, vacuum, carbon monoxide, water, and nitrogen on silica fracture. In the case of the first three environments the fracture mechanism exhibiting the lowest energy was ring contraction. However, moisture enhanced fracture is predicted to occur by ring opening through hydrolysis.

Clearly the distribution of three-member rings in the silica structure will have an important effect on the crack growth rate. However, it is not necessary that the entire crack front move simultaneously through the structure. When portions of the crack front encounter weak rings, so called "kinks" can form (17). The energy of these "kinks" is such that they can move parallel to the front and so cause the crack to move forward.

SUMMARY

The primary mechanism of crack growth in silica is one in which the stress enhanced reaction of the environmental species, e.g. H_2O either in liquid or vapor form, with the highly strained Si-O bonds at the crack tip causes the crack to extend. This reaction is modified, but not substantially changed, by glass modifiers.

Environments other than water which can donate both electrons and protons to the Si-O bond, e.g. NH_3, ammonia, and N_2H_4 (hydrazine), also will enhance crack growth rates. No corrosion products need be formed. The stress acts to polarize the Si-O bond, decrease the bonding electrons on both the Si and O, decrease the stearic hindrance for a molecule of the environment to reach the crack tip, and in general, decrease the activation barrier for the process. Certain aspects of the stressed silica structure appear to be most vulnerable to attack by water, namely three-member silica rings. The point is made that cracks can move forward at "kink" sites where rupture is more likely, with the remaining portions of the crack front following.

There remain, however, a number of unanswered questions about the environmentally enhanced fracture process in silica as well as all other materials. For example:

- What is the real nature of a crack tip? Fracture mechanics equations are not applicable as one approaches the non-linear elastic regime near a crack tip, so the value of the crack tip stress becomes uncertain. It is this crack tip stress that drives the reactions, so comparing materials with different crystal structures may not give a uniform answer.
- Although we use both exponential and power law expressions for crack velocity as a function of stress intensity factor (K_I), no one expression can be identified as fundamental. Because of the wide range of crack velocity data, fitting expressions to data will not provide a guide for the choice of an equation.
- We have no way at present of predicting, a-priori, the position or the slope of V-K_I curves from knowledge of the composition or structure of a material.
- Although both abinitio and semi-empirical models are useful guides to the fracture process, they ignore the effects of thermal vibrations.

I hope that we can find ways to provide answers to these questions in the near future.

ACKNOWLEDGMENTS

First I would like to acknowledge all of my collaborators who have made my contributions in this field possible. In this regard I must mention, Edwin Fuller, David Greenspan, Terry Michalske, Jack Mecholsky, Katherine Simmons, Sheldon Wiederhorn, Grady White, and Winnie Wong-Ng. Helpful conversations with Jon West are gratefully acknowledged. To my graduate school advisor, and mentor, Larry Hench I give my thanks and gratitude for teaching me the true meaning of conducting research.

REFERENCES

[1]E. Chevandier and G. Wertheim, Note Sur L'Élasticité et sur Cohésion des Différentes Espéces de Verre, *Ann. de Chim. et de Phys., 3 serie*, t. XIX (Février

1847) 129-138

[2]W. Fairbairn and T. Tate, "On the Resistance of Glass Globes and Cylinders to Collapse from External Pressure; and on the Tensile and Compressive Strength of Various Kinds of Glass," *Phil. Trans. of the* Royal *Society of London*, **149** 213-47 (1859).

[3]C. Brodman, "Einige Beobachtungen über die Festigkeit von Glasstäben," Vorgelegt von W. Voigt in der Sitzung vom 3, 1894.

[4]L. Grenet, Bull. Soc. Encour. Ind. Nat. 4, 839 (1899). Partial English Translation, F. W. Preston, *J Ceramic. Soc.* **18**, 219 (1935).

[5]A. A. Griffith, "Phenomena of Rupture and Flow Solids," *Trans. Roy. Soc. (London)*, A221, 163-198 (1920).

[6]E. Orowan, "The Fatigue of Glass Under Stress," *Nature*, Cavendish Laboratory, Cambridge, No. 3906, 341-343, 1944

[7]C. Gurney, "Delayed Fracture in Glass," *Proc. Phys. Soc. (London)* **59**, 169-85 (1947).

[8]D. A. Stuart and O. L. Anderson, "Dependence of Ultimate Strength of Glass Under Constant Load on Temperature, Ambiuent Atmosphere, and Time," *J. Am. Ceram. Soc.*, **36** [12] 416-424 (1953).

[9]W.B. Hillig and R.J. Charles, pp. 682-705 in *High Strength Materials.* Edited by V.F. Zackey, Wiley & Sons, New York, 1965.

[10]G. R. Irwin and J. A. Kies, *Weld. J. Res. Suppl.*, **33**, 1954, p. 193s.

[11]G. R. Irwin, *Handbuch der Physik*, Vol. **VI**, Springer, Berline, 1958, p. 551.Introduction to Fracture Mechanics, Chapter One pp. 1-17

[12]S. M. Wiederhorn, "Influence of Water Vapor on Crack Propagation in Soda-Lime Glass," *J. Am. Ceram. Soc.*, **50** [8] 407-14 (1967).

[13]S. M. Wiederhorn, S. W. Freiman, E. R. Fuller, Jr., and C. J. Simmons, "Effects of Water and Other Dielectrics on Crack Growth," *J. Mater. Sci.*, **17**, 3460-78 (1982).

[14]S. W. Freiman, "Effects of Alcohols on Crack Propagation in Glass," *ibid.*, **57** [8] 350-53 (1974.)

[15]L.L. Hench, "Physical Chemistry of Glass Surfaces," Proceedings of the XIth congress on glass, Prague, 1977. D. M. Sanders and L. L. Hench, "Mechanisms of Glass Corrosion," *J. Am. Ceram. Soc.*, **56**, [7], 373-377 (1973).

[16]T. A. Michalske and S. W. Freiman, "A Molecular Mechanism for Stress Corrosion in Vitreous Silica," *J. Am. Ceram. Soc.*, **66**, [4], 284-288 (1983).

[17]T. A. Michalske and B. C. Bunker, "Slow Fracture Model Based on Strained Silicate Structures," *J. Appl. Phys.* **56** [10], 2686-2693 (1984).

[18]S. W. Freiman, G.S. White, and E. R. Fuller Jr., "Subcritical Crack Growth in Soda-Lime Silica Glass," *J. Am. Ceram. Soc.*, **68** [2] 108-13 (1985).

[19]G. S. White, D. C. Greenspan, and S. W. Freiman, "Corrosion and Crack

Growth in 33% Na_2O-67% SiO_2 and 33% Li_2O-67% SiO_2 Glasses," *J. Am. Ceram. Soc.*, **69** [1] 38-44 (1986).

[20]C.C. J. Roothaan, "New Developments in Molecular Orbital Theory," *Rev. Mod. Phys.*, **23**, 69-89 (1951). J.A. Pople and D. L. Beveridge, *Approximate Molecular Orbital Theory*, McGraw-Hill, New York, 1970.

[21]W. Wong-Ng, G. S. White, and S. W. Freiman, "Application of Molecular Orbital Calculations to Fracture Mechanics: Effect of Applied Strain on Charge Distribution in Silica," *J. Am. Ceram. Soc.*, **75** [11] 3097-3102 (1992).

[22]R. S. Mulliken, "Electronic Population Analysis on LCAO-MO Molecular Wave Functions. II. Overlap Populations, Bond Orders, and Covalent Bond Energies," *J. Chem. Phys.*, **23**, 1841-46 (1955).

[23]C. G. Lindsay, G. S. White, S. W. Freiman, and W. Wong-Ng, "Molecular Orbital Study of an Environmentally Enhanced Crack Growth Process in Silica," *J. Am. Ceram. Soc.*, **77** [8] 2179-2187 (1994).

[24]G.S. White and W. Wong-Ng, "Molecular Orbital Calculations of Environmentally Enhanced Fracture in Silica", in Fracture Mechanics of Ceramics, Vol. 12, Fatigue, Composites, and High Temperature Behavior, R.C. Bradt, D.P.H. Hasselman, D. Munz, M. Sakai, and V. Shevehenko, eds, Plenum Press, New York and London (1996) p. 219.

[25]J. K. West and L. L. Hench, "Silica fracture Part I: A ring contraction model," *J. of Materials Science*, **29**, 3601-3606 (1994).

[26]J. K. West and L. L. Hench, "Silica fracture Part II A ring opening model via hydrolysis," *J. of Materials Science*, **29** 5808-5816 (1994).

[27]J. K. West and L. L. Hench, "Silica fracture Part III Five- and six-fold ring contraction models," *J. of Materials Science,* **30** 6281-6287 (1995).

[28]J.K. West and L. L. Hench, "The effect of environment on silica fracture: vacuum, carbon monoxide, water and nitrogen," *Philosophical Magazine A*, **77**, [1], 85-113, (1998).

THE QUANTUM AND FRACTAL NATURE OF BRITTLE FRACTURE

J. J. Mecholsky, Jr.
Department of Materials Science & Engineering
University of Florida,
Gainesville, FL 32611

Jon K. West
Engineering Innovations, Inc.
4300 NW 23rd Ave, Suite 362
Gainesville, FL 32606

Larry L. Hench
Department of Materials
Imperial College of Science, Technology and Medicine,
London SW7 2BP, UK

ABSTRACT

This paper addresses the apparently disparate views of fracture in brittle materials with the purpose of demonstrating the interrelationship of the fracture process at all length scales. Experimental measurements have shown that the fractal dimension is directly proportional to the fracture energy, γ_c, during the fracture process, i.e., $\gamma_c = \frac{1}{2} E a_0 D^*$ where E is the elastic modulus, a_0 is a structural parameter and D* is the fractal dimensional increment. Molecular dynamics modeling shows that the fractal nature of the fracture surface can be explained by the molecular motion induced during fracture. Quantum mechanical molecular orbital calculations suggests that the most likely reaction pathway to brittle fracture is a series of ring contractions, which are related to a_0 and form the basis for further propagation. Thus, we have identified a reasonable atomic level foundation for the structural parameter, a_0. Based on a comparison of atomic and molecular modeling with macroscopically measured values of D*, we can calculate a_0. The value of a_0 can be related to the strained structure of the materials whether amorphous or crystalline. As a result of the combination of experimentation and modeling, we suggest that the brittle fracture process is a percolation of a series of ring contractions along the crack front, which result in the observed fracture surfaces for several brittle materials.

INTRODUCTION

What is brittle fracture? More importantly, how do we attempt to answer the question? Numerous investigators have attempted to determine the nature of brittle fracture.[1-4] Most agree that the key to understanding rests with the nature of the atomic bond. No one has really been able to identify the exact nature of the bond rupture process in fast fracture. In addition, once the first bond breaks, it is not clear as to the nature of the sequence of events that have become known as the fracture process. We propose that the nature of the fracture process can be determined by the analysis of a combination of techniques both analytical and experimental. First, we present several fractographic observations of fracture, then we analyze previous results of molecular dynamics calculations. Finally, we incorporate a semi-empirical molecular orbital model to determine the details of bond "fracture" at the atomic level. The three approaches are related by fractal geometry. This paper is written to suggest a model of fracture in amorphous silica, single crystal silicon, zinc sulfide and alumina based on specific fracture surface observations, molecular dynamics results and molecular orbital theory and to propose a method for experimentally verifying the model. In addition, we present a suggested methodology for comparing atomic, molecular and macroscopic length scales in the fracture process.

OBSERVATIONS OF FRACTURE AND FRACTAL GEOMETRY

Fractography has been used for years to identify the cause of failure.[1,2] It has been used occasionally to aid in the understanding of the fracture process. However, the potential for determining the sequence of events during the fracture process from fractographic data is great and should continue to be utilized. There are some very useful and comprehensive summaries of fractography in the literature.[3-5] We summarize here the salient features.

Observations of the fracture surface can relate to us something of the process of brittle fracture. There are characteristic features that have been observed on brittle fracture surfaces known by various names depending on the field of study. For example, in the study of inorganic glasses, there are regions of crack propagation that have become known as mirror, mist and hackle followed by macroscopic crack branching (Figure 1). In organic polymers there are mirror regions, mackerel markings and parabolic markings[6]. Despite the difference in names, these features have similar origins.

Ravi-Chandar and Knauss[7] noted the self-similarity of fracture surfaces on PMMA although they did not use the term "self-similar." Poncelet[8] showed that the "mirror" region on glass surfaces was not smooth, but rather showed features which appeared to be similar in geometry to mist features. The characteristic features on brittle fracture surfaces contain information about the stress at failure, the size and type of defect causing fracture and the nature of the fracture process[9]. These characteristic features appear on brittle polymers, inorganic glasses, glass ceramics, polycrystalline ceramics and single crystals[3]. Single crystal silicon has features reminiscent of the hackle regions in glasses[10]. The hackle features are self-similar to other features on the surface and are scale invariant, as well, i.e., magnified regions appear the same as smaller magnified regions. Of course, macroscopic crack branching can occur as well. Sakai *et al.*[11] and Mecholsky *et al.*[12]showed that the branching traces can be represented by a fractal pattern.

It is apparent from observations of the fracture surface over the years that there are features which appear self-similar and scale invariant[1 3]. The significance of this scaling behavior is that by using fractography, fracture mechanics principles, and fractal geometry, these features can be related to other properties such as fracture toughness, elastic modulus and atomic bond lengths. This knowledge can help describe the creation of the surface during bond rupture and crack propagation.

Observations of the fracture surface have shown that linear elastic fracture mechanics (LEFM) is a continuum approach that provides a predictive capability and quantitatively explains the Griffith approach to fracture[14-16]. Griffith hypothesized that the observed low strengths in brittle materials were due to the existence of flaws, i.e., cracks[17]. The analysis is based on the fundamental equation of fracture mechanics which relates the stress at failure, σ_f, and the size of the fracture initiating (sharp) crack, c, to the toughness of the material in the form of the critical stress intensity factor, K_{IC}, or critical fracture energy, γ_c :

$$K_{IC} = Y\,\sigma\,(c)^{1/2} = \sqrt{2E\gamma} \tag{1}$$

where Y is a geometric and load factor which accounts for the shape of the fracture- initiating crack and the loading condition, and E is the elastic modulus.

The stress at failure can be related to the radius of the mirror-mist, r_1, mist-hackle, r_2, or crack branching, r_3, boundaries measured from the origin of fracture through a fracture mechanics-type equation :[3,18]

$$K_{bj} = Y'\,\sigma\,r_j^{1/2} \tag{2}$$

where K_{bj} are the crack branching stress intensities associated with each of the boundaries, r_j, respectively, i.e., j = 1,2 or 3, and Y′ is a geometric and load factor similar to that in Eq. 1, but can have a different value. Notice that the existence of Eqs (1) and (2) implies that the crack-size-to-mirror size ratios are constant, i.e., $c/r_j = A_j$. This relationship is quite useful, because if an origin is difficult to identify, the equivalent semi-circular crack size can be determined from the size of the mirror boundary.[9,19] It is really the effective size, c, that affects strength and fracture behavior.

Studies in brittle fracture have suggested that the nature of the tortuosity that results from the fracture process can be explained using fractal geometric techniques[13,20-23]. In other words, this fact implies that fracture is a fractal process, which in turn implies that there is a self-similar component to the fracture process, as observed[10,24], that fracture is scale invariant, also observed[24], and that the fracture process and fracture surface can be described by a fractal dimension. Several studies have identified fractal dimensions for fracture surfaces[22,24-27]. Atomic force microscope studies have shown that the fracture surface is not smooth at the atomic level for many materials and has the same tortuosity at the atomic level as at the macroscopic level[28,29]. The sum total knowledge concerning the fracture process is descriptive and useful, but does not answer the fundamental question as to the nature of bond "fracture" and subsequent fracture surface structure. In order to more fully understand the fracture process, selective modeling is needed.

The atomic structure at the crack tip ultimately leads to the fractal nature of the fracture surface[20,21]. A relationship between basic materials constants, fractal geometry, and fracture topography has been shown to fit many experimental data [1,13,21,22].

$$\gamma_c = \gamma_0 + \frac{a_o D^* E}{2} \tag{3}$$

where
γ_c = fracture energy
γ_0 = fracture energy corresponding to $D^* = 0$.
D^* = fractal dimensional increment
E = Young's modulus
a_o = structural materials constant

All of these variables are independently measurable except a_o. The interpretation of a_0 is not complete. However, in Table I we see that a_0 appears to be related to the strained structure during fracture.

Table I. Characteristic a_0 Values for Classes of Materials

Material Class	D*	a_0 (Å)	Interpretation
Single Crystals	0.04 – 0.1	1 – 10	stretched bonds
Glasses	0.8 – 0.12	10 – 20	free volume
Glass Ceramics	0.5 – 0.33	20 – 60	glass-crystal clusters

MD MODELING AND EXPERIMENTS

Modeling affords us the luxury of performing "experiments" at the atomic scale in a relatively easy manner. Actual experiments in which observations can be made at the atomic scale are extremely tedious and in most cases cannot be controlled in all aspects, i.e., control of stress level, crack size, environment, as well as precise measurements. Thus, model "experiments" are very convenient if there is a way of verifying the results with actual experimental data. Previous molecular dynamics models of fracture in amorphous silica[30] and single crystal silicon[31] have shown that fracture can be reasonably explained by the location of the free volume in the structure. The simulations were able to be verified experimentally because fracture surfaces were constructed using the models and the fractal dimension determined[31]. The fractal simulation of the fractured surfaces matched experimentally determined fractal dimensions for the same materials (Table II). Thus, these comparisons provided confidence that the modeling was, at least, approximately correct. However, even with this encouragement, the actual mechanism of bond rupture was still unknown because molecular dynamics models the forces between atoms and ignores the electronic details. Molecular orbital modeling can provide the precise arrangement of atomic configurations accounting for the atomic structure. Problems in the past with modeling is that there has been no easy way of experimentally verifying the proposed results.

Table II. Fractal Dimensions for Experimental & Simulated Fracture Surfaces

Material	Slit-Island Measurement	AFM Measurement	MD Simulation
silica glass	2.10 ± 0.02	2.11 ± 0.01	2.11 ± 0.02
silicon			
{100}	2.16 ± 0.04	2.19 ± 0.02	2.16 ± 0.06
{110}	2.10 ± 0.04	2.11 ± 0.02	2.11 ± 0.05
{111}	2.06 ± 0.02	2.05 ± 0.01	2.09 ± 0.04

MO MODELING AND EXPERIMENTS

The way we attempt to answer the question: "What is brittle fracture?" is by modeling the fracture process at the atomic level using molecular orbital theory, at the molecular level by incorporating the molecular dynamics results and at the macroscopic scale using previous fracture mechanics and finite element analyses. The key requirement in this modeling is that there are techniques available for experimentally verifying the results. Fractal analysis provides the tool for this comparison. We present here a methodology by which fracture can be described at all levels of observation from the atomic to the macroscopic.

A quantitative model for the initial step in the fracture process in brittle materials has been developed using AM1 (Austin Method 1) semi-empirical molecular orbital theory [32-35]. A set of silica rings were initially optimized to find the ground state geometries, and heats of formation. The set of rings included a 3-member, 4-, 5-, and 6-member ring wherein the silica tetrahedra was the unit member. Each tetrahedra in the ring consisted of two non-bridging oxygens (NBO) terminated with hydrogens forming silanols and two bridging oxygens connecting neighboring silica tetrahedra [36,37].

Each of these rings were then strained in 0.01 Å increments to observe their response to the attempt at breaking a bridging Si-O bond. A very unexpected result occurred. Each of the rings condensed or contracted into the next smaller ring with a single silica tetrahedra being extracted in the direction of the strain (cf. Figure 2). Note that there is not a smooth transition from one state to another, but rather a "jump" from one configuration to another (lower energy) configuration. AM1 UHF PRECISE SADDLE (transition state) calculations on this type of bond reconfiguration event clearly showed that formation of a pentacoordinate silica as the transition state. The energy barrier to this type of bond conformation was found to be only +26.1 kcal/mol for the 4-member ring in a vacuum which is approximately one quarter that of a Si-O bond. The 3-member ring has the highest barrier to bond reconfiguration via ring contraction at +62.4 kcal/mol and the 5-membered ring has the lowest at +17.0 kcal/mol. The 6-member ring has an intermediate value of +40.6 kcal/mol. The effective or average energy barrier to bond conformation was estimated using the Bell and Dean[38] distribution of ring sizes in amorphous silica (a-silica). Thus, a weighted average barrier could be determined for bond reconfiguration in vacuum.

In Figure 2, the diameter of the basic ring structure before bond reconformation is designated as "a" while the strain fraction after fracture via ring contraction is defined as "ε". It is then hypothesized that the atomic level materials constant is the ratio of a/ ε. Using the weighted average values (based on the Bell and Dean distribution) yield a/ ε as 11.1 while a_o, determined experimentally from equation 3, is 10.0 for a-silica.The value of a/ ε was then determined for the fracture of single rings via ring contraction (not weighted averages) for alumina, silicon, and zinc sulfide. Figure 3 shows the relationship between these theoretical, a/ ε, values (indicated as calculated a_0) as a function of the materials constant a_o. The slope is slightly greater than 1 with a correlation coefficient of 89.4%. The results support the conjecture that the fractal nature of fracture surfaces originates from the atomic structure at the crack tip.

FRACTURE PROCESS

With the application of molecular orbital calculations, previous determinations from MD simulations[30,31] and determinations of experimental fractal analyses[20-22,39-41,43], we are prepared to suggest the mechanism and process of brittle fracture. When external strain is applied to the tip of a crack, individual groups of atoms are strained in such a way, based on their structure and bond energies, as to minimize the energy necessary to change their configuration to a more stable structure. This change leads to a creation of space, i.e., free volume, at the tip of the crack known as crack advancement. This advancement is not random and follows a path dictated by the least energy path afforded by the time scale. Collectively these advancements form a discontinuous crack front, which is related to the atomic structure and atomic vibrations and propagates out of the plane of the crack with increasing energy and increasing deviations from the "plane" of the crack. The crack keeps advancing as long as there is energy supplied to increase the fracture surface. The energy is supplied so fast that the stability of the vibrations of the molecules become chaotic and eventually macroscopic crack branching occurs. The distance to crack branching is controlled by the near field strain at the tip of the crack[42] and is proportional to the crack size[20]. The energy stored before fracture is proportional to the stress squared at fracture, which, in turn, is inversely proportional to the crack size. Thus, for small cracks the distance to branching is small and there are more branches than for a large crack (for the same size of material). Brittle fracture is a series of events related to the free volume of the structure, which due to vibrations and lowest energy configurations develops into a pattern. The pattern that develops is started at the tip of the crack on the atomic scale with ring contractions based on minimum energy. These ring contractions occur individually all along the crack "front". In some locations the contractions stabilize and in other locations two or more rings supplement each other and become larger in a fractal manner, i.e., they percolate through the structure. These enlarged groups, in turn, annihilate and supplement other similar groups to subside and grow, respectively. The groups that grow do so in a self-similar manner and produce the structure we observe as the fracture surface. At some point, more energy is available than necessary and the crack branches based on energy levels and geometric atomic structure.

At the atomic level we can use quantum mechanics to describe the fracture process as a ring contraction process all along the crack front dictated by minimum energy and availability of free volume. The models of this process can be verified experimentally by comparing a/ε to a_o. On the molecular scale, MD modeling describes the creation of the fracture surface and is compared to experimental observations by a comparison of the fractal dimensions. Finally, on the macroscopic scale, continuum (fracture) mechanics can be used to identify many macroscopic quantities to be verified using fracture mechanics, fractography and fractal analysis, i.e., Eq. 3, cf. Figure 4.

CONCLUSIONS

The mechanism and process of brittle fracture has been suggested based on a combination of fracture surface observations, molecular dynamics results and molecular orbital calculations. The relationship between fracture toughness, fractal dimension and atomic (fracture) structure can be related at many length scales. Future fracture studies should include both modeling (analysis) and experimental tests in an interactive manner. Fractal geometry offers a very useful tool for comparing models of fracture with experimental results and in analyzing experimental details.

ACKNOWELEDGEMENTS

The authors acknowledge very useful discussions with Dr. Dann Passoja as well as thank him, Dr. George Onoda and Dr. Steve Freiman for reading the manuscript and making useful suggestions. The authors also gratefully acknowledge the support of the Air Force Office of Scientific Research (Grant No. F49620-95-0382), the National Institutes of Health (NIH Grant No. DE09307-07), the Quantum Theory project at UF and CAChe Scientific, Oxford Molecular Group, Inc., Beaverton, Oregon.

REFERENCES

[1] J. J. Mecholsky, Jr., "Quantitative Analysis of Fracture Surfaces", Cer Trans, Am. Cer. Soc. (1992).

[2] e.g., F. W. Preston, "A Study of the Rupture of Glass," J. Soc.of Glass Tech. 10 234-269 (1926).

[3]R. W. Rice, "Ceramic Fracture Features, Observations, Mechanisms and Uses," in Fractography of Ceramic and Metal Failures, ASTM STP 827, J. J. Mecholsky, Jr. And S. R. Powell, Jr., eds. American Soc. For Testing and Materials, Phila. PA, 1 - 100 (1984).

[4]V. D. Frechette, *Fractography of Glass and Ceramics*, The American Ceramic Soc. (1990).

[5] G. D. Quinn, J. J. Swab and M. J. Slavin, "A Proposed Standard Practice for Fractographic Analysis of Monolithic Advanced Ceramics," MTL TR 90-57, Army Materials Laboratory, NTIS Access No. ADA-231989 (1990).

[6] L. Plangsangmas, J. J. Mecholsky, Jr. and A. B. Brennan, "Determination of Fracture Toughness of Epoxy Using Fractography," J. Appl. Polymer Science72, 257-268 (1999).

[7] K. Ravi-Chandar and W. G. Knauss, "An Experimental Investigation into Dynamic Fracture: II. Microstructural Aspects," Inter. J. Fracture, 28 65-80 (1984).

[8] E. F. Poncelet, "The Markings on Fracture Surfaces," Trans. Soc. Glass Tech., 52 279-288 (1958).

[9] J. J. Mecholsky, Jr. "Quantitative Fractography: An Assessment" in Ceramic Trans. 17, Fractography of Glasses and Ceramics II. Frechette and Varner eds., Am. Cer. Soc., Westerville, OH, 413-451 (1991).

[10]Y. L. Tsai and J. J. Mecholsky, Jr., "Fractal Fracture of Single Crystal Silicon," J. Mater. Res. 6, 1248 - 1263 (1991).

[11]T. Sakai, M. Ramulu, A. Ghoshand R. C. Bradt, "A Fractal Approach to Crack Branching (Bifurcation) in Glass, in Ceramic Transaction 17 op. Cit. (1991).

[12]J.J. Mecholsky, Jr., R. Linhart, B. D. Kwitkin and R. W. Rice "On the Fractal Nature Crack Branching In MgF_2" J. Materials Res.13 ,11 (1998).

[13] J. J. Mecholsky, Jr., T. J. Mackin and D. E. Passoja, "Self-similar Crack propagation in Brittle Materials," Adv. In Ceramics 22: Fractography of Ceramics and Glasses Varner and Frechette, eds.,The Am. Ceramic Soc (1988).

[14] P. N. Randall, "Plain Strain Crack Toughness Testing of High Strength Metallic Materials", in ASTM STP 410, pp. 88-126 (1967).

[15]ASTM Annual Book of Standards, E616, p.674, The American Society of Testing and Materials, 1916 Race St., Phila. PA 19103 (1984).

[16]I. Bar-on, Applied Fracture Mechanics, pp. 645-651 in Engineered

Materials Handbook (Ceramics and Glasses) ASM International (1991).

[17] A.A. Griffith, Trans. Roy. Soc. London, A221 (1921) 163.

[18]S. W. Freiman, J. J. Mecholsky, Jr., and P. F. Becher, "Fractography: A Quantitative Measure of the Fracture Process," in Ceramic Transactions 17, op. Cit., 55-78 (1991).

[19]D. A. Krohn and D. P. H. Hasselman. J. ACerS 57 440 (1974).

[20] J. J. Mecholsky, Jr. And S. W. Freiman, " Relationship Between Fractography and Fractal Geometry,"J.ACerS. Soc.74[12]3136-38 (1991).

[21]J.J. Mecholsky, Jr., "Fractography, Fracture Mechanics and Fractal Geometry: An Integration," Ceramic Trans., 64, Fractography of Glasses and Ceramics III, Varner, Frechette and Quinn, eds, pp. 385-393, 1996.

[22]J. C. Russ, "Fractal Surfaces," Plenum Press, NY, 1996.

[23]J. J. Mecholsky, Jr. , D. E. Passoja and K. C. Fineberg-Ringel, "Quantitative Analysis of Brittle Fracture Surfaces Using Fractal Geometry, J. Am. Ceramic Society (1990).

[24]B. B. Mandelbrot, D. E. Passoja and J. Paullay, Fractal Character of Metals Nature 308:721-722.

[25]B. Dubuc, J. F. Quiniou, C. Roques-Carmes, C. Tricot and S. W. Zucker, "Evaluating the Fractal Dimensions of Profiles", Phys. Rev. A39(3) 1500-1512(1989).

[26]J. J. Mecholsky and T. J. Mackin Fractal Analysis of Fracture in Ocala Chert J. Mater. Sci Ltrs. 7, 1145-47 (1988).

[27]P. A. Scott, T. Engelder and J. J. Mecholsky, Jr., "The Correlation Between Fracture-toughness Anisotropy and Crack Surface Morphology in the Ithica Formation, Appalachian Basin," in Fault Mechanics and Transport Property of Rocks, ed. B. Evans and T. Wong, Academic Press NY 341-370 (1992).

[28]D. M. Kulawansa, L. C. Jensen, S. C. Langford and J. T. Dickinson. STM Observations Of The Mirror Region Of Silicate Glass Fracture Surfaces. J. Mater. Res. 9 [2] 476-485 (1994).

[29]E. Guilloteau, H. Arribart and F. Creuzet. Fractography Of Glass At The Nanometer Scale.Mater.Res. Soc. Symp. Proc.409,365-370 (1996).

[30]J. H. Simmons, T. P. Swiler and R. Ochoa, J. Non-Crystalline Solids 134, 179-182 (1991).

[31]Y.L. Tsai, T.P. Swiler, J.H. Simmons and J. J. Mecholsky, Jr., "Fractal analysis of Molecular dynamics Simulations of Brittle Fracture," in Computaional Modeling of Materials and Processing, Eds., J.H. Simmons, E.R. Fuller, Jr., A.L. Dragoo and E.J. Garboczi, Ceramic Transactions V., 69, American Ceramic Soc., Columbus, OH, (1995).

[32]MOPAC (Computer Program) Version 3.8, CaChe Scientific, Beaverton, Oregon.

[33]Jon K. West and Larry L. Hench, "Silica Fracture: Part I: A Ring Contraction Model," J. Materials Science, 29 (1994) 3601-3606.

[34]Jon K. West and Larry L. Hench, "Silica Fracture: Part III: 5-Fold and 6 Fold Silica Ring Contraction Models,"J.Mat'ls Sci.,30 (1995)6281-6287.

[35]J. K. West and L. L. Hench, "The Effect of Environment on Silica Fracture:Vacuum,CarbonMonoxide,WaterandNitrogen,"Phil.Mag. (1997).

[36]J. K. West and S. Wallace, "Interactions of water with trisiloxane rings. Theoretical analysis", J. Noncryst. Solids, 152 (1993) 109-117.

[37]L.L. Hench and J.K. West, "Molecular Orbital Models of Silica", Annual Reviews of Materials Science, 25 (1995) 137-68.

[38]R.J. Bell and P. Dean, Phil. Mag., 25 (1972) 1381.

[39]C. Y. Poon, R. S. Sayles and T. A. Jones," Surface Measurement and Fractal Characterization of Naturally Fractured Rocks, J. Phys. D:Appl. Phys. 25 1269-1275(1992).

[40]A. V. Neimark, "Characterization of Self affinity in the global region" Phys. Rev B 50 [20]435-438 (1994)

[41]E. Bouchaud, G. Lapasset, J. Planes and S. Naveos, "Statistics of branched fracture surfaces" Phys. Rev. B (1995).

[42]H. P. Kirchner and J. C. Conway, Jr.,"Fracture Mechanics of Crack Branching in Ceramics," in Fractography of Glass and Ceramics, Advances in Ceramics, Vol. 22, The American Ceramics Society, Westerville, OH, pp. 187-213 (1988).

[43]D. E. Passoja, Fundamental Relationships Between Energy and Geometry in Fracture. Adv. In Ceramics 22 Fractography of Glasses and Ceramics, ACerS (1988).

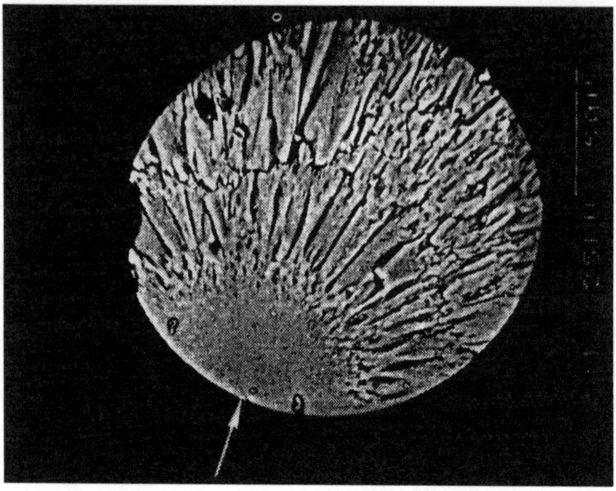

Figure 1 – Typical Fracture Surface Showing Mirror, Mist and Hackle Regions. The arrow indicates the location of the fracture-initiating crack.

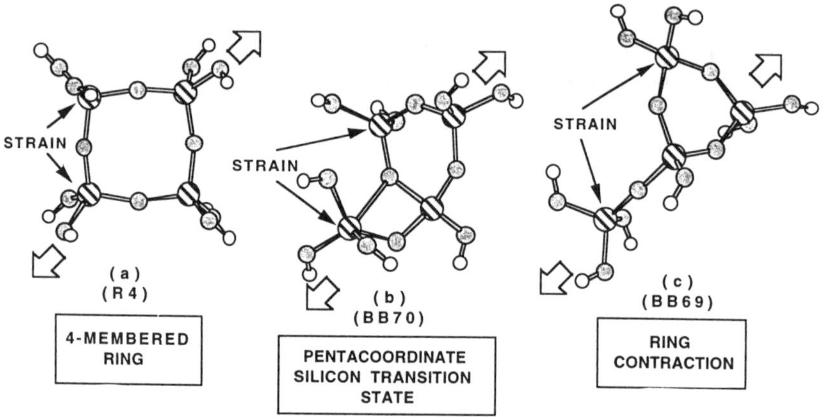

Figure 2 – Suggested Model for the Strained Structure That Leads to Reconformation During the Fracture Process. The original "ring" structure size is "a" and the strained bond change from (a) to (c) is given by c'– c. The energy barrier in the pentacoordinate transition state represents the minimum energy of all the possible configurations.

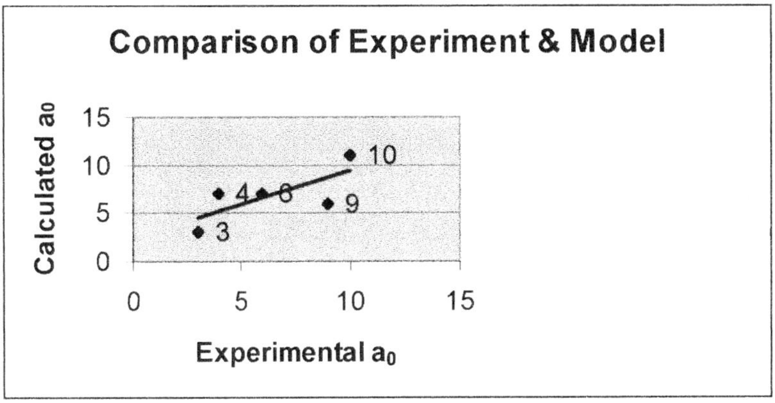

Figure 3 - Linear Relationship Between Experimental and Calculated Values of a_0. Numbers in figure indicate values of a_0 in Å; i.e., silica glass = 10 Å; Si (110){100} = 6 Å; Si (100){110} = 4 Å ; Si (111){110}=6 Å; alumina = 3 Å; and zinc sulphide = 9 Å.

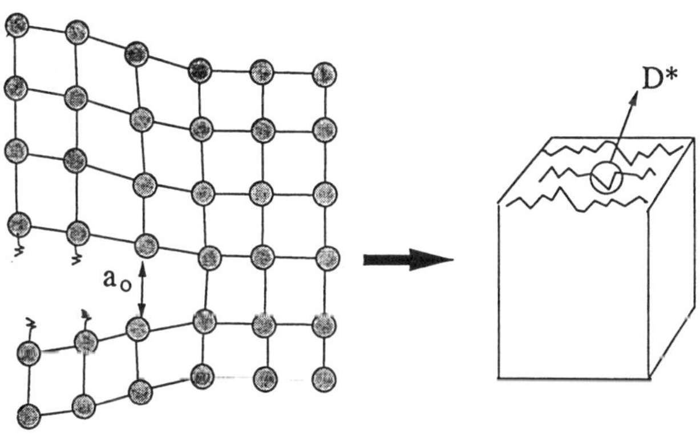

$$\gamma_f = \gamma_o + Ea_oD^*/2$$

Figure 4 – Schematic of Bond Breaking Process at Different Length Scales. The structure on the left represents the atomic scale, characterized by a_0. The structure on the right represents the macroscopic process, characterized by D* . Eq. 3 is shown.

MICROWAVE ABSORPTION BY MATERIALS: THEORY AND APPLICATION

J.K. West
Engineering Innovations, Inc.
P.O. Box 147050
Gainesville, FL 32614-7050

D.E. Clark
Department of Materials Science and Engineering
University of Florida
Gainesville, FL 32611

ABSTRACT

The absorption of microwave energy by ceramic materials is an elusive topic. The advantages of microwave processing indicate that this is one of the most important new directions in materials science, however, the theory behind the physical mechanisms has been lagging. This study combines quantum molecular orbital theory with the theory of the coupled oscillator to indicate a possible microwave absorption mechanism. The AM1 semi-empirical quantum molecular orbital (MO) models of water, silica, and silicon carbide are presented to support the hypothesis that coupled molecular vibrations absorb microwave energy directly. The Quantum coupled Oscillator Model appears to predict the primary microwave absorption features of these materials.

INTRODUCTION

Materials processing with microwaves has become very important.[1] However, the mechanism of microwave absorption in a dielectric solid are uncertain at best. Most atomic and ionic polarization mechanisms are at too high a frequency to couple directly with the microwave energy.[2] Resonance cannot be the mechanism for direct absorption of microwaves by infrared vibrations. All infrared vibrations, oscillations and rotations are at frequencies thousands of times higher than microwave energy. The absorption of energy by any oscillator from an external source is basically zero if the difference in frequencies (driver and driven) is greater than 3 or 4 times. As a result, in this study we have directed our attention to oscillations that are associated with specific structures that should have identical vibrations but do not. They are degenerate vibration states. Microwave absorption by these degenerate states may be modeled by the secondary frequency of a coupled oscillator. The coupled oscillator could be formed from two earby defect structures on the surface of a particle or in the grain boundaries of ceramics. Microwaves may resonate with the secondary frequency of coupled oscillators which can be 1000's of times lower than the infrared frequencies.

The frequencies of the secondary vibrations (w^- in Figure 1) of coupled oscillators are typically orders of magnitude lower than the primary frequencies (w_1 and w_2). This effect could conceivably shift the secondary frequency into the microwawve region of the electromagnetic spectrum. In order to find prospective sites for coupling to occur, molecular models were constructed for water, steam, silica, and silicon carbide. For these structures, the infrared (vibrational) spectra were calculated and a computer search for coupling effects was made using a microwave operator, $\theta_{\mu\omega}$. The result of this search yielded a qualitative match to known microwave absorption features for these materials.

The coupling between defects occurs through the bulk material and only slightly perturbs the natural frequency of the defect. The authors acknowledge that this hypothesis must be tested thoroughly and that this study only represents the initial findings of this model.

THE COUPLED OSCILLATOR

A schematic of the mechanically coupled oscillation[3], the basis of the proposed model for microwave absorption, is shown in Figure 1. The primary frequency of the mass, m_1, is given by w_1 and the coupled frequency, w^-, for energy transfer between the two masses. The solution of the two simultaneous second order differential equations yields two solutions or modes w^+ and w^- where (Simon pp. 194)[3]

$$X_1 = \tfrac{1}{2} A (\cos w_1 t + \cos w_2 t) \tag{1}$$

and

$$X_2 = \tfrac{1}{2} A (\cos w_1 t - \cos w_2 t) \tag{2}$$

with two identical oscillators loosely coupled where

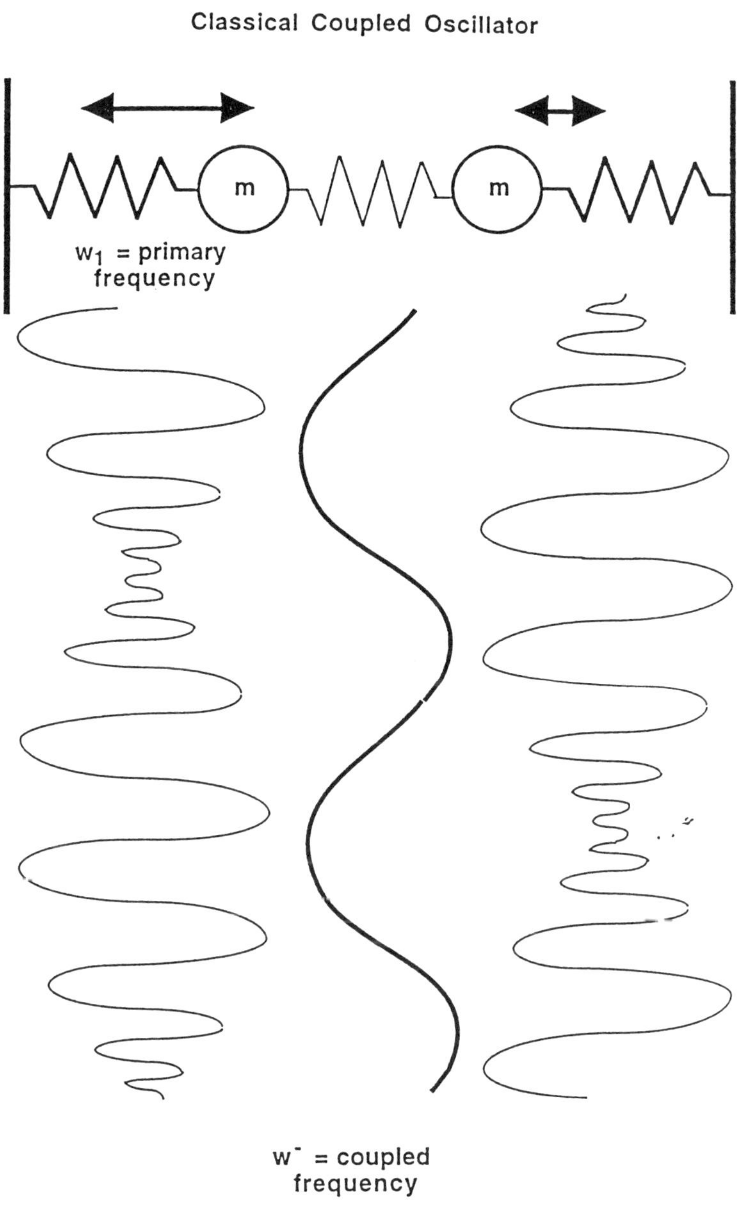

Figure 1
Coupled Harmonic Oscillator

$$w^+ >> w^- \tag{3}$$

Then,

$$w^+ = (w_1 + w_2)/2 \tag{4}$$

$$w^- = (w_1 - w_2)/2 \tag{5}$$

Solving for w^- we obtain

$$2w^- = w_1 - w_2 = \Delta \tag{6}$$

which is the frequency that may allow direct microwave absorption for which

$$X_1 = A \cos (w^- t) \cos (w^+ t) \tag{7}$$

and

$$X_2 = -A \sin (w^- t) \sin (w^+ t) \tag{8}$$

THE MICROWAVE OPERATOR

The formal expression of this hypothesis is the proposed existence of a microwave operator, $\theta_{\mu\omega}$. From Schrodinger s equation,

$$H_{vib}\, \Psi = E_{vib}\, 2 \tag{9}$$

where H_{vib} is the vibrational Hamiltonian, Ψ is the wavefunction and E_{vib}, is the vibrational eigenvalue matrix, it is proposed that

$$\theta_{\mu\omega}\, H_{vib}\, \Psi = \theta_{\mu\omega}\, E_{vib}\, \Psi \tag{10}$$

yields

$$= E_{\mu\omega} \Psi \tag{11}$$

where $E_{\mu\omega}$.is the eigenvalue matrix for microwave absorption by a material. $E_{\mu\omega}$.has the form: (equation 12)

$$E^j_{\mu\omega} = \frac{1}{2}\left(E^{j+1}_{vib} - E^j_{vib}\right) \tag{12}$$

and the oscillator strengths, $X_{\mu\omega}$, are estimated using: (equation 13)

$$X^j_{\mu\omega} = X^j_{vib} \tag{13}$$

for identical oscillators of equal masses being weakly coupled through the bulk of a materials or via hydrogen bonding-like interactions.

In order to evaluate $\theta_{\mu\omega}$ and $X_{\mu\omega}$, the vibrational eigenvalues and oscillator strengths, E_{vib} and X must be determined for each of the materials being studied. In this study, semi-empirical quantum molecular orbital methods were used.

MOLECULAR ORBITAL METHODS

We developed the molecular models using a CAChe work system, Oxford Molecular Group, Beaverton, OR, that provides an interface to several semi-empirical methods including ZINDO[4] (Professor M.C.Zerner's Intermediate Neglect of Differential Overlap program) through a Macintosh environment as well as MOPAC[5] that provides an interface to MNDO, AM1 and PM3.

The semi-empirical Austin Method (AM1)[6-9] used in this study is able to calculate molecular structures, heats of formation, transition state energies, and IR spectra. A comparison of the geometries and spectra of silica rings using both AM1 and PM3 shows that both models represent the geometries and infrared spectra better than previous MNDO IR spectra and geometries.[10,12] AM1 appears to represent the bulk distribution of silica rings slightly better than PM3[12] while PM3 calculates slightly better bond lengths for the C-H. AM1 calculates a closer C=C bond length, Table I.

As a result, the geometry for the pentasiloxane chain used as the first structure in this study was optimized using AM1 Unrestricted Hartree-Fock (UHF). This UHF AM1 optimized geometry was then submitted through MOPAC to a normal mode analysis (FORCE) calculation to determine the IR spectrum and vibrational eigenvalues.

Both AM1 and PM3 methods yield force constants between all neighboring atoms which are converted to a vibrational spectrum using a FORCE, normal mode analysis, calculation within MOPAC. For details see Stewart[12,14]. Thus, the frequency of vibrational transitions are predicted by the calculations that compute the force necessary to distort atom pairs in the molecular cluster from their equilibrium geometries. The intensities of the infrared active transitions (oscillator strength of the changing dipole) are plotted. The vibrational spectra obtained in this manner are idealized, gas phase spectra and, therefore, are only an approximation. Molecular transitions which are forbidden by symmetry in the MO models are present in experimental spectra of the condensed phase. Anharmonicity effects, overtones, or combination transitions are also absent from the theoretical spectra and therefore will be absent from the microwave spectra. Thus, the calculated IR and microwave spectra are sharp, discrete lines whereas experimental modes are broadened by rotational levels, instrument optics, molecular motions, and other solid state effects. Some of these effects are included in the calculated spectra by introducing a Gaussian-like distribution for

Table I
Comparison of Bond Lengths (Å)

MOLECULE	BOND	EXPER.	AB INITIO	ZINDO	AM 1	PM 3
Nitrogen	N-N	1.094	1.104	1.149	1.106	1.098
Carbon Monoxide	C-O	1.128	1.101	1.196	1.171	1.135
Ethylene	C-H C=C	1.086 1.338	1.08 1.31	1.11 1.31	1.098 1.326	1.086 1.322
Silica (4-ring)	Si-O Si-OH	1.62	1.60	1.689 1.701	1.70 1.73	1.65 1.69

each vibrational mode with a half width of 40 cm^{-1} and a transmittance of 20% for the most intense oscillation. In the microwave spectra the half width of 2 GHz was used.

It is important to note that absolute errors in these quantum calculations can at times be large, especially in the IR peak locations. These calculations have been very successful, however, in showing trends, directions, and shifts in energies or wavelengths[9-15]. In this light, we are looking for changes in structure and differences in energy that correspond to our observations. The coupling that we are seeking occurs between two identical structures that should have the same vibrational frequency. If they have different frequencies then the bulk or connection between the two oscillators has modified either one or both. The differences in vibrational energy, not absolute values is the important feature in our analysis of microwave absorptions.

RESULTS OF THE MODELING

The optimized AM1 PRECISE geometry of a pentasiloxane chain is shown in Figure 2. There are in general two bridging oxygen bonds and two non-bridging oxygen bonds per silica tetrahedra. The non-bridging oxygens are terminated with hydrogens forming silanols. The calculated value of the heat of formation was -1297.7 kcal/mol. More structural details and the complete IR and UV spectra for this molecule are presented elsewhere[15]. Figure 2 also shows a small range in the IR from 100 cm^{1} to 50 cm^{-1} where the Si-OH rocking vibrations occur.

Two absorptions that lie within 1.74 cm^{-1} of one another are labeled w_1 and w_2. These peaks lie at 64.23 cm^{-1} and 62.49 cm^{-1}, respectively. As indicated in Figure 2, these vibrations correspond to silanols that are separated by three silica tetrahedra. All adjacent silanols show a greater separation of their vibrational modes than these two. Half the differencc between w_1 and w_2 would represent a coupled resonant frequency of 26 Gigahertz.

That means that microwaves of 26 GHz would couple with these vibrations. As the material is heated, the coupling force constant should be reduced. That should reduce the difference between w_1 and w_2 and lower the coupling frequency, w^-.

The modes in the 50-100 cm^{-1} range are all Si-OH rocking vibrations. They should be identical but vary with location on the C5 molecule. The modes with the closest frequencies are Si-OH structures on adjacent silicon atoms but on non-adjacent sites. The ncxt closest modes are Si-OH structures on adjacent silicon atoms. Thus, the closer silanols have a stronger coupling and higher coupling frequencies w^- for microwave absorption.

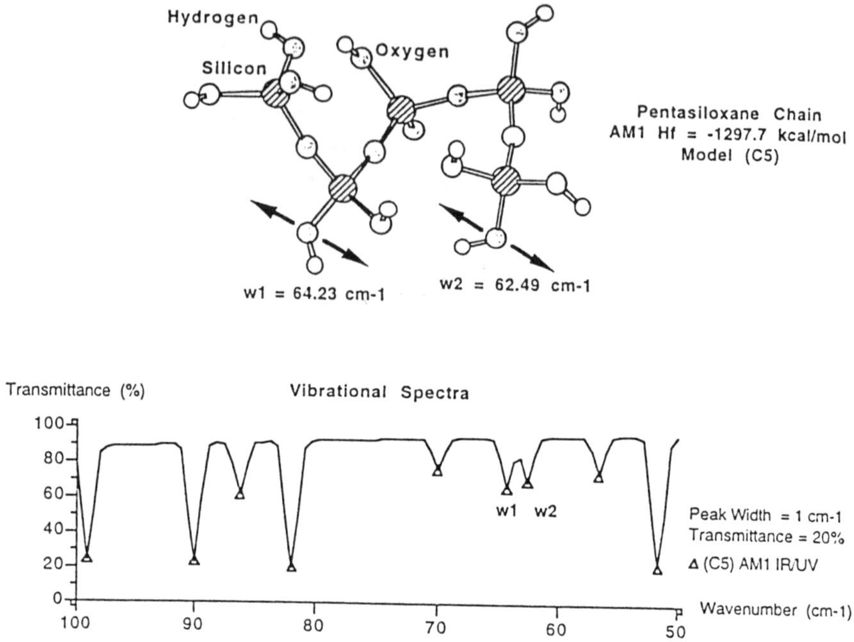

$$\Delta = w1 - w2$$

$$\Delta = 2w-$$

$$w- = \frac{\Delta}{2} \left(30 \text{ "GHz/cm}^{-1}\text{"}\right) = 26 \text{ GHz}$$

Figure 2
Pentasiloxane Chain with Theoretical IR Spectra

This type of mechanism would suggest that grain boundaries and defect structures are the most likely sites for structures such as the pentasiloxane chain. Similar studies on the disiloxane (dimer) show similar results.

MICROWAVE ABSORPTION IN WATER

The entire food industry has made use of microwave energy in one way or another because water absorbs at 2.45 GHz. The mechanism has (until now) remained a mystery and the debate over the "microwave effect" on materials continues[16]. Using the simple model of the coupled oscillator and the Quantum MO spectra of three hydrogen bonded (coupled) water molecules, a theoretical microwave spectra was calculated and is shown in Figure 3. The main theoretical absorption features are found at 2.4, 6.15, 9.45, 19.65, and 42.9 GHz. The absorption at 2.4 GHz is the feature that is the primary microwave heating band used in industry and in the home cooking units. However, the band is not very active at room temperature because the O-H Rocking vibration eigenvalues exist at 1900 cm^{-1} and are only slightly occupied. The H-O-H bending vibrations near 100 cm^{-1} are highly occupied at room temperature and lead to a microwave absorption centered around 6.2 GHz. the width of the peak at 6.2 GHz probably overlaps the 2.4 GHz region and absorption occurs and should increase as the temperature increases because higher temperature states are being occupied. The theoretical features correspond with experimental dielectric measurements on liquid water. Figure 4 shows the experimental Cole-Cole plot for liquid water[17] and the theoretical predictions for microwave absorption plotted as a function of decreasing frequency. The main experimental absorption features are found at 0.58, 1.7, 3.2, 4.6, 9.2, 23.7, and 34.9 GHz. The highest microwave absorptions occur at the highest imaginary dielectric, $\in''$, corresponding to 9.2 and 23.7 GHz. The theoretical model predicts corresponding peaks for the highest microwave absorptions at 9.45 and 19.65 GHz. This is reasonably close to the theoretical microwave absorption calculated for a coupled oscillator.

The same approach yielded a transparent window in the 2.4 GHz band for water vapor (3 isolated water molecules). High frequency absorption still exist for the vapor state as shown in Figure 5 but the oscillator strengths are reduced by more than 50% (see the 20 GHz peak). In a similar analysis on the population of higher energy states, this model of water vapor has no occupied vibrational states at room temperature. This explains why RADAR works and is not absorbed by water vapor (gas) in our atmosphere.

Figures 6 and 7 show the theoretical microwave spectra for 2-member and 5-member silica rings. The structural member of a silica ring is the silica tetrahedra with two bridging oxygen (BO) bonds and two non-bridging oxygen

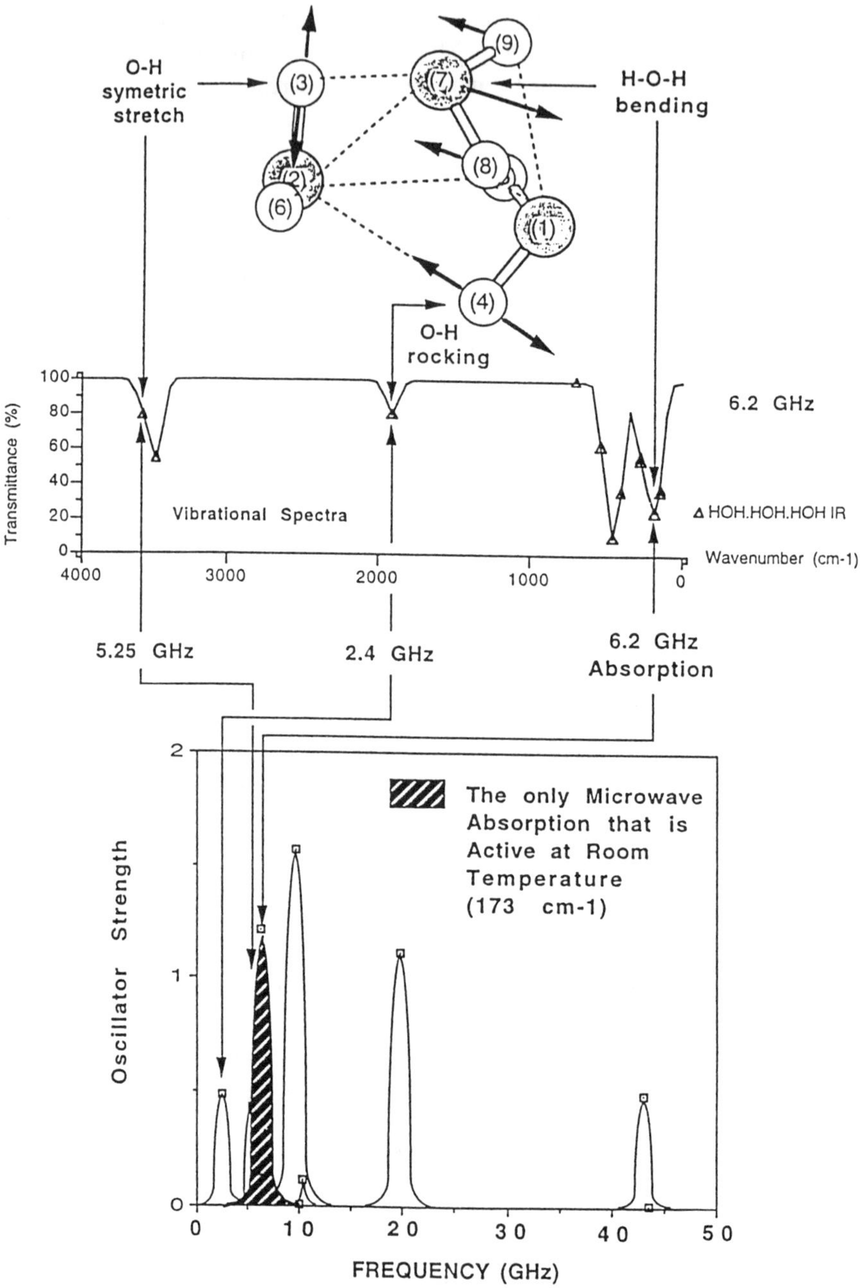

Figure 3
Calculated IR Spectrum for Liquid Water

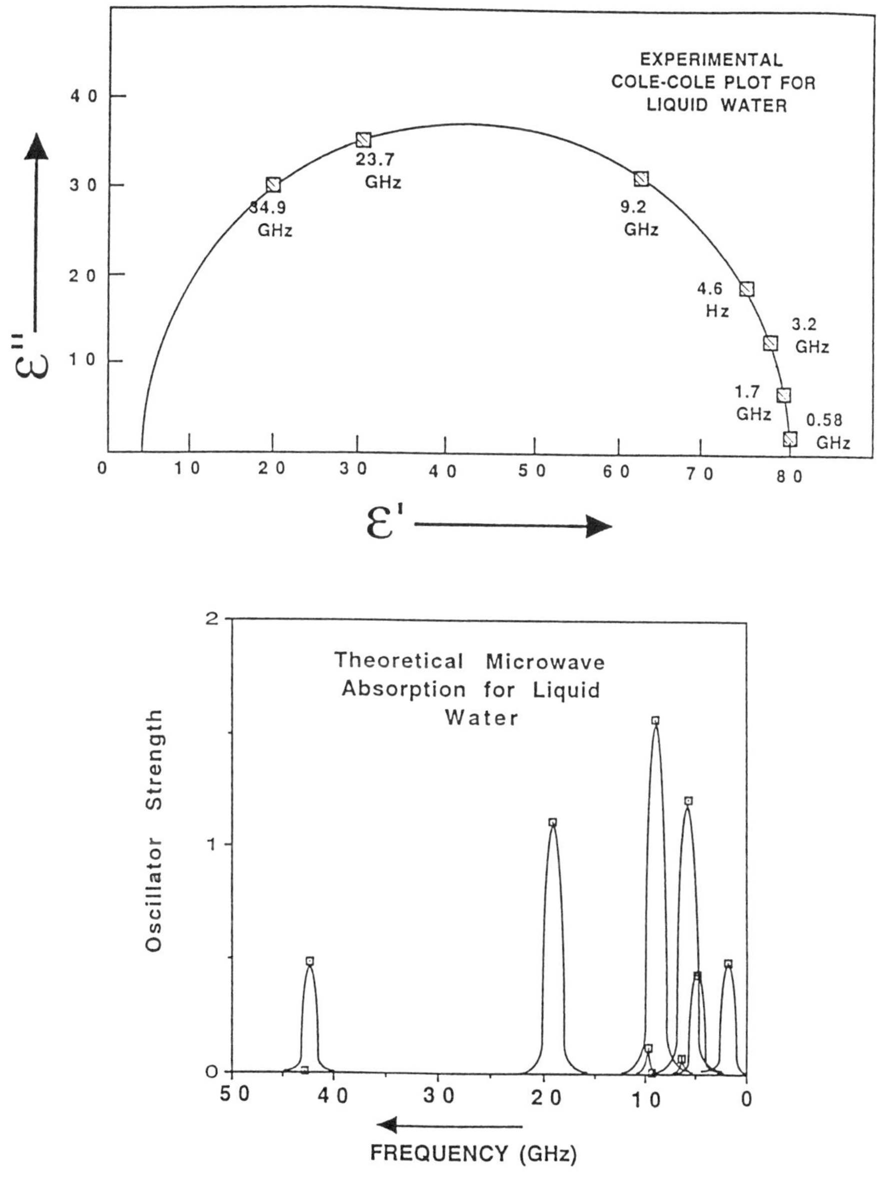

Figure 4
Calculated Microwave Absorption Spectrum for Liquid Water
Compared with the Experimental Cole-Cole Plot for Liquid Water

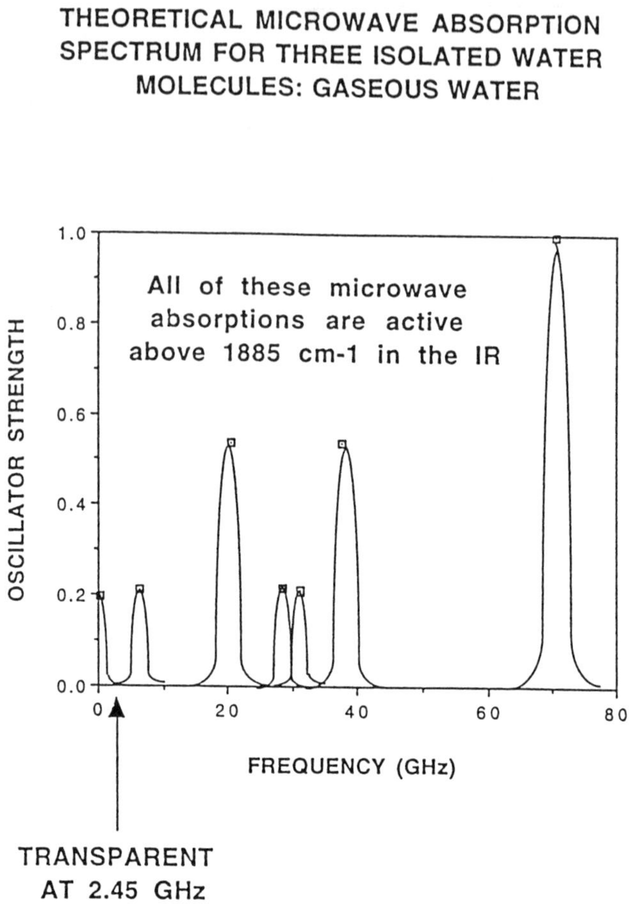

Figure 5
Calculated Microwave Absorption Spectrum for Gaseous Water

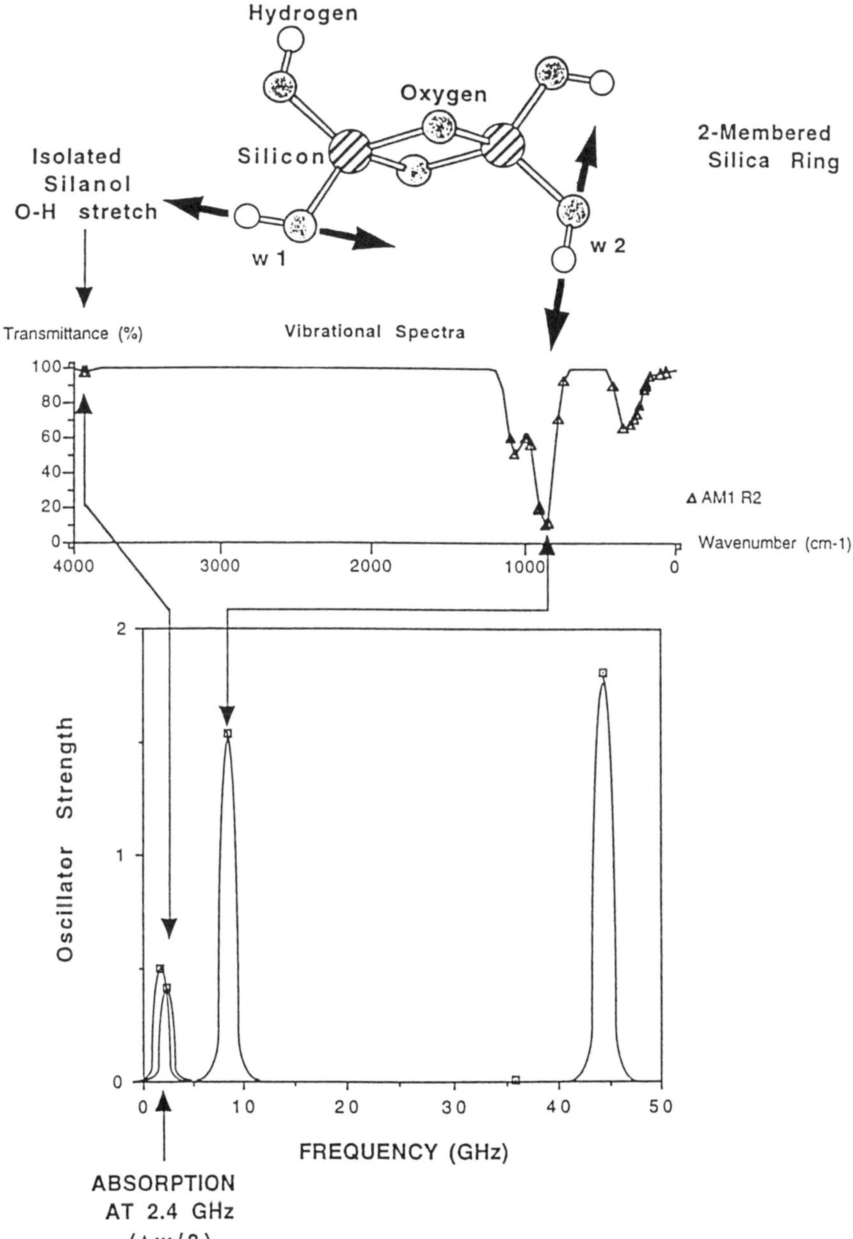

Figure 6
Calculated Microwave Absorption Spectrum for a 2-Member Silica Ring

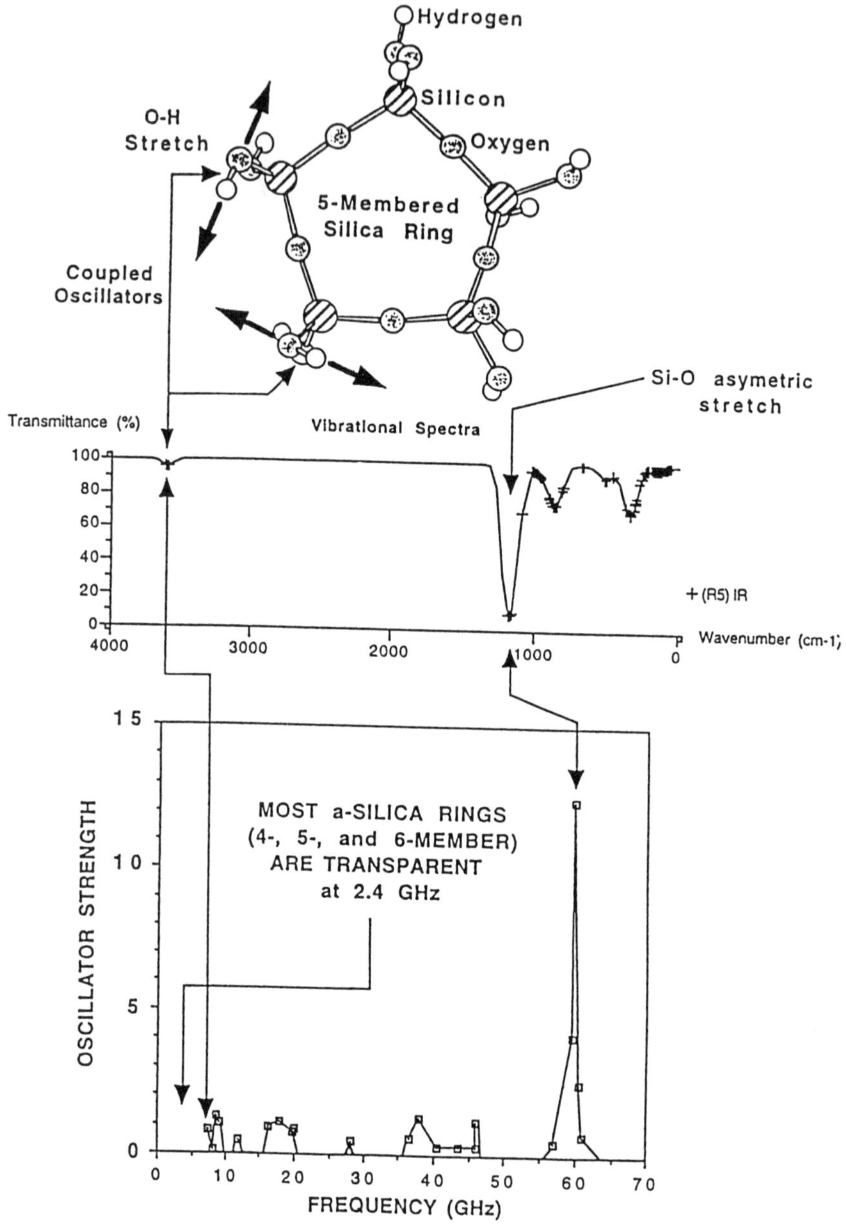

Figure 7
Calculated Microwave Absorption Spectrum for a
5-Member Silica Ring

(NBO) bonds terminated with hydrogens forming silanols. The IR spectra were calculated for each ring then a computer search of each vibrational mode was made to find coupled oscillators. The microwave frequencies were then extracted along with the oscillator strength of the primary vibrations. Figure 6 shows an absorption for 2-membered rings at 2.4 GHz. This indicates that porous sol-gel silica should absorb microwave energy because of the high surface concentration of small rings. In bulk silica, small rings do not exist[18]. Figure 7 shows that a 5-member silica ring is transparent in the 2.4 GHz region. All other silica rings (3-, 4-, and 6-member) show similar theoretical microwave spectra. This matches observational evidence very well for silica glass that we all know is good for use in home microwave ovens[1].

In the case of a 3-membered ring of silicon carbide, a significant absorption band is found in the 2.4 GHz region as shown in Figure 8. Silicon carbide is a known microwave energy enhancer and the absorption mechanism has both real and imaginary components. Silicon carbide is used in small quantities to initiate microwave heating of other less absorptive ceramics[1].

Figure 9 shows the results of contamination of a structure. If a soluble salt (LiCl) is added to water, microwave heating decreases[19]. The temperature of pure water increases more rapidly than does that of the salt water. This effect has also been reported for NaCl salt water and KCl containing foods[20]. However, if the frequency is increased above 2.45 GHz, the heating rate increases[19]. This effect is predicted by the theoretical model. At 2.45 GHz, LiCl salt water shows a decrease in the oscillator strength for the 2.4 GHz region. The model also shows an increased oscillator strength for bands above and below this frequency. This is what is observed experimentally for microwave absorption in LiCl salt water[19].

Finally, in Figure 10, a model for water in the crystal form of ice is shown. The model is a six-sided structure with water molecules forming the corners. The IR spectra is significantly different from the liquid water model shown in Figure 3. The coupling between water molecules has shifted the microwave interaction from the H-O-H rocking mode (Figure 3) for liquid water to the O-H stretch (Figure 10) for water ice. The oscillator strength is also reduced from 0.49 for liquid water (Figure 9) to 0.29 for ice (Figure 10). These theoretical results have been obtained without the use of parameters or coefficients in the operation on the AM1 IR eigenvalue matrix. These results are confirmed by industry and experiment[21] where dielectric properties of frozen foods change significantly when thawed and aabsorb less microwave energy when frozen[21].

MICROWAVE TEMPERATURE EFFECT

There is a temperature effect of microwave absorption where the energy transfer increases as the temperature of the material increases. This may be

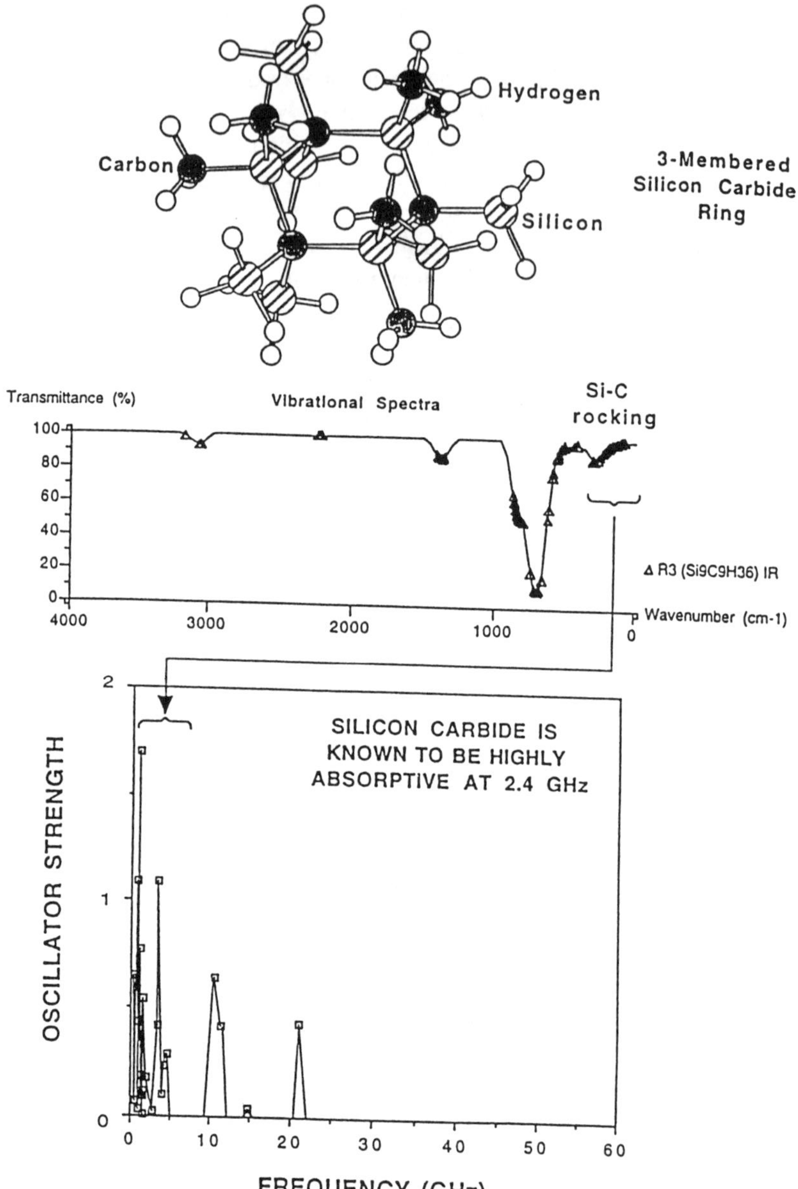

Figure 8
Calculated Microwave Absorption Spectrum for a
3-Member Silicon Carbide (SiC) Ring

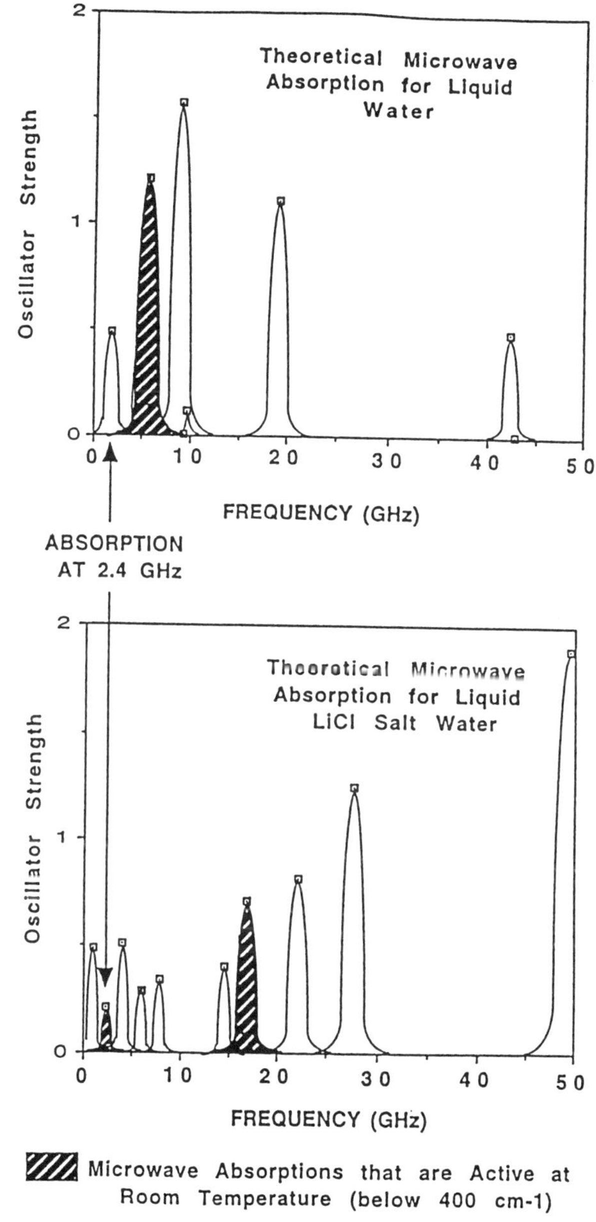

Figure 9
Calculated Microwave Absorption for Liquid Water and LiCl Salt Water

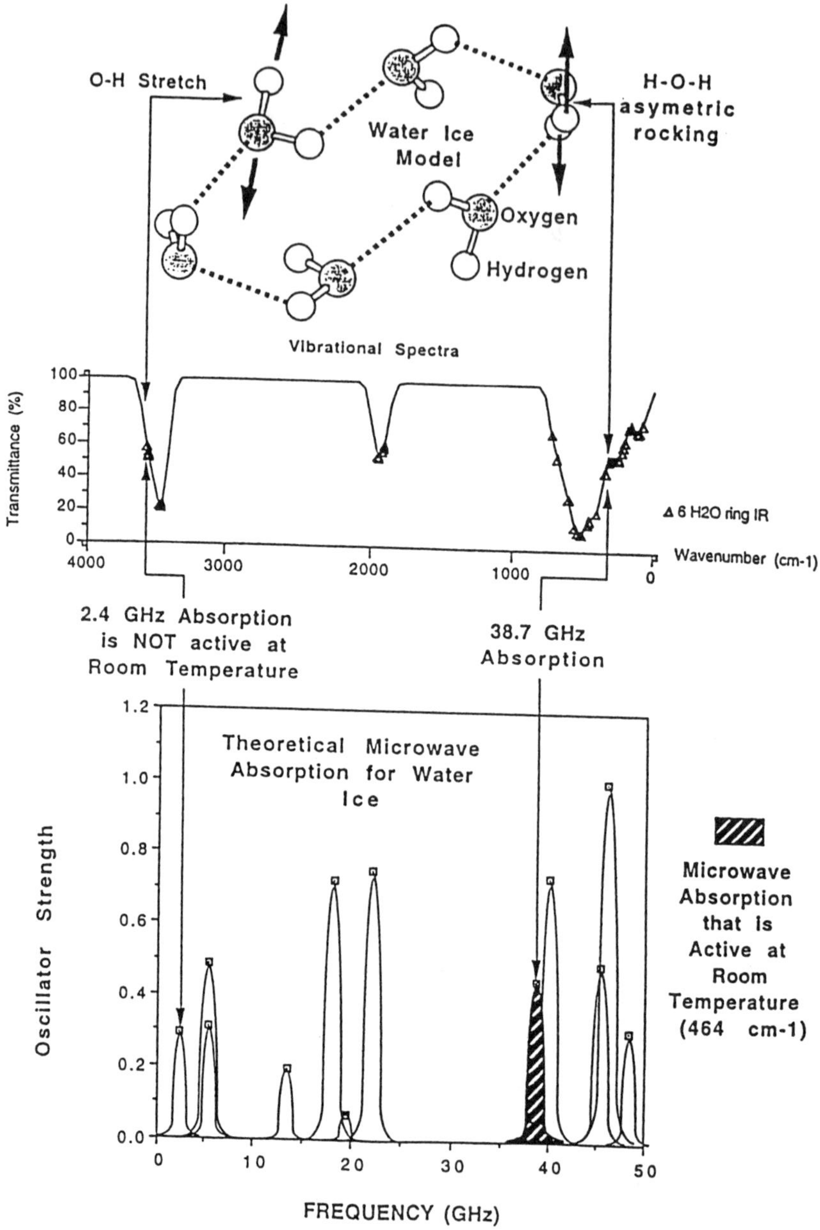

Figure 10
Calculated Microwave Absorption for "Crystal" Water (Ice) and the Infrared Spectrum

explained using the Quantum Coupled Oscillator (QCO) Model. Remembering that photons are bosons and therefore their energies fall into a (Gaussian-like) Maxwell-Boltzman Distribution and they are not constrained by Pauli Exclusion. There will be maximum population of vibrational states centered around the temperature of the material. Because of the distribution of energies, there will always be a small number of populated states at higher and lower energies (wavenumbers) in the infrared spectrum. As the temperature is increased, the lower energy states remain populated and the population of higher energy states increases. The microwave absorption would then be increased because the higher energy vibrational states become more populated and, therefore, the chances for degeneracy and the coupled oscillator effect increase. The total number of excited vibrational states would increase with temperature which would increase the number of populated coupled oscillator states. Microwave absorption would increase with increasing temperature for nearly all materials.

CONCLUSIONS

The AM1 IR molecular orbital models of water ice, liquid water, water vapor, salt water, silica rings, and silicon carbide indicates that microwave energy could be absorbed directly by resonance through coupled oscillators. The coupled oscillator mechanism and the microwave operation, $\theta_{\mu\omega}$, correctly identify major features of microwave absorption for all of these materials. Future work will concentrate on identifying the phase changes and characterization of defect structures that could absorb microwave energy directly via coupled oscillators. New developing variable frequency microwave sources could be targeted toward verification of theoretical spectra based upon quantum molecular orbital theory and the coupled oscillator mechanism.

ACKNOWLEDGMENTS

The authors gratefully acknowledge the support of the The National Science Foundation Grant No.s 9526040 and 9413890, Electric Power Research Institute (EPRI) Grant No. 96011809, the Quantum Theory Project, University of Florida and Oxford Molecular Group (CAChe Scientific), Beaverton, Oregon, and Engineering Innovations, Inc. (EII), Gainesville, Florida.

REFERENCES

1. D.E. Clark, "Microwave Processing of Materials,"Annu. Rev. Maters. Sci., 26, 299- 331 (1996).
2. L.L. Hench and J.K. West, Principles of Electronic Ceramics, John Wiley & Sons, New York, 1990.

3. K.R. Symon, Mechanics, 2nd Edition, Addison-Wesley Publishing Co. Inc., Reading MA, 1960.
4. ZINDO Version 3.0, CAChe Scientific, Inc., Beaverton, OR.
5. MOPAC Version 6.1, CAChe Scientific, Inc., Beaverton, OR.
6. M.J.S. Dewar, E.G. Zoebisch, E.F. Healy, and J.P. Stewart, "Am1: A New General Purpose Quantum Mechanical Molecular Model," J. Am. Chem. Soc.,107, 3902-3909 (1985).
7. M.J.S. Dewar and C. Jie, "Organometallics," J. Am. Chem. Soc., 6, 1486-1490 (1987).
8. J.A. Pople and D.L. Beaveridge, Approximate Molecular Orbital Theory, McGraw-Hill Book Company, New York, 1970.
9. M.C. Zerner, "Semi-Empirical Molecular Orbital Methods," in Review of Computational Chemistry II, K.B. Lipkowitz and D.B. Boyd (eds.), VCH Publishing, N.Y., 1991, pp. 313-366.
10. J.K. West and L.L. Hench, "A Pm3 Molecular Orbital Model of Silica Rings and Their Vibrational Spectra," J. Non-Crystal. Solids, 180, 11-16 (1994).
11. T. Chia, J.K. West and L.L. Hench, "A Molecular Orbital Model of Gel-Silica IR Spectra," in Better Ceramics Through Chemistry VI, A.K. Cheetham, C.J. Brinker, M.L. Mecartney and C. Sanchez (eds.), Vol. 346, Materials Research Society, Pittsburgh, PA, 1994, pp. 727-732.
12. J.K. West and L.L. Hench, "Molecular Orbital Models of Silica Rings and Their Vibrational Spectra," J. Amer. Ceram. Soc., 78[4] 1093-96 (1995).
13. J.J.P. Stewart, "Optimization of Parameters For Semi-empirical Methods I. Methods," J. Comput. Chem. 10[2] 209-221 (1989)
14. J.J.P. Stewart, "Reply To Comments on a Comparison of AM1 With Recently Developed PM3 Method," J. Comput. Chem. 11[4] 543 (1990).
15. J.K. West and L.L. Hench, "The UV-Visible Spectrum in Porous Type VI Silica: Application and Theory," J. Non-Crystal. Solids 195 45-53 (1995).
16. D.L. Johnson, "Temperature Gradients, Thermal Stability, and Hot Spots in Microwave Heating of Ceramics", in Ceramic Transactions, Vol. 80, D.E. Clark, W.H. Sutton, and David A. Lewis, Eds., (1997)
17. J.B. Hasted, Aqueous Dielectrics, Chapman and Hall, London, 1973.
18. R.J. Bell and P. Dean, "The Structure of Vitreous Silica: Validity of the Random Network Theory, Phil. Mag., Vol. 25, (1972) pg 1381
19. J.M. Moore, Private Communication
20. J.C. Anantheswaran and L. Liu, "Effect of Viscosity and Salt Concentration on Microwave Heating of Model Non-Newtonian Liquid Foods in a Cylindrical Container", Journal of Microwave Power and Electromagnetic Energy, Vol. 29, No. 2, (1994), pgs. 119-126

21. A.K. Datta, "Electromagnetics, Heat Transfer, and Misture Tansfer in Mcrowave Hating of Solid and Liquid Foods," in Ceramic Transactions, Vol. 80, D.E. Clark, W.H. Sutton and D.A. Lewis, (eds.), (1997)

SURFACE MODIFICATION OF GLASSES USING MICROWAVE ENERGY

Duangduen Atong, David E. Clark, and Diane C. Folz
Department of Materials Science and Engineering
University of Florida
Gainesville, FL 32611

ABSTRACT

Surface modification of a sodium aluminosilicate glass using a slurry ion-exchange was performed. This was conducted in both a conventional furnace and microwave oven to draw parallels between the two processing methods. Exchange of a larger potassium ion for a smaller sodium ion in the glass was investigated. Microhardness measurements, Fourier Transform Infrared Reflection Spectroscopy (FT-IRRS), and Electron microscope analysis (EMPA) were performed to evaluate the effect of microwave radiation on the ion exchanged glass. Results showed an increase of hardness of the glass samples treated using microwave energy when compared to conventionally processed samples. From the FT-IRRS analysis, microwave ion exchanged samples exhibited very minor spectra difference as compared to the as-received, annealed, and conventionally ion exchange processed glasses. However, EMPA revealed a much larger depth of surface alteration for the microwave ion exchanged glass.

INTRODUCTION

Microwaves represent a major part of the electromagnetic radiation spectrum that has been identified as a national resource managed by the Federal Communications Commission (FCC). The use of microwave energy can be divided into two categories: (1) signal generation and long distance transmission which is important in radio and television broadcasting, communications and RADAR, and (2) signal generation and interaction which has been used primarily in food cooking and processing. Although the first category represents a large and growing market, it is the second category that has the greatest potential for <u>new applications</u>, growth and future benefits for society in such applications as waste remediation, electronic chip manufacture, mining and minerals processing (coal, oil, gold), medical treatments. In these cases, the microwave energy is used to interact specifically with the materials to detect and characterize flaws, cause heating, induce chemical reactions and, in some instances, produce substantial alterations.

The possibility of microwave processing ceramics was known in the late 1950s and was investigated on a limited basis in the 1960s by Tinga et al [1], Levinson [2] and Bennett et al [3]. In 1975, while investigating microwave drying of high alumina castables, Sutton [4] observed that in addition to removing water, the microwaves also heated the ceramic, in some cases to very high temperatures. The preliminary work in polymers was focused on using microwave energy to cure thick cross-sections of rubber. Because rubber is a poor thermal conductor, long curing times in conventional ovens are required. With use of microwave energy, however, rubber can be pre-heated rapidly and uniformly, thereby shortening the final curing times required. This work led to the first industrial application of microwave energy in polymer processing.

Most of the microwave investigations have been and still are performed at 2.45 GHz. This has resulted in some creative engineering in the development of microwave susceptors, microwave/conventional heating hybrids, and absorbing additions to overcome the room temperature, low absorption of many materials. Scientists in the Ukraine and Russia used high frequency (37 and 84 GHz) microwaves for very rapid heating and processing. This has resulted in many applications such as plastic coating of oil piping, glazing of ceramics, soldering and brazing, treatment of polymers, composites and semiconductors and the synthesis of materials [5]. Researchers at Los Alamos and Oak Ridge National Laboratories used both 2.45 GHz and higher frequencies (28 and 60 GHz) to prepare ceramic-glass seals; sinter Al_2O_3, ZrO_2, B_4C and TiB_2; heat treat Si_3N_4; fabricate composites; and form ceramic joints. The labs also developed unique microwave susceptors, hybrid systems and casketing techniques to produce materials that were difficult to sinter (B_4C, TiB_2) and poorly absorbing (Al_2O_3 at 2.45 GHz). Scientifically, one of the most interesting discoveries was reported by researchers [6] from Oak Ridge National Laboratory who demonstrated that Al_2O_3 could be sintered at much lower temperatures with microwave energy compared with electric resistance heating. This research indicated that a non-thermal "microwave effect" reduced the apparent activation energy for sintering. Other investigators have observed similar microwave effects (i.e., reduced processing temperatures and times, and altered reaction pathways and products) in glasses[7-10], polymers [11-12] and organic compounds [13], and these results have created much interest and controversy in the microwave research community.

With proper understanding and control, many technically important materials can be heated rapidly, uniformly, selectively, less expensively and with greater control than is possible with conventional methods. The **unique selective and internal heating** phenomena associated with microwave energy can lead to products and processes that cannot be achieved using conventional methods.

Another significant advantage of microwave processing observed by researchers is that of **uniformity** in processes and products. At the University of Florida, researchers are using a combination of microwave and radiant energy, microwave hybrid heating (MHH), to obtain an extremely uniform temperature throughout materials [14]. In this case, a microwave-absorbing material incorporated into a refractory matrix is placed around the samples in the microwave cavity. Using this method, heat is transferred to the material's surface using radiant heat and to the interior of the sample by microwave energy. The typical thermal gradients observed in conventional processing due to surface heating and thermal conduction from the surface to the sample interior is minimized or eliminated. Figure 1 is a plot of the porosity versus sample thickness for an alumina sample processed using the MHH method.

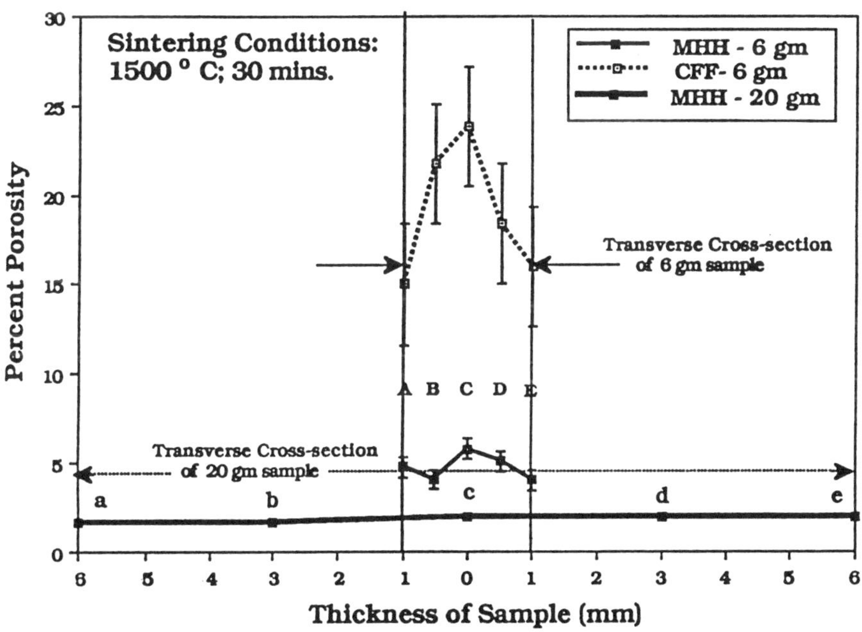

Figure 1. Porosity versus sample thickness for alumina samples (6 g and 20 g) processed using conventional fast-firing (CFF) and microwave hybrid heating (MHH) methods.

The conventionally processed (fast-firing method) 6 g sample shows a pronounced porosity gradient across the cross-section, while the same size sample processed using MHH shows much more uniformity. It is notable that, when the sample size was increased to 20 g, the larger sample processed using MHH exhibited almost no change in porosity from the surface to the interior, an indication of the thermal uniformity during processing. The conventionally fast-fired sample did not survive processing, exploding in the furnace due to extreme thermal gradients.

As these proceedings are dedicated to surface-active processes in materials, the work presented detailed in this paper discusses the use of microwave energy to alter the surface of glasses via the ion exchange process.

Glass is intrinsically a strong material with a theoretical strength in the range of 10^6 psi. However, due to the presence of surface flaws, glass strength is limited to a few thousand psi (8 to 10 ksi) and its failure is always tensile in nature. The strength can be increased significantly by incorporating compressive stresses into the glass surface. The ion exchange process is a surface modification technique based on a chemical technique [15-22]. It is used to develop a pattern of surface compression and tension within the glass. Some applications for chemically strengthened glass include aircraft windshields, high strength transmitting lenses, or dental porcelain. Surface modification can also be important in application other than strength. In addition to glass, it is also possible to modify ceramic surfaces using ion exchange. Other potential applications for this technology include electronic circuits, optical materials and devices, bearings, and products where wear is important.

Chemical strengthening of glass is based on ion exchange by diffusion. The glass is immersed in a molten salt bath at a temperature well below its softening point (Figure 2). When ion exchange takes place, smaller ions are replaced in the surface by larger ions from the molten salt. The surface of the glass is thus altered and a high compressive stress is generated, commonly 50 to 100 ksi. Current methods of surface modification require long times in conventional ovens due to slow ion exchange reactions. Uniformity and depth of the modified surface are also factors in quality control of surface strengthened glasses. The cost of the process is increased significantly due to the lengthy heat treatments required to achieve a strong, uniform product. By using microwave energy as a heat source for ion exchange, the rate of exchange can be enhanced and deeper penetration of the alkali ions into a glass matrix can be achieved[7-8].

This study was undertaken to evaluate the effect of microwave radiation on the ion exchange reaction of potassium (K^+) for soduim (Na^+) of a commercial glass. Chemical/heat treatments on the glass were conduct in both conventional furnace and microwave oven to draw parallels between the two processing methods.

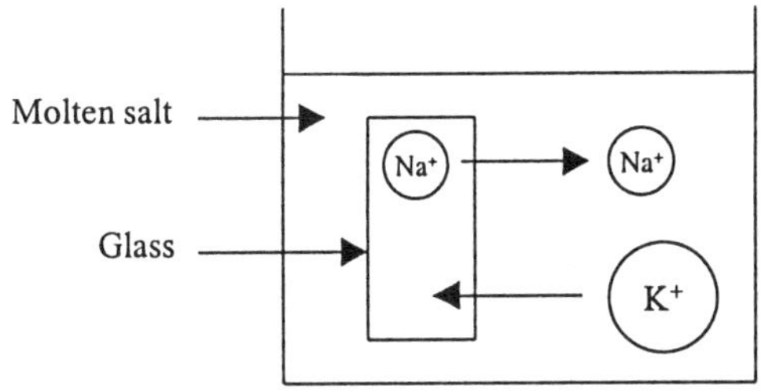

Figure 2. Ion exchange process (exchange of K^+ for Na^+ ions)

EXPERIMENTAL

The glass used for this study was a sodium aluminosilicate type. The glasses were cut to 1 inch x 1 inch using a diamond scribe. Once cut, samples were individually sonicated in acetone for a period of 15 minutes. The samples were then dried in an oven at 90 °C for 30 minutes prior to the ion exchange process.

Ion exchange was carried out using potassium nitrate salts. The clean glass sample was dipped in a pyrex crucible containing salt and was isothermally heated at 400 °C or 500 °C for 30 minutes or 1 hour in both a conventional box furnace and a microwave oven. After ion exchange, the gel-coated glass samples were removed from the furnaces, air cooled, and then sonicated in acetone for 15 minutes. The samples were dried in the same manner as described earlier and stored in a desiccator until characterization and testing were performed.

Two heat treatments were performed on the glass samples. The first series of samples were ion-exchanged in a conventional box furnace or a microwave oven. A second series were subjected to a thermal heat treatment without being subjected to ion-exchange processing under the same condition as the first series. The second series were performed to ensure that any changes in properties of the ion exchanged samples (series 1) were mainly due to the ion exchange effect, and not because of a thermal heat treatment effect. Conventional heat treatments were carried out in a Lindberg box furnace. The microwave heat treatments were carried out in Raytheon Radarline Multimode Microwave Oven Model QMP 2101A-6 operating at 2.45 GHz capable of delivering either 800 or 1600 watts. The samples were placed in a insulated chamber lined with a silicon carbide (SiC) susceptor.This type of heating process is referred to as microwave hybrid heating (MHH). An inconel shielded, K-type thermocouple was positioned at the top of the sample to monitor slurry temperature.

A glass transition temperature measurement was carried out using Differential Thermal Analysis (DTA) at a rate of 10 °C/min. Vicker's microhardness measurements also were performed on the glasses samples using a Leco DM-400 microindenter at a load of 1 kg for a period of 10 seconds. Ten indentations were performed on each sample. FT-IRRS analysis of all samples were conducted using the Nicolet Impact 410 Fourier Transform Infrared Spectrometer. The FT-IRRS in conjunction with a reflactance stage was used to determine the infrared reflectance spectra of the glass samples.

RESULTS AND DISCUSSION

The strengthening process was carried out above the melting point of potassium nitrate (337 °C) but below the annealing point of glass so that the stresses introduced on the glass surface were not removed by relaxation. The transition temperature (T_g) of the glass is about 900°C (Figure 3). However the upper use temperature of potassium nitrate was limited by the onset of the nitrate decomposition which is 525 °C. Therefore all the ion exchange reactions were carried out in the 400-500 °C temperature range.

When ion exchange was carried out below the annealing point of the glass, compressive stresses built up at the surface. Microhardness results are illustrated in Figure 4-5. The ion exchanged glass samples exhibited an increase in hardness compared to the as received and the annealed glasses subjected to the same chemical/heat treatment (Figure 4). The hardness was higher for the microwave processed samples when compared to that of conventionally ion exchanged samples. This result is significant with increasing time and temperature exposure

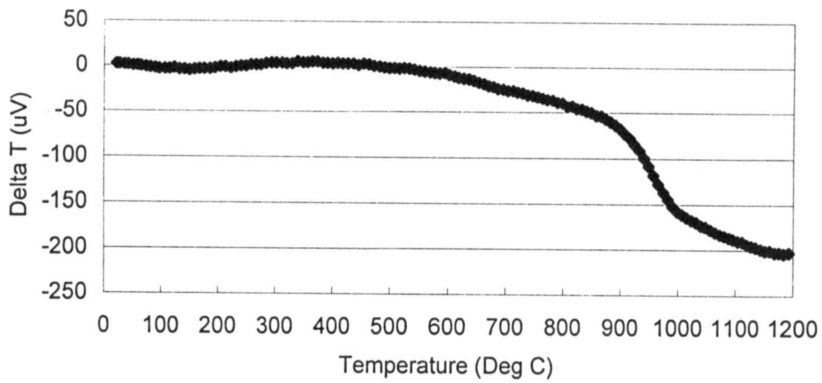

Figure 3. DTA analysis of the sodium aluminosilicate glass.

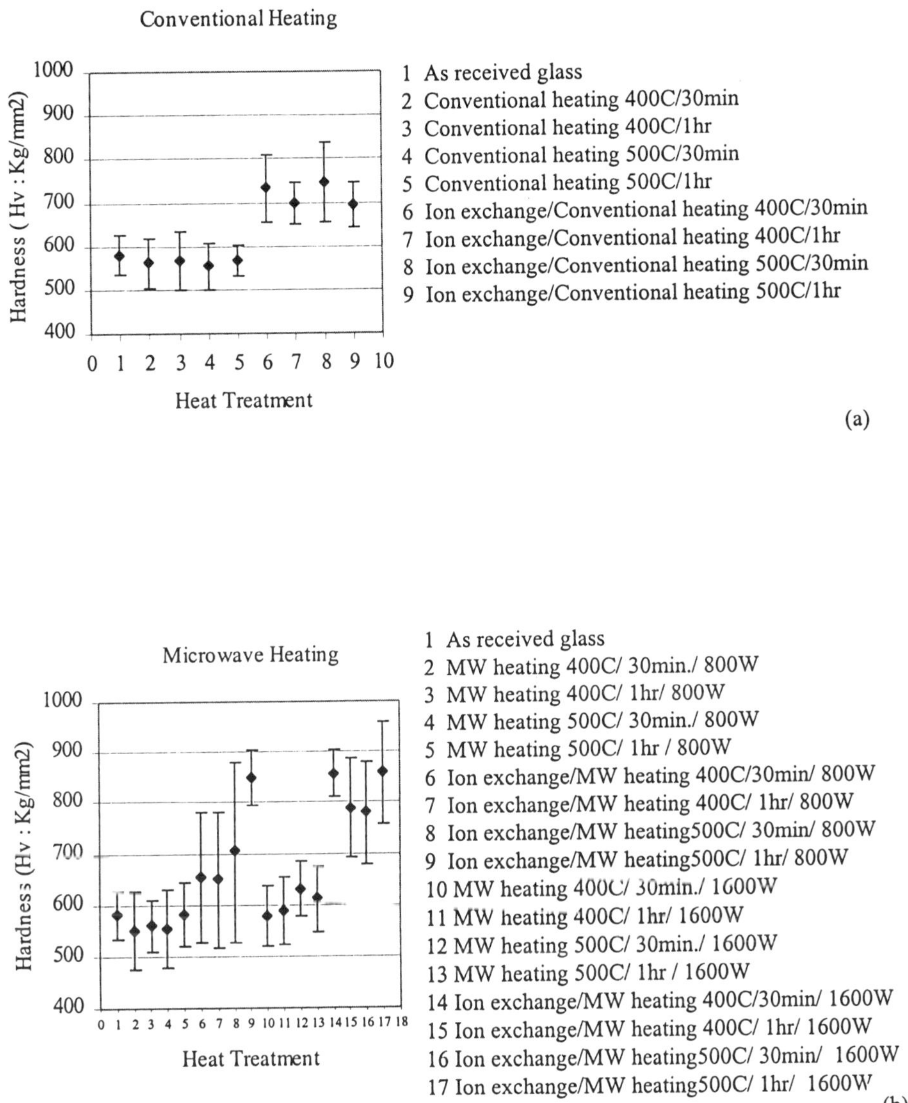

Figure 4. The Vicker's hardness of the glass subjected to different heat/ chemical treatment: (a) Conventional heating, (b) Microwave heating

to ion exchange process (Figure 5). Since the glass hardness was increased as a result of ion exchange in the microwave field, no structural relaxation appears to occur upon microwave irradiation. Table 1 summarizes the microhardness results.

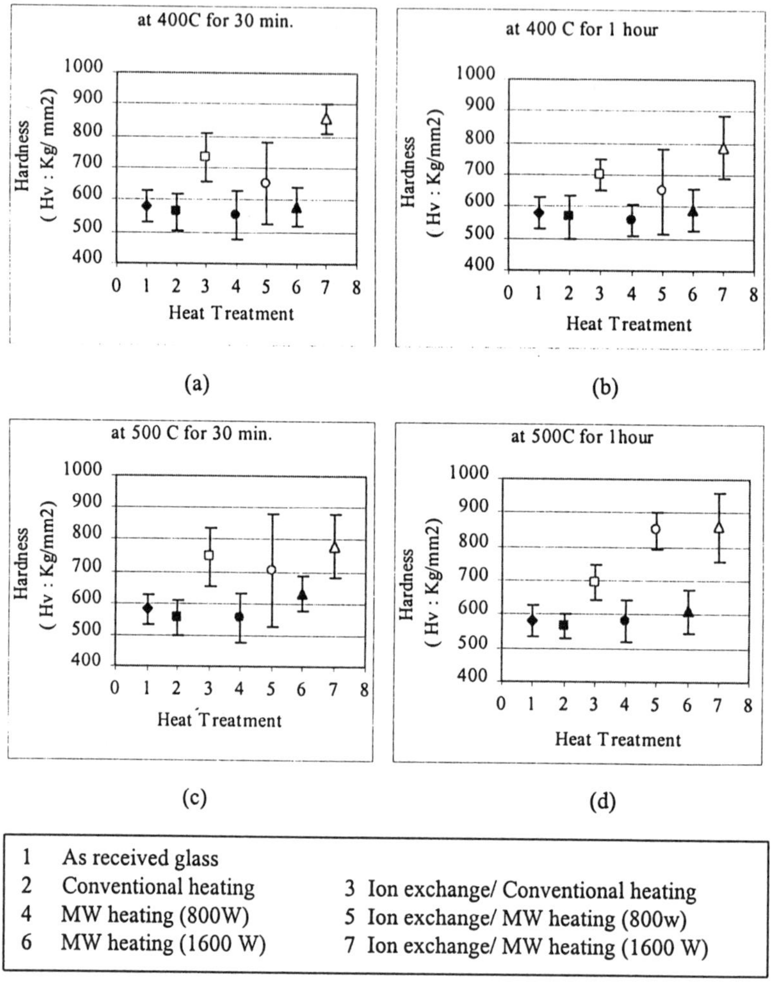

Figure 5. The Vicker's hardness of the glass subjected to different heat/ chemical treatment at: (a) 400°C, 30min, (b) 400°C, 1 hr, (c) 500°C, 30min, (d) 500°C1 hr

Table 1. The Vicker's Hardness of the Glass subjected to different Heat/Chemical Treatments

Heat and /or Chemical Treatment	**Microhardness (Hv: Kg/mm²)**	**Standard deviation**
As received glass	581	3
Conventional Heat	562	6
Conventional Ion- exchange	718	26
Microwave Heat	582	28
Microwave Ion- exchange	767	88

The FTIR spectra of all glass samples were obtained (Figure 6). Infrared spectroscopy is based upon the interaction between the infrared radiation and the

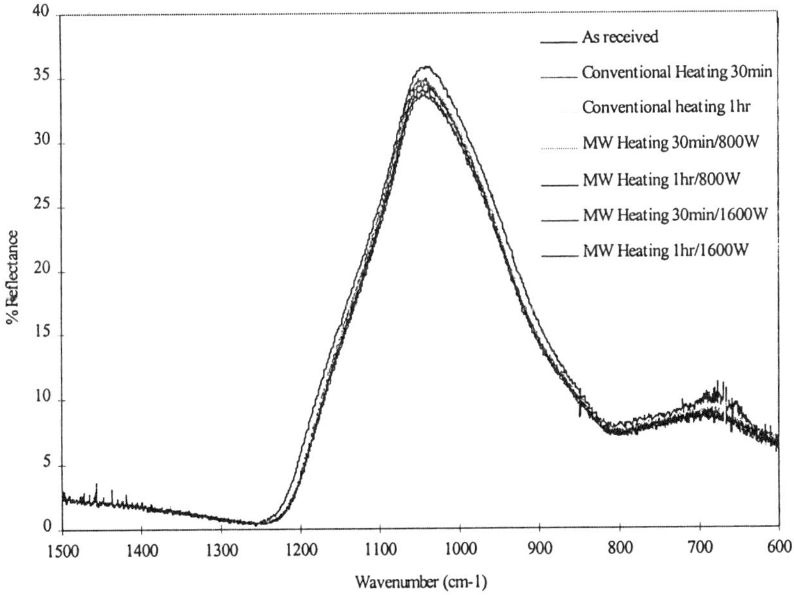

Figure 6. FTIR spectra of the corning glass before and after ion exchange under conventional and microwave heating at 500°C

vibrations of dipole moments of atomic bonds. The vibrations of silicon oxygen bonds (Si-O-Si) and non-bridging oxygen bonds (Si-O-R) (R represents an alkali ion, either Na^+ or K^+)were of interest in this work. The stretch vibrations of Si-O-Si bonds give rise to a peak center at about 1050 cm^{-1}. The peak which usually occur at ~950 cm^{-1} due to non-bridging oxygen was not observed for this glass.

The main purpose for using FTIR was as a quality control tool. All of the pretreated samples yielded identical spectra within the experimental error range of the instrument. Only very minor spectra differences were observed between the pretreated and ion exchanged glasses and there were insufficient to draw any significant conclusion. Therefore the surface chemistry of the sample before and after treatment were evaluated using Electron Microprobe Analysis. The results of the glasses ion exchanged by conventional and microwave method are shown in Figure 7. It can be seen that when all other experimental parameters are equivalent, the microwave field enhances the extent of ion exchange. That is, K^+ penetrates deeper in the presence of the microwave field.

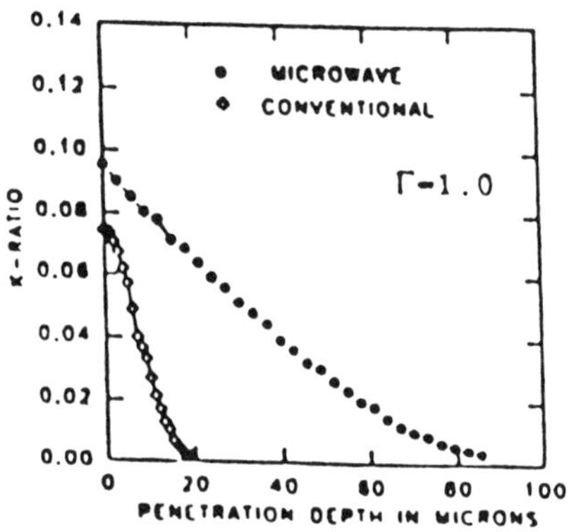

Figure 7. Penetration depths of potassium into the Γ=1 glass composition as a result of Na^+ <--> K^+ interdiffusion at 400°C for 30 minutes, Γis the Al/Na ratio [7].

SUMMARY

The feasibility of using microwave energy at 2.45 GHz to surface modify commercial glasses by ion exchange was demonstrated. Ion exchange of the glass was performed in molten KNO_3 salts in both the microwave oven at 800 and 1600 watts and the conventional box furnace at 400 °C and 500 °C for 30 minutes and 1 hour. Microwave ion exchanged glass samples exhibited an increase in hardness compared to the annealed glasses when subjected to the same thermal heat treatment. The mean hardness for the microwave ion exchanged glass samples was higher when compared to that of conventionally ion exchanged samples. Although FT-IRRS appears to be a good quality control tool, it was unable to distinguish difference between the pretreated and ion exchanged glasses. However Electron Microprobe Analysis data show that the surface of glass was significantly more altered in the presence of microwave field.

ACKNOWLEDGEMENT

The authors wish to express their sincere gratitude to EPRI for funding this work. The authors acknowledge Z. Fathi for the permission to use portions of his previous work.

REFERENCES

1. C.R.James, W.R.Tinga, and WAG,Voss, "Energy Conversion in Closed Microwave Cavities,", Microwave Power Engineering, Academic Press, New York, pp. 28-37,1968.
2. M.L.Levinson, " Method of Firing Ceramic Articles Utilizing Microwave Energy," US Patent No. 3,585,258, 1971.
3. C.E.G. Bennett, N.A.McKinnon, and L.S.Williams, "Sintering in Gas Discharges,", Nature, 217, pp. 1287-88, 1968.
4. W.H.Sutton, "Microwave Firing of High Alumina Castables," Microwave Processing of Materials, Mater. Res. Soc. Symp. Proc., Pittsburgh: Mater. Res. Soc., pp. 287-95, 1988.
5. B.E.Paton, V.E.Salyarevich, and M.G.Slusarezuk, "Gyrotron Processing of Materials,", MRS Bull. XVII (14): 58-63, 1993.
6. M.A.Janney and H.Kimrey, "Microwave Sintering of Alumina at 28 GHz,", Ceramic Powder Science II, Ceram. Trans., Westerville, OH: Am. Ceram. Soc., 1:919-24, 1988.
7. Z. Fathi, I. Ahmad, J.H. Simmons, A.R. Lodding, and D.E.Clark, " Surface Modification of Sodium Aluminosilicate Glasses Using Microwave Energy," D.E.Clark, F.D. Gac and W.H. Sutton eds., Ceramic Transactions, Vol 21, 623-629, 1991
8. Z. Fathi, D.C. Folz, R. Hutcheon, and D.E. Clark, " Surface Modification of Sodium Aluminosilicate Glasses Using Microwave Energy II," D.E. Clark, W.R. Tinga, J.R. Laia eds., Ceramic Transactions, Vol. 36, 333-340, 1993.
9. W.C. LaCourse and M.Akhtar, " Process for Strengthening Glass," US Patent No. 4,872,896,1989

10. J.Sehgal and L.D.Pye, " Strengthening Glass Articles withElectromagnetic Radiation and Resulting Product," US Patent No. 5,102,833, 1992.
11. W.I.Lee and G.S.Springer, "Microwave Curing of Composites," J. Composite Mater., 18:387-409,1984.
12. J.Wei and M.C.Hawley, "Utilization of Microwaves in Processing of Polymer Composites - Past, Present and Future," Proc. Am. Chem. Soc. Div. Polym. Mater. Sci. Eng., 72:10-12,1995.
13. S. Stinson ,"Use of Microwave Heating to Speed Organic Reactions Continues to Grow," C&EN (Northeast News Bureau), May 1996, pp. 45-46.
14. A.De, I.Ahmad, E.D. Whitney and D.E.Clark, " Microwave (Hybrid) Heating of Alumina at 2.45 GHz, D.E.Clark, F.D. Gac and W.H. Sutton eds., Ceramic Transactions, Vol 21, 319-328, 1991.
15. H. Miska, "Understanding the basics of chemically strengthened glass," Materials Engineering, June 1976.
16. J. Zarzycki, Glasses and the vitreous state, Cambridge University Press, 1991
17. R.F. Bartholomew, "Ion exchange," Engineered Materials Handbook, Volume 4, Ceramics and Glasses, The Materials Information Society, 1991
18. F.V. Tooley, The Handbook of Glass Manufacture," Volume 2, 3rd Edition, Ashlee Publishing Co., Inc, 1984
19. W. Vogel, Chemistry of Glass, The American Ceramic Society, Inc., Columbus, Ohio, 1985
20. R.F. Bartholomew and H.M. Garfinkel, Chemical Strengthening of Glass, Glass Science and Technology, Volume 5, D.R. Uhlmann and N.J. Kreidel, Ed., Academic Press, 1980
21. F.M. Ernsberger, Techniques of Strengthening Glasses, Glass Science and Technology, Volume 5, D.R. Uhlmann and N.J. Kreidel, Ed., Academic Press, 1980
22. H.M. Garfinkel, Strengthening Glass by Ion-Exchange, Glass Ind., Vol 50, p28-31 and 74-76.

MICROWAVE CRYSTALLIZATION OF GLASS

A. Boonyapiwat, D. C. Folz, and D. E. Clark
Department of Materials Science and Engineering
University of Florida
Gainesville, FL 32611

ABSTRACT

Microwave crystallization of $Li_2O{\bullet}2SiO_2$ was investigated. The crystal growth rate in microwave crystallization was greater than in conventional crystallization. The microwave heating rate of glass was compared to that of the fully crystallized glass-ceramics. Dielectric properties as a function of temperature of the glass-ceramic also were investigated. The glassy phase exhibited higher dielectric constant and losses than those of the fully crystallized glass-ceramics. For a nucleated glass sample, the dielectric properties changed with the phase transformation. Results from the analysis of microwave power distribution at the interface between the glass and crystal suggested that a multimode microwave applicator was suitable for uniform crystal growth.

INTRODUCTION

It has been observed that microwave processing can enhance the processing kinetics in several ceramic and glass systems [1,2]. These enhanced processing kinetics are thought to be the result of microwave-enhanced ionic transport. Since the phase transformation in glass-ceramic processing involves ionic transport across the interface between the glassy phase and the crystalline phase, the use of microwaves may offer advantages over conventional processing. Preliminary research has shown that the kinetics of glass crystallization may be accelerated in a microwave field [3,4].

Incorporation of a microwave processing unit into a glass-ceramic processing line would not be complicated. A possible heating schedule and the

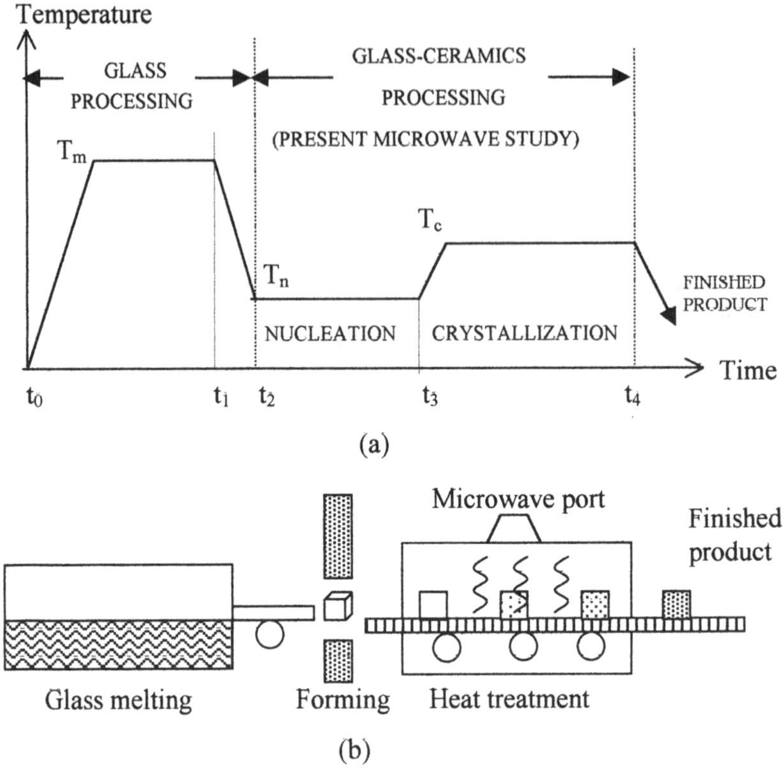

Figure 1. Glass-ceramic processing: (a) heating schedule, and (b) microwave heat treatment unit incorporated into a glass-ceramic process line.

processing line with the incorporation of a microwave processing unit are shown in Figure 1. After the glass is melted at T_m, it is cooled between t_1 and t_2. During this time, the glass is formed into its final shape. The as-formed glass products are subjected to heat treatment in the temperature range of the phase transformation to glass-ceramics. In the conventional process, the conversion of glass to a glass-ceramic takes place in a conventional furnace or lehr. This unit can be replaced with a microwave heating unit, as shown in Figure 1 (b). No other sophisticated modifications to the glass-ceramic processing line are needed. There is no need for a microwave susceptor or the use of microwave hybrid heating (MHH) which generally is required to raise the temperature to a point where glasses and ceramics begin to absorb microwaves. Since the temperature of the as-formed products is still high, direct microwave heating is possible in glass-ceramic processing.

The goal of this study was to investigate the effects of microwave energy on glass-ceramic processing. The microwave effect on the transformation kinetics was determined through a comparison of crystal growth with microwave heating and conventional heating. The study of microwave interaction with different phases in glass-ceramics was based on the dielectric property analysis and a comparison of the heating rate during direct microwave irradiation.

The analysis of microwave power absorption near the interface between glass and crystal also was performed. Since a curved interface between materials of different dielectric properties can cause an uneven electric field distribution along the interface [5], one may suspect that the interface between crystal and glassy matrix can cause uneven electric field distributions around the crystal during microwave processing. This uneven electric field distribution can lead to uneven power absorption in the glass around the crystal. The information from microwave power distribution analysis near the interface between the glass and the crystal can be used to suggest a practical microwave processing method for glass-ceramics.

EXPERIMENTAL

Glass Preparation and Heat Treatment

The $Li_2O{\bullet}2SiO_2$ glass-ceramic system was examined in this study. A nucleating agent was not incorporated into the glass composition. The glass produced by melting the primary constituents, 5.0 μ Min-U-Sil SiO_2 and reagent grade Li_2CO_3, was obtained from Specialty Glass Company (Oldsmar, FL). After the glass composition was melted and homogenized in a fused silica crucible for 24 hours at 1350°C, the glass was wet-fritted in deionized water, then milled with alumina grinding media and sieved to –325 mesh.

The glass frit powder was remelted to produce the required specimen configuration. The frit was melted in a covered platinum crucible at 1400°C and was held at that temperature for 6 hours. Rods of 14 mm diameter and 60 mm length were cast by pouring the molten glass into a graphite mold. After casting, the rods were annealed in an electric furnace at 450°C for 1 hour before being furnace-cooled to room temperature. Glass specimens were produced by cutting the rods using a low-speed diamond saw.

Phase Transformation Kinetic Study

Two methods of heat treatment were employed in this study: microwave hybrid heating (MHH) and conventional heating. The reason for using MHH rather than direct microwave irradiation (even though this glass can be directly heated by microwaves) was that MHH allowed the precise temperature measurement using a thermocouple. (Based on previous experience, temperature measurements during direct microwave heating require a special temperature probe.) The details of the experimental set up for MHH and conventional heating were described elsewhere [4].

Quantitative stereology was used to evaluate the microstructure of the samples [4].

Heating Rate Study

A heating rate study was conducted to evaluate the possibility of using the direct microwave heating technique in glass-ceramic processing. This study was based on monitoring the increase in temperature of a sample during direct microwave heating (a microwave susceptor was not used). A variable frequency microwave oven, a Variwave® Model LT 1500 from Lambda Technologies, Inc. (Raleigh, NC) was chosen for this study because it was equipped with a special temperature probe that allowed monitoring the temperature during direct microwave heating. The Variwave® contains a travelling wave tube capable of sweeping microwaves over frequencies ranging from 6.5 to 18 GHz. The power of the oven is tunable from about 0 to 150 watts, depending on the selected frequency range. The cavity dimensions of the oven are 0.25 x 0.25 x 0.30 m^3. The temperature of the specimen was monitored by an optical fiber-based thermal monitoring system manufactured by Luxtron Fiber Optics (Santa Clara, CA). This system is based on the thermolunimesence property of the material at the tip of the optical probe. Since the maximum operating temperature of the optical probe is 300°C, the heating rate experiments were conducted over a range from room temperature to 250°C.

Dielectric Properties Measurement

Dielectric property measurements were performed at AECL Research (Chalk River, Ontario, Canada). The measurements were based on the cavity shift or cavity perturbation technique [6]. The measurement was done by first heating the sample conventionally to a specified temperature, and then rapidly moving the sample into a measurement cavity. The measurement was then performed before significant cooling occurred. After measurement, the sample was brought back to the furnace and was heat treated to another specified temperature. The dielectric property measurements were performed on samples while heating from room temperature to 600°C and during cooling-down to room temperature.

Quasistatic analysis is used in this study. This technique is based on the assumption that the applied electric field is electrostatic, even though the real applied electric field may not be electrostatic [5]. This technique can be used in the case where the area of interest is small compared to the wavelength of the applied electric field [5]. Because the dimension of the area of interest (a crystal and the nearby surrounding glassy phase in glass-ceramics which is in the micron range) is much smaller than the wavelength of the microwave in use (millimeter range), quasistatic analysis can be used.

Since $Li_2O \bullet 2SiO_2$ crystals have a prolate spheroidal shape, the analysis is performed in prolate spheroidal coordinates. The details of derivation of equations and the analysis are described elsewhere [7].

RESULTS AND DISCUSSION

The crystal radius as a function of time is shown in Figure 2. Crystal growth can be evaluated from the slope of the crystal radius versus time. While the results of crystal growth from conventional heating agreed with the results from Deubener et al [9], the crystal growth rate observed in microwave heating was greater than in conventional heating. Moreover, microwave heating results compared to conventional heating appeared to have shorter induction times. These results agreed with previous work [4]. The induction time was thought to be the result of very slow crystal growth rates (size-dependent growth rate) when the crystal sizes were in the range of critical nuclei [8,9]. The induction time for crystal growth is always observed in the single-stage heat treatment [8]. The results from this study agreed with the theory discussed above since the induction time only occurred for the as-cast and annealed samples (single-stage heat treatment). Samples that were heat treated at 470°C to develop stable nuclei did not exhibit an induction time. Both microwave and conventional heating showed similar results with regard to induction times.

The starting nuclei/crystal size or the volume fraction of crystals did not affect the crystal growth rate in microwave heating. However, the as-cast samples appeared to have a slightly faster growth rate than the samples having stable nuclei. For conventional heating, the crystal growth rate did not show an effect from pretreatment of the sample. Heat treatment (470°C) that was in the glass transition range (450-480°C) [10] may have resulted in a more dense glass structure. For the same composition, microwave absorption by the material having a dense structure tended to be lower than that of the material having a more open structure.

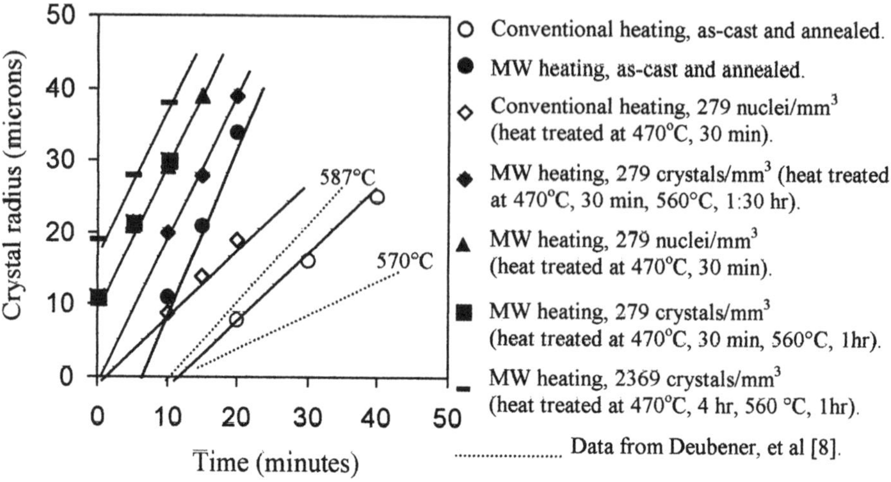

Figure 2. Crystal size versus processing time for conventional and microwave crystallization at 580°C.

The microwave heating rate of glass as compared to the fully crystallized glass-ceramic is shown in Figure 3. The glass sample was heated from room temperature without using a microwave susceptor or insulation, while the fully crystallized sample did not heat at all. Higher microwave power resulted in higher heating rates in glass samples. Thus microwave processing of glass-ceramics was a self-limiting process since heating stopped when the glass was fully converted to a glass-ceramic.

The dielectric properties of the $Li_2O{\bullet}2SiO_2$ glass-ceramic system are shown in Figure 4. The glassy phase of $Li_2O{\bullet}2SiO_2$ had higher dielectric constant and dielectric loss values than the crystalline phase of the same composition. The values of the dielectric properties as a function of temperature for the glass and the fully crystallized glass-ceramics measured during heating were the same as the values measured during cooling. However, the pre-nucleated glass exhibited hysteresis behavior. The dielectric properties of this sample were the same values as those of the glass up to the temperature where crystallization occurred. As the sample was cooling down, the dielectric properties of this sample followed those of the crystalline phase. This behavior resulted from the growth of the nuclei into crystals. Microstructural analysis of samples performed after measuring the dielectric property confirmed this behavior.

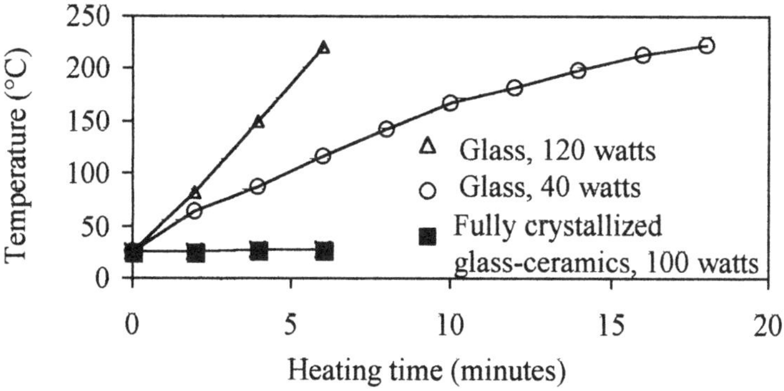

Figure 3. Temperature of the sample heated by direct microwave irradiation as a function of time (frequency = 9.5 GHz).

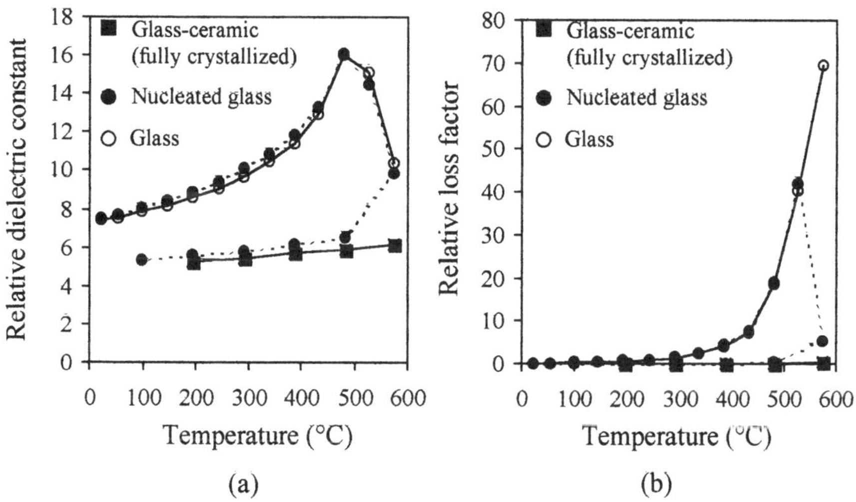

Figure 4. Dielectric properties of the $Li_2O{\bullet}2SiO_2$ glass-ceramic system at 2.467 GHz as a function of temperature: (a) relative dielectric constant, and (b) relative loss factor.

The fact that the dielectric losses of the glassy phase had higher values than those of the crystalline phase can be used to explain the heating rate in Figure 3. The relative dielectric losses of the glassy phase at room temperature were four times higher than those of the crystalline phase (0.152 for glass and 0.038 for crystal). The microwave power absorption in a material (per unit volume) can be expressed as [11]

$$\frac{P}{V} = \omega \varepsilon_0 \varepsilon_{eff}^2 E_{rms}^2 \tag{1}$$

where: P = power absorb,
V = volume,
ε_0 = dielectric constant of free space,
ε_{eff} = effective dielectric losses, and
E_{rms} = root mean square of the internal electric field.

As predicted by equation (1), the microwave absorption in the glassy phase at room temperature is about 16 times that of the crystalline phase.

A schematic illustrating the direction of the electric field and the positions where the microwave power absorption was evaluated is shown in Figure 5. The calculated distributions of microwave power absorption near the glass/crystal interface for two cases, applied electric field direction parallel and perpendicular to the major axis of crystal, are shown in Figure 6. The behavior of microwave power absorption near the interface in both cases were similar. The microwave power absorption at the position close to the glass/crystal interface varied with the angle of the direction from the applied electric field. The microwave power absorption became uniform as the distance from the glass/crystal interface increased. Generally, the lowest microwave power absorption was at the interface having angle 0° from the direction of the applied electric field, whereas the highest absorption was at 90° to the applied electric field. The microwave power absorption in glass was higher than that in the crystal (ranging from about 40 to 230 times), as seen in Figure 6. This resulted from the glassy phase having much higher dielectric losses than the crystalline phase.

It is suggest that the results shown in Figure 6 can be used to predict the behavior of crystal growth during microwave processing. Glass-ceramics processed in a microwave applicator having a single electric field direction (such as single-mode or single-path type applicators) tend to exhibit oriented crystal growth, while the glass-ceramics processed in an applicator having many electric field directions (multimode applicator) may tend to exhibit uniform crystal growth. All the experiments in this study were performed in multimode applicator types, so the oriented crystal growth was not observed.

SUMMARY

- The $Li_2O{\bullet}2SiO_2$ glass can be heated directly using microwaves, while the crystalline phase of the same composition cannot be heated. Using higher applied microwave power resulted in higher heating rates in glass samples.

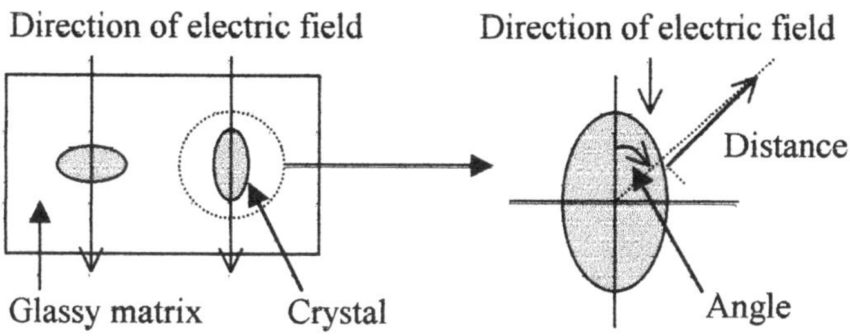

Figure 5. Schematic illustrates the direction of electric field and the positions where the microwave power absorption are evaluated. The case that the applied electric field parallel to the major axis of crystal is shown in detail..

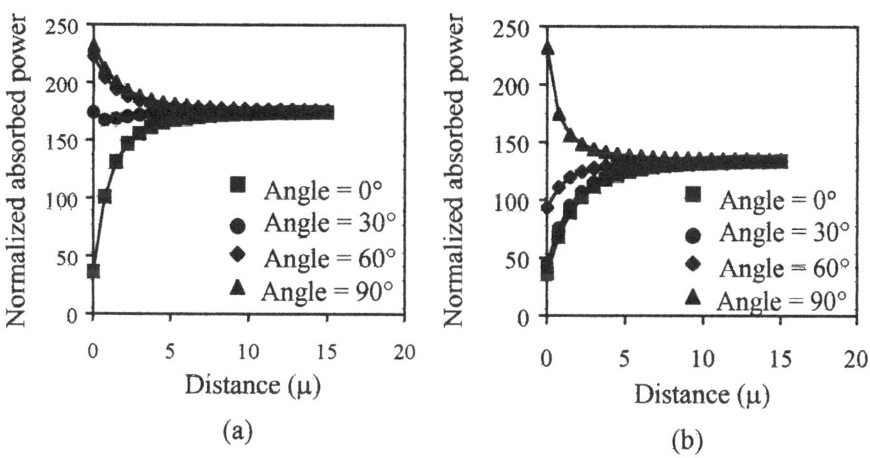

Figure 6. Normalized absorbed power (to the power absorbed by crystal) as a function of distance from the crystal surface, at several directions (crystal size: major axis length = 10μ, minor axis length = 6.1μ, temperature = 550°C): (a) applied electric field parallel to major axis, and (b) applied electric field perpendicular to major axis.

- The starting size and the number of nuclei or crystals did not effect the crystal growth rate during microwave heating under the conditions in this study.

- Crystal growth rates in a microwave field are higher than those produced using conventional heat treatments.

- The values of the dielectric constant and the dielectric losses of the glassy phase are higher than those of the crystalline phase.

- The values of the dielectric constant and the dielectric losses change during the phase transformation from glass to glass-ceramics.

- The microwave power distribution near the glass/crystal interface evaluated by the quasistatic method leads to the perception that microwave processing of glass-ceramics in an applicator having a single electric field direction may yield oriented crystal growth while processing in a multimode applicator yields uniform crystal growth.

ACKNOWLEDGEMENT

The authors wish to express their gratitude to R. M. Hutcheon for dielectric property measurements and EPRI for financial support (contract No. 08060-12A2).

REFERENCES

1. W. H. Sutton, Am. Ceram. Soc. Bull., 68, p. 376-386 (1989).
2. D. E. Clark and W. H. Sutton, Ann. Rev. Mater. Sci., 26, p. 299-331 (1996).
3. A. D. Cozzi, Z. Fathi, R. L. Schulz, and D. E. Clark, Ceram. Eng. Sci. Proc., 14(9), 856-862 (1993).
4. A. Boonyapiwat and D. E. Clark, p. 505-514, in Microwaves: Theory and Application in Materials Processing, edited by D. E. Clark, D. C. Folz, S. J. Oda, and R. Silberglitt, The American Ceramic Society, Westerville, Ohio (1995).
5. S. Ramo, J. R. Whinnery, and T. Van Duzer, Field and Waves in Communication Electrinics, John Wiley & Sons, New York (1994).
6. F. P. Adams, M. S. deJong, and R. M. Hutcheon, J. Microwave power Electromagn Energy, 27, 131-135 (1992),
7. A. Boonyapiwat, PhD Dissertation, University of Florida, to be published.
8. J. Deubener, R. Bruckner, and M. Sternitzke, J. Non-Cryst. Solids, 163, 1-12 (1993).
9. K. F. Kelton and A. L. Greer, J. Non-Cryst. Solids, 79, 295-309 (1986).
10. J. Y. Thompson, PhD Dissertation, University of Florida (1995).
11. A. C. Metaxas and R. J. Meredith, Industrial Microwave Heating, Peter Peregrinus, London (1988).

THUNDER – AN ULTRA-HIGH DISPLACEMENT PIEZOELECTRIC ACTUATOR

B.M. Copeland[1], J.D. Buckley[2], R.G. Bryant[2], R.L. Fox[1] and R.F. Hellbaum[3]
NASA Langley Research Center
Mail Stop 390[1], 226[2], 197[3], Hampton VA 23681-0001

ABSTACT

THUNDER is an acronym for "THin layer composite UNimorph piezoelectric Driver and sEnsoR". It is a new ulrta-high displacement composite actuator developed at NASA Langley Research Center. THUNDER, significantly improves State-of-the-art piezoelectric technology and provides inordinately large displacement. This paper discusses the material fabrication technology, the effects of processing and the initial trends that occur with several sets of wafers under varying conditions.

INTRODUCTION

Piezoelectric materials include a class of materials called ferroelectrics [1]. The ferroelectric material discussed in this paper is ceramic based. Some ionic crystalline materials have unit cells that do not have a center of symmetry (fig.1) and as a result of this condition their unit cells contain a small electric dipole. An industrially important ceramic in this class of material is Lead Zirconate Titanate technically called PZT. These ceramic materials exhibit, over a range of temperatures, a spontaneous electric polarization that can be reversed or reoriented by application of an electric field. The phenomenon experienced by this ferroelectric ceramic PZT is called the piezoelectric effect [2]. The piezoelectric effect is an electromechanical effect by which a mechanical force on a ferroelectric material can produce an electric response and an electric potential can produce a mechanical distortion (fig. 2).

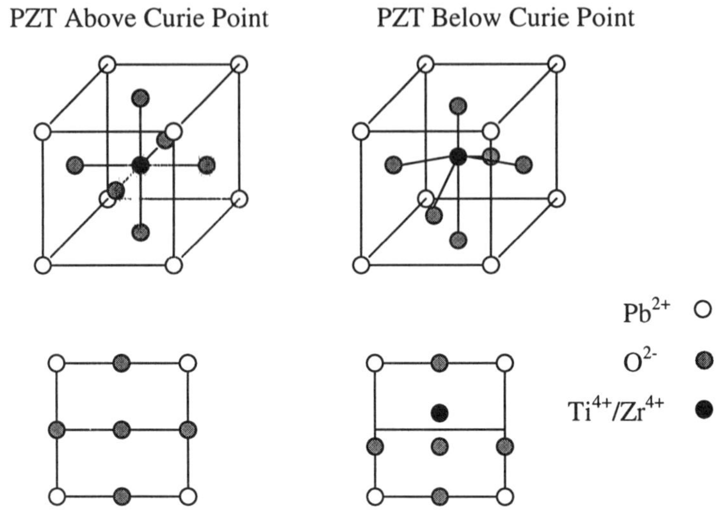

Figure 1. Schematic Representation of PZT Above and Below the Curie Temperature

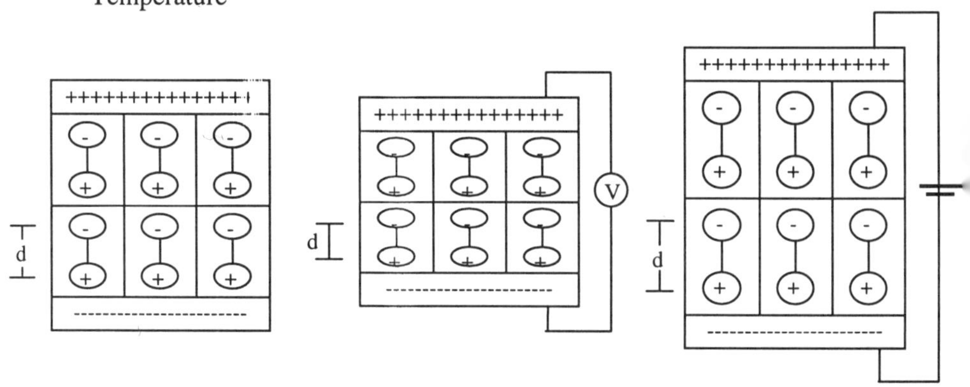

Figure 2. (a) Schematic illustration of electric dipoles within a piezoelectric material.
(b) Compressive stresses on material cause a voltage difference to develop due to change in electric dipoles.
(c) Applied voltage across ends of sample causes dimensional change and changes the electric dipole moment [1,2].

THUNDER is an acronym for "THin layer composite UNimorph piezoelectric Driver and sEnsoR." THUNDER is a new ultra-high displacement actuator developed by the National Aeronautics and Space Administration's

Langley Research Center. THUNDER, significantly improves the state-of -the-art of piezoelectric technology and provides inordinately large mechanical displacement [3,4]. As actuators, typical unimorph piezoelectric devices have relatively small movement when voltage is applied and also generate small voltages under mechanical force (Table 1) [5]. A typical piezoelectric device subjected to an electric force will move about 0.012 mm/volt for a disk approximately 25 mm. in diameter by 0.15 mm. thick. Stacks of piezoelectric materials, mechanically joined in series and electrically driven in parallel, can increase movement at the same voltage required of a single disk. The total displacement is approximately equal to the sum of each disk in the stack [5].

New high displacement, pre-stressed piezoelectric devices have been researched and developed at NASA Langley Research Center (LaRC). These include RAINBOW™ (Reduced And Internally Biased Oxide Wafers), developed at Clemson University under a NASA grant [5] and THUNDER in the process of being developed at NASA LaRC. These devices use two different processes to achieve their increased mechanical output. The RAINBOW™ uses a thermal chemical process that results in reducing some of the metal oxide in the PZT wafer to a metal that results in creating an internal prestressed structure [5]. The THUNDER actuator uses a ultrahigh performance hot melt adhesive, LaRC™-SI, to bond metal foils to PZT ceramic at elevated temperature to create a prestressed condition when cooled to room temperature [3,4]. These high displacement solid state actuators have the potential to be used in place of solenoids, active springs for vibration and jitter suppression, pump and valve diaphragms, accelerometers and cams to create small high frequency stepper motors.

The objectives of this paper are to present materials and process technology and to discuss the initial trends that occur with several sets of wafers under varying conditions.

Table I. Ceramic Actuator Technologies

Type	Configuration	Max Stress* (Kg/cm^2)	Actuator Movement w/Voltage	Actuator Behavior (P or E)	Actuator Strain (%)
Monolithic (d_{33} mode)	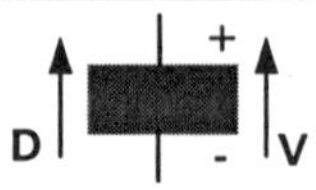	400	Expansion	P	0.40
Monolithic (d_{31} mode)	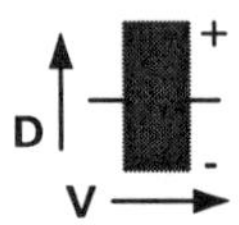	400	Contraction	P	- 0.15
Monolithic (s_{11} mode)	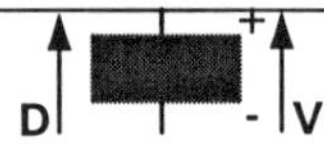	400	Expansion	E	0.24
Monolithic (s_{12} mode)	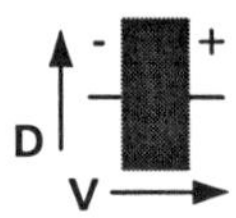	400	Contraction	E	-0.08
Composite Structure (d_{33} mode) (flextensional)	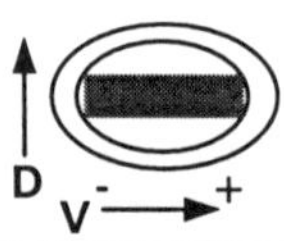	100	Contraction	P	-1.0
Composite Structure (d_{33}/d_{31}) (Moonie)	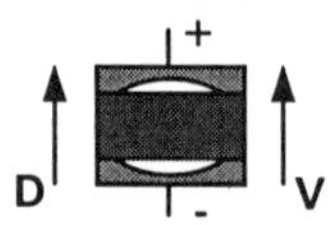	5	Expansion	P	1.2
Unimorph (bender)	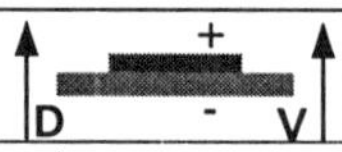	0.1	Expansion/ Contraction	P/E	10
Bimorph (bender)	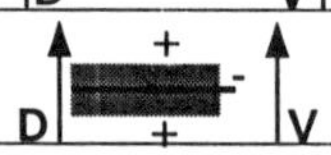	0.1	Expansion/ Contraction	P/E	20
RAINBOW™ Monomorph (bender)	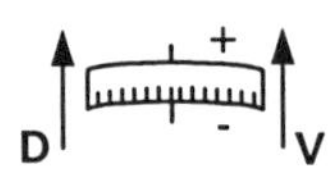	5	Expansion/ Contraction	P/E	500

Notes: V = Voltage; D = Actuator Displacement; P = Piezoelectric; E = Electrostrictor; Strain values at + or – 10kV/cm; * = Maximum generated stress

MATERIALS AND PROCESSING PROCEDURE

Materials and Equipment

The materials used to manufacture THUNDER devices are a piezoelectric ceramic wafer, metallic backing material and an adhesive in spray and film form. The wafers used in this study were a composition of lead zirconate titanate ($Pb(Zr_{0.52}Ti_{0.48})O_2$) PZT with dimensions of 6.35 cm x 3.81 cm x 0.018 cm Two foil materials were used, including aluminum (0.03 mm commercial grade aluminum foil) and 1020 steel (0.08mm thick). The construction of the piezoelectric devices requires equipment and facilities that include a spray gun, a convection oven, high temperature vacuum fixtures and a 400°C pressure autoclave.

Processing

Four basic steps were involved in manufacturing the piezoelectric devices: (1) Spray coating of the PZT ceramic wafers, (2) Construction of the component layers, (3) Assembly of the layers and (4) Bonding and thermal processing in an autoclave.

1. Spray Coating of the Ceramics: LaRC™-SI /N-methyl-pyrolidinone (NMP) 10% (w) solution (acquired from Imitec, Inc.) was sprayed on to the ceramic using an air brush. Both sides of the PZT ceramic wafer were cleaned by a process of washing with acetone, followed by methanol and lastly with trichloroethane. Two coats of the solution (~0.005 mm. per coat) were sprayed on each side of the PZT ceramic and dried to a tack free state in an oven for 30 min. at 60°C.

2. Construction of the Layers: The top foil layer was 0.03 mm aluminum textured with sandpaper, cleaned using the procedure discussed in (1) and sprayed with LaRC™-SI solution. This aluminum layer protects the device while serving as an electrical contact for the PZT wafer. The materials (foils and LaRC™-SI film, with a 3% offset) were cut to the size of the PZT. The 1020 steel back layer was fabricated in the same manner as the top aluminum layer except for texturization.

3. Assembly: The materials were assembled in the order shown in figure 3 and thermally processed to produce the composite piezoelectric THUNDER device. The order of the layers, starting from the bottom of the device was the 1020 steel (prestress layer), LaRC™-SI film, PZT ceramic, LaRC™-SI film, and top textured metallic aluminum foil.

4. Bonding Process in the Autoclave: An autoclave plate with fixturing was prepared for final thermal processing of the composite wafer. Figure 4 illustrates a side view of the wafer. First a vacuum of ~10 mm of mercury was pulled on the autoclave plate to seal the fixtured composite wafer to the plate. The plate with the wafer was placed in the autoclave heated to 160°C at 5°C/min at which time the autoclave was pressurized to 0.51 MPa. Temperature was increased to 320°C at 5°C/min and the pressure was increased to 1.05 MPa and held for 30 minutes. Cooling was accomplished at 5°C/min and pressure was released at 180°C. The fixture was removed when the autoclave cooled to room temperature. Figure 5 shows the time, temperature and pressure profile used in the autoclave processing of the THUNDER device.

Figure 3. Wafer Stack Configuration (without curvature).

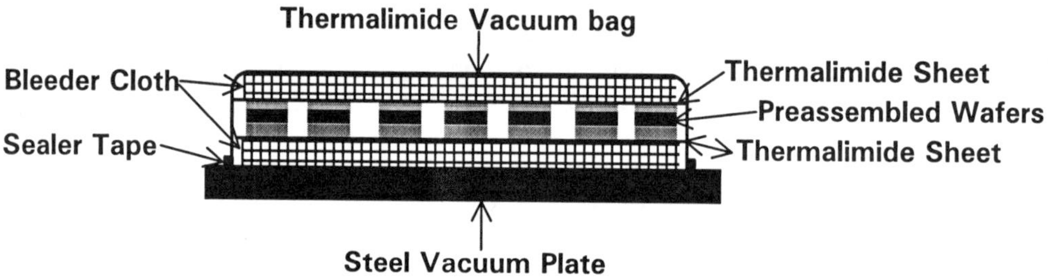

Figure 4. Side View of Vacuum Bagged Wafers for Autoclave Processing.

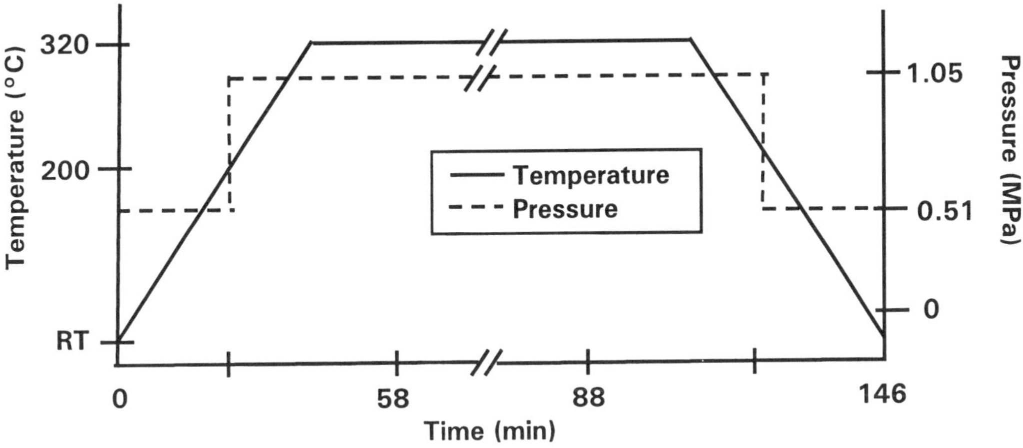

Figure 5. Time, Temperature, Pressure Profile for Autoclave Process.

A processed wafer is shown in figure 6. The wafer's curvature results from the prestressed condition caused by the differences in the coefficients of thermal expansion of the various bonded layers that make up the composite laminate.

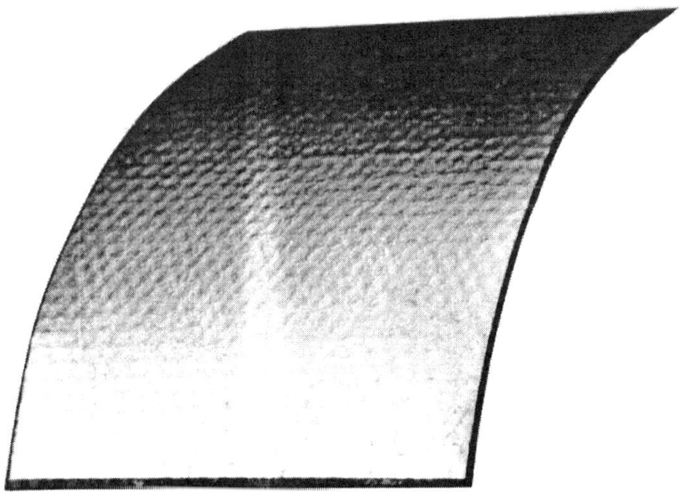

Figure 6. Manufactured composite piezoceramic THUNDER actuator.

CHARACTERIZATION

Baseline characterization was performed by affixing the THUNDER actuators at opposite ends using a compliant mounting system consisting of a hinged fixture on linear slide bearings. Measurements were then taken using approximately 50 to 800 volts (sine) peak to peak with a lower limit of –150 volts above 300 volts peak to peak. Frequencies ranged from 1 to 100 Hz in 25 Hz steps under a 50 g line-load placed on center. The results were averaged over several cycles at each frequency and voltage and the center displacements were recorded.

PRELIMINARY RESULTS

The convention that was chosen to describe the lay-up of these actuators is as follows: Top Layer Thickness/Ceramic Thickness/Bottom Layer Thickness : Top Layer Material/Ceramic Type/Bottom Layer Material. Thus, 1/8/3 Al/PZT-5A/Steel is a lay-up with 1 mil (~0.04 mm) aluminum, 8 mils (~0.3 mm) PZT-5A and 3 mil (0.12 mm) steel. The area dimensions are given, the adhesive thickness is omitted (typically 1 mil) and "X" is used to represent a variable. Several THUNDER wafers that varied only in length were compliantly mounted at each end. The center displacements were examined using different sinusoidal peak-to-peak voltages at 1, 25, 50, 75 and 100 Hz (Figures 7-9).

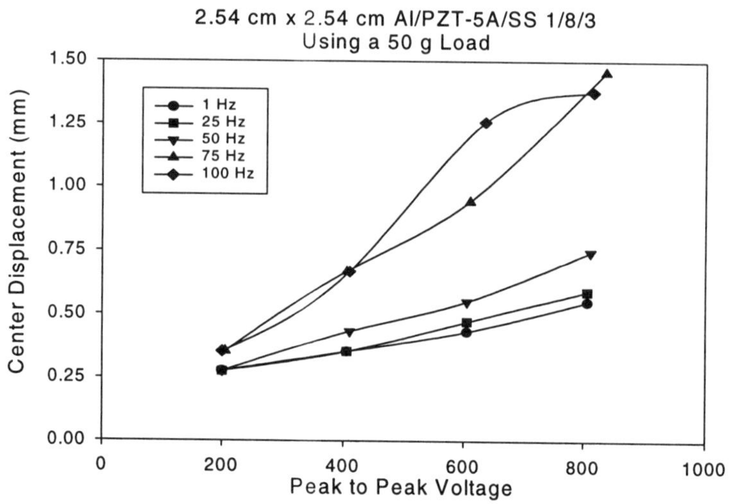

Figure 7. Center Displacement vs. Sinusoidal Voltage at Different Frequencies

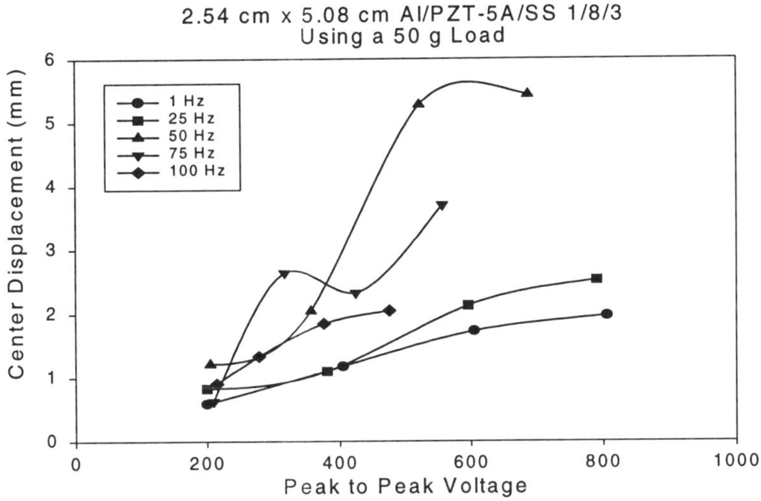

Figure 8. Center Displacement vs. Sinusoidal Voltage at Different Frequencies

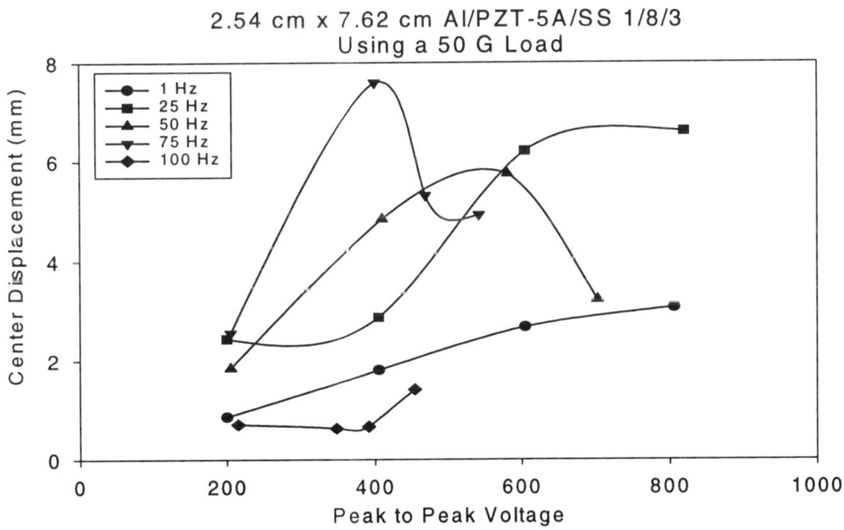

Figure 9. Center Displacement vs. Sinusoidal Voltage at Different Frequencies

As shown on Figure 7, as the voltage increases, the center displacement increases and as the frequency increases, the displacement decreases. In a similar manner, as the length of the wafer increases, the displacement increases at a given voltage and frequency. However, some of these trends are not readily apparent for the wafers examined in Figures 8 and 9. There are several explanations for this behavior. The first is that as the THUNDER actuators increase in length, they begin to behave like vibrating plates, which develop nodes at different frequencies. Also, the some of the wafers may be operating near their mechanical resonance frequency generating inordinately large displacements. Lastly, as these actuators increase in length and afford higher displacements, the system in which they are affixed to, the slide and the mechanical LVDT, may be generating artifacts in the data caused by mechanical inertia.

CONCLUSIONS

THUNDER has improved the state-of-the-art of piezoelectric devices by producing stack-like displacements with a single actuator. THUNDER provides a comparatively large mechanical output displacement when measured against typical piezoelectric specimens. This is primarily the result of the in-situ mechanical prestress. Preliminary test results show that a 7 cm. rectangular specimen will bend approximately 7.5 mm on center depending on thickness and applied voltage. Test results also show displacements ~50 times greater than the device thickness at low frequencies. These piezoelectric devices have shown load displacement capabilities exceeding 10 kilograms and have been operated to over 400 million cycles without failure. These highly efficient electrically insulated actuators can be operated under a variety of environmental conditions.

REFERENCES

1. Van Vlack, L.H. : Elements of Material Science and Engineering 6th Edition, Addison-Wesley Publishing Co. 1989, pp 484-485.

2. Jaffe, B., Cook, W. R. and Jaffe, H.: Piezoelectric Ceramics, Academic Press, New York, 1971.

3. Gregory, R., Bryant, R., Fox, R. L., Hellbaum, R. and Jalink, A.: Manufacture, Characterization and Application of High-Performance Composite Piezo-ceramics, Technology 2007 Conference NASA Langley Research Center Hampton, VA. Sept. 1997.

4. Mossi, K. M., Selby, G. V. and Bryant, R. G.: Materials Letters, 35, 39, (1998).

5. Haertling, G.: Materials ,Processes, Properties and Applications of Ferroelectric Ceramic Materials Workshop, NASA Langley Research Center Hampton, VA. Sept. 1993.

Nuclear Waste Glasses and Corrosion

NUCLEAR WASTE GLASSES- Suitability, Surface Studies and Stability

George G. Wicks
Senior Advisory Scientist
Westinghouse Savannah River Technology Center
Aiken, SC USA

ABSTRACT

Every major country involved with long-term management of high-level radioactive waste (HLW) has either selected or is considering glass as the matrix of choice for immobilizing and ultimately, disposing of the potentially hazardous, high-level radioactive material. There are many reasons why glass is preferred. Among the most important considerations are the ability of glass structures to accommodate and immobilize the many different types of radionuclides present in HLW, and to produce a product that not only has excellent technical properties, but also possesses good processing features [1]. Good processability allows the glass to be fabricated with relative ease even under difficult remote-handling conditions necessary for vitrification of highly radioactive material. The single most important property of the waste glass produced is its ability to retain hazardous species within the glass structure and this is reflected by its excellent chemical durability and corrosion resistance to a wide range of environmental conditions [1, 2].

INTRODUCTION

In addition to immobilization of HLW, glass matrices are also being considered for isolation of many other types of hazardous materials, both radioactive as well as non-radioactive. This includes vitrification of various actinides resulting from clean-up operations and the legacy of the cold war. Other types of wastes being considered for immobilization into glasses include transuranic wastes, mixed wastes, contaminated soils, asbestos, incinerator ashes, medical wastes, electronic circuitry, weapons parts, and a variety of other potential hazardous materials or components. The ability of glass structures to incorporate and then immobilize many different elements into durable, high integrity, waste glass products is a direct function of the unique random network structure of the glassy state.

WASTE DISPOSAL STRATEGY

The strategy that has been adopted for long term management of HLW involves removing the waste from temporary storage facilities, chemically processing the waste to reduce volume when appropriate, and then immobilizing the potentially hazardous radionuclides into a solid, non dispersible form-borosilicate glass. The waste glass forms produced can then be temporarily stored on-site at the vitrification facility, but eventually, will be shipped off-site to carefully selected geologic repositories. Once at the geologic site, the forms will be placed in deep burial as one element of a multibarrier isolation system directed at permanently disposing of the radioactive material. It is important that the glass systems chosen have the ability to incorporate and immobilize HLW constituents and that the waste glass produced will perform well during fabrication, interim storage, transportation, and final geologic disposal.

WASTE GLASS STRUCTURE

For immobilization of HLW, borosilicate glass systems are preferred world-wide. In the case of Savannah River Site (SRS) defense HLW, waste glass forms are produced from approximately 70% glass forming chemicals added to about 30% waste constituents. The glass components are generally added to the liquid waste stream in the form of a multi-component pre-melted glass or frit prior to waste glass melting operations, which produce solidified radioactive products [3]. Although there are many individual elements that are found in nuclear waste glass systems, these components can play only one of three basic roles in the glass structure; *network formers, intermediates* or *modifiers* [4].

Constituents such as silica and boric oxide are generally added to the waste stream as major components of the glass frit. The silicon and boron atoms are NETWORK FORMERS and are located in the center of oxygen polyhedra in the configuration of tetrahedra or triangles. These polyhedra are then tied together by sharing corners, generally in accordance with Zachariasen's rules, which then make up the 'framework or skeleton' of the random network structure of the solidified waste glass form. Another structural role that both glass frit and waste elements can play is that of INTERMEDIATES, which is exemplified by major components found in the waste such as alumina. These components can replace the network formers and still retain the framework structure of the glass. Other cations can move to the singly-bonded oxygen ions that are created, for charge neutrality. The final role that components can play is the most prevalent, that of MODIFIERS. In this case, important waste components such as cesium and strontium, along with alkali and alkali earth constituents, are located within the holes of the random network structure, and can also be associated with nearby singly-bonded oxygen ions. In Figure 1, a model depicting a simplified waste glass random network structure is shown. An important point to note is that both glass and waste components become an integral part of the random network structure of the glass. Components are incorporated by primary and/or secondary bonding which helps explain why glass is able to retain radionuclides so well during leaching and why different elements can leach at different rates [5].

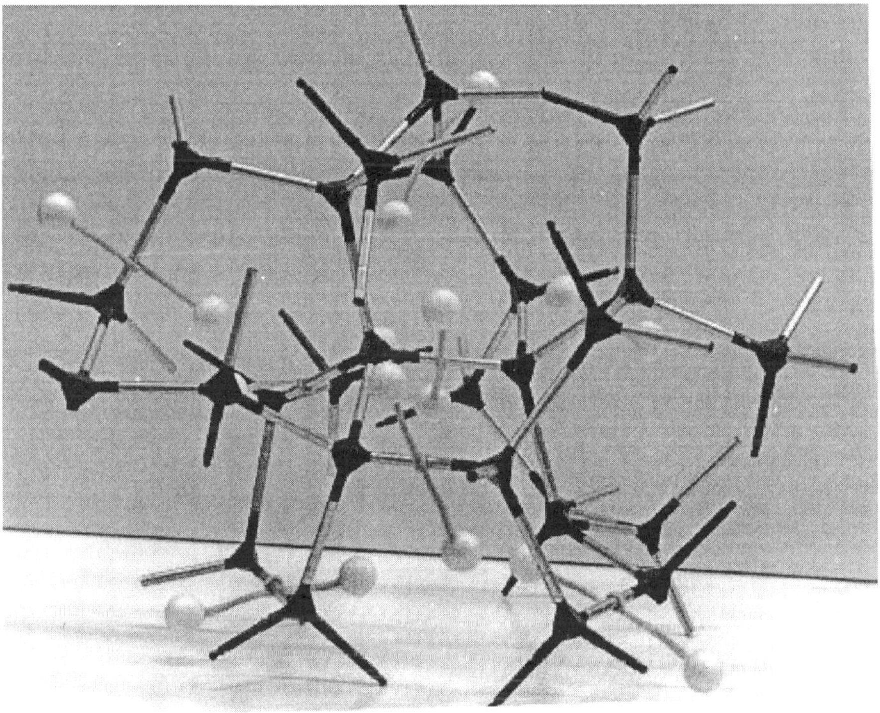

Figure 1. Simplified Random Network Structure of Nuclear Waste Glasses

GLASS SUITABILITY FACTORS

There are a variety of factors which contribute to the suitability of glasses for immobilization of hazardous materials. These considerations generally fall into two major technical categories. First, involves PROCESSING CONSIDERATIONS, which include the ease of being able to produce waste forms even under difficult remote conditions, and second, TECHNICAL PERFORMANCE FEATURES, which can be related to the structure and composition of the vitrified waste forms. Technical performance features are in five major areas of interest; (a) flexibility/ waste compatibility (b) mechanical integrity, (c) thermal stability, (d) radiation effects and (e) chemical durability. Chemical durability is generally considered the most important technical property of the final waste glass form, but there are also other important considerations of a less technical nature. Consideration of all factors are essential in the development of high integrity, cost-effective waste forms and subsequent systems, designed to permanently manage or dispose of hazardous materials. Following is a brief description of glass suitability factors for immobilization of potentially hazardous materials, with an emphasis on SRS high level waste glasses:

• Processing Considerations

Vitrification Facilities & Practical Operating Experience. The only HLW immobilization facilities in the world are glass-making or vitrification operations. This includes France and their successful production facilities at Marcoule and LaHague. Other vitrification plants in production world-wide include facilities located in Belgium, England as well as the United States, along with additional vitrification plants located or being planned in other countries. In the United States, the first major waste vitrification plant is the almost one billion dollar Defense Waste Processing Facility located in Aiken, South Carolina, which is currently in production vitrifying hazardous constituents in more than 30 million gallons of HLW.

• Technical Performance

Flexibility. Glass has demonstrated the ability to accommodate not only the 40 or more different elements that are found in HLW waste streams, but also large variations in waste composition. The reason for this feature is a result of the relatively open random network structure that characterizes glass systems and its ability to accommodate elements or radionuclides of different sizes, charge, and characteristics, as well as differing amounts of these constituents [1].

Thermal Stability. Waste glass products possess good thermal stability. Upon cooling from the melt or from self-heating due to radionuclide decay, waste glasses can phase separate or crystallize [6]. Many different studies have been performed to assess the effects of these structural and chemical changes on the performance of the product [7]. In one series of studies, Savannah River Site waste glass was purposely devitrified [8], even though extensive devitrification is not expected for this system. From the resulting time-temperature-transformation [TTT] curves, the phases formed were identified and leaching tests showed that even in this 'worst-case' scenario, the effects of the crystalline phases that formed on chemical durability were not significant.

Mechanical Integrity. Waste glass products also possess good mechanical integrity. Glasses are relatively brittle materials and cracking can occur due to stresses induced during fabrication or from accident scenarios during handling, transportation or storage operations. A variety of mechanical tests have been performed ranging from laboratory-scale studies to drop tests of full-size canisters containing simulated waste glass, at speeds up to 80 mph [9,10]. In one set of drop tests performed by Battelle Pacific Northwest Laboratory on SRS canisters filled with simulated waste glass, several interesting observations were noted. First, glass forms fracture into relatively large chunks inside the canisters and fracture was generally localized to the area of impact. Second, the amount of increased surface area produced was low along with the amount of resulting fines or small particles. The waste glass product possessed more than adequate mechanical stability under anticipated as well as accident conditions.

Radiation Stability. As radionuclides are incorporated into glass structures, a significant radiation field can be produced. The glass can be irradiated by alpha and beta particles, gamma rays and neutrons that result from decaying radionuclides. The effects that these components have on important properties of waste forms have been assessed for a variety of parameters, including chemical and mechanical integrity, stored energy, helium accumulation, density changes and radiolysis [11,12]. Based on all existing data, waste glass forms perform very well under all of the radiation conditions expected during all stages of solidification and isolation of the HLW. In experiments performed on Savannah River Site waste glass doped internally with curium-244, an intense alpha emitter considered by many to be the most detrimental radiation effect, only very minor effects were noted [13]. The glass remained intact after being subjected to these extreme conditions (simulating more than one million years of storage) and its leaching rate, even with this very high radiation field, was seen to be within a factor two of the undoped glass.

Chemical Durability. The most important and most studied property of solidified waste glass forms is its chemical durability. This provides a measure of how well the waste glass will retain radionuclides under anticipated as well as accident scenarios. Nowhere is this more important than in the final resting place of the waste, the geologic repository. The two most important observations made after evaluating the chemical durability of a variety of waste glass systems under many different repository conditions is that 1) *leaching of glass is very low when subjected to realistic scenarios and conditions* and 2) *not only is the chemical durability of the waste glass systems good, but durability can actually improve with time.* These observations are consistent with our understanding of how and why glass structures incorporate and immobilize hazardous components, and the subsequent behavior these systems exhibit when subjected to extreme conditions. These data have been observed not only in laboratory studies and repository interactions testing, but have also been confirmed in actual field or in-situ testing of the simulated HLW waste forms in various geological settings [14-19].

Actinides and especially plutonium, also incorporate very well into glass structures. Due to this effect, as well as other important considerations, Pu-glasses are expected to have even better chemical durability than the already internationally accepted HLW glasses. Recently, this has been confirmed for a special group of borosilicate Pu-glasses known as Löffler compositions [20].

In order to assess, understand and ultimately, predict the long term reliability of waste glass systems, chemical durability has been assessed as a function of many important parameters that would be encountered during each stage of the solidification, transportation, interim and permanent disposal scenarios [21]. The parameters are related to the chemistry and structure of the glass systems studied. These important variables affecting the chemical durability of waste glasses include time, temperature, solution pH, Eh, composition (waste, glass, leachate and homogeneity), devitrification, waste loading, surface area of sample to leachant ratio [SA/V], flowrate, pressure, surface finish, glass cracking and fines, radiation effects, geology, hydrology, and package components [canister metal, possible overpack, potential backfills]. Based on all data currently available, the chemical durability of waste glasses should be extremely good when subjected to realistic values of these parameters.

SURFACE STUDIES AND STABILITY OF NUCLEAR WASTE GLASSES

• Integrated Study Approach

A typical approach for assessing glass durability is to immerse a glass in a corrosive medium under defined conditions and then to either measure weight loss or analyze the solution, to detect elements leached from the glass. Weight loss measurements are often the simplest to perform while solution analysis provides additional information on the leaching behavior of specific elements of interest. These techniques are very useful and continue to be used effectively in many studies today. However, this approach has limitations, since possible alteration products or back reactions forming on glass surfaces are generally not detected. This can be especially relevant for more complex glass systems, which is what characterizes nuclear waste glasses. These reactions and the products that may form can have significant effects on measurements and subsequent interpretation of data.

In the 1960's and 1970's, investigators such as Hench and Clark of the University of Florida began routinely combining solution analysis with bulk and surface studies of leached waste glasses, to obtain a more complete picture of the leaching process [22]. Over recent years, new and more sophisticated surface analytical tools have been developed along with more accurate solution analyses. Recognizing that the analytical tools used to study nuclear glass systems all have limitations along with advantages, an integrated study approach was used which combined and overlapped each of these techniques. This provides important information on the chemistry and structure of species of interest on and in leached surface layers and therefore, provides a more complete picture of the leaching process and corrosion mechanisms involved. The integrated study approach and the various analytical tools used to characterize surface layers and interfaces in nuclear glasses are depicted schematically in Figure 2, along with their sampling depths [22,23].

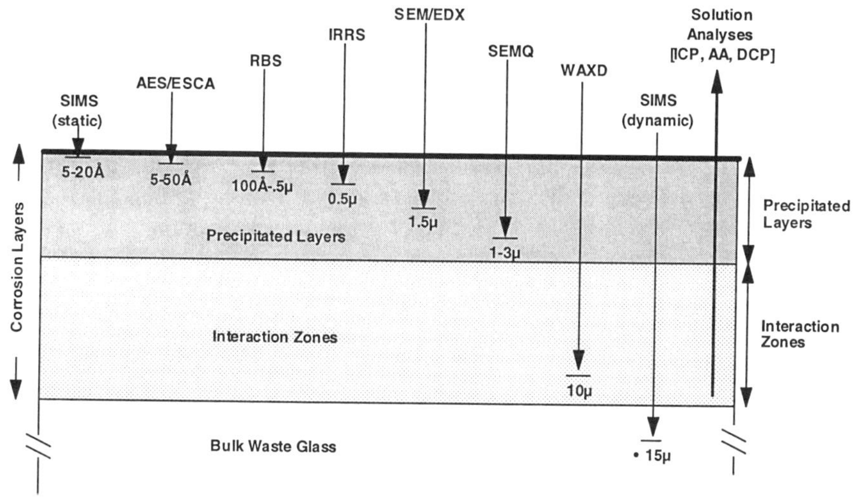

Figure 2. Analytical Tools Used for Integrated Study Approach of Leached Glasses

• Nuclear Waste Glass Surface Layers and Stability

As a result of the numerous surface studies performed by the nuclear waste management community on U.S. and international waste glasses, surface layers have been observed on a wide array of nuclear glass systems and under a wide range of conditions. These layers are generally complex and often, contain a variety of crystalline and non-crystalline phases as well as a series of sublayers, which can be characterized and more fully understood using the integrated study approach. The chemistry, morphology, density and general characteristics of the layers are a strong function of the chemistry of the glass and leaching conditions. The leached glass surface layers can be very important as protective layers, resulting is an already durable glass product exhibiting improved durability with time. While some systems exhibit significant benefits of protective surface layers under specific conditions, other systems and other conditions can produce surface layers that provide much less protection to the glass underneath and in some cases, no protection at all. Surface layers observed in various nuclear waste glass systems all fall within the 5 cases as defined and described by Hench [24].

An example of layers formed and elemental distributions for simulated SRS HLW, after being leached under various conditions and leachants, is shown in Figure 3. An interesting feature of these layers is that for this system, they are generally enriched in major waste components (ex. Fe) while being deficient in major frit components (ex. Si) that are added to the waste to produce the glass forms. Surface layers, with similar as well as unique features, have also been observed and characterized for this same system leached not only in deionized water, but also in tuff, granite, shale, salt and basalt groundwaters. Leached layers and improved glass behavior have also been observed for leaching of the glass in the presence of lead, while deleterious effects have been noted for leaching in the presence of carbon steel, a potential package component.

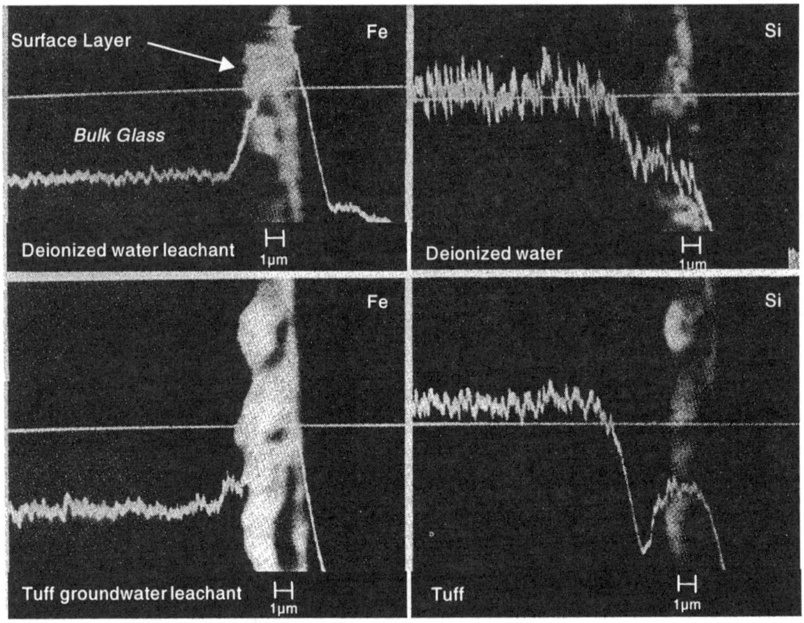

Figure 3. Leached Layers on SRS Waste Glass

Based on these and many other studies that have been performed over the past 30 years, both qualitative and quantitative glass leaching models have been developed to explain the observed leaching behavior and the effects of the surface interactions that have been characterized. In one approach, a simplified 3-stage corrosion process based on glass structural changes has been postulated for nuclear waste glasses involving interdiffusion, matrix dissolution and the effects of surface layer formation [25]. This concept and other modeling efforts are described in detail elsewhere [26-30].

SUMMARY

The random network structure of glass provides a unique matrix that is particularly suited for incorporation and subsequent immobilization of potentially hazardous radioactive and non-radioactive materials. Based on all data currently available, the performance of a wide variety of waste glass systems and their ability to retain hazardous species is excellent, when tested under realistic conditions, as determined by many studies performed by many different investigators in many different countries. In addition to possessing outstanding chemical durability, the durability of nuclear waste glasses generally improves with increasing time. This behavior has been observed not only in laboratory tests, but also in actual field experiments conducted with simulated HLW glasses buried in Sweden, Belgium, England, and the United States.

REFERENCES

1 Wicks, G.G., "Nuclear Waste Glasses", in Treatise in Materials Science and Technology- GLASS IV, (M. Tomozawa and R.H. Doremus ed) Vol. 26, pp. 57-118, (1985).

2 Wicks, G.G., "Nuclear Waste Glasses: Corrosion Behavior and Field Tests, in Corrosion of Glass, Ceramics and Ceramic Superconductors, ed. by D.E. Clark and B.K. Zoitos, pp. 218-268 (1992).

3 Wicks, G.G. and Bickford, D.F., "Doing Something About High Level Nuclear Waste", MIT Technology Review, p. 51 Nov/Dec (1989).

4 Kingery, W.D., Bowen, H.K. and Uhlmann, D.R., Introduction to Ceramics, 2nd ed., (1976).

5 Wicks, G.G., "Structure of Glasses", in Encyclopedia of Materials Science and Engineering, Vol. 3, p. 2020, Pergamon (1986)

6 Tomozawa, M., Singer, G.M, Oka, Y., and Warden, J.T., in Ceramics in Nuclear Waste Management (T.D. Chikalla and J.E. Mendel, ed.) CONF-790420, p. 193 (1979).

7 Turcotte, R.P. and Wald, J.W. in Scientific Basis for Nuclear Waste Management (C. Northrup ed.), Vol. 2, p. 141 (1980).

8 Bickford, D.F. and Jantzen, C.M., in Scientific Basis for Nuclear Waste Management VII (G.L. McVay ed.), Vol. 26, p. 557 (1984).
9 Smith, T.H and Ross, W.A., "Impact Testing of Simulated HLW in Canisters", BNWL-1903, Pacific Northwest Laboratory (1975).
10 Perez, J.M. and Westsik, J.H. "Effects of Cracks on Glass Leaching", ORNL Conference on Leachability of Radioactive Solids, p. 35 (1980).
11 Weber, W.J. and Roberts, F.P., Nucl. Tech. 60: p. 178 (1983).
12 Matzke, Hj. and vanGeel, J., in Disposal of Weapon Plutonium (E.R. Merz and C.E. Walter, NATO Series1: Disarmament Technologies, 4, p.93 (1996).
13 Bibler, N.E., in Scientific Basis for Nuclear Waste Management (S.V. Topp ed.), Vol. 6, p. 681 (1982).
14 T. McMenamin ed., Testing of HLW Forms Under Repository Conditions, Commission of European Communities, EUR 12 017 EN (1989).
15 T. McMenamin ed., In-Situ Testing of Radioactive Waste Forms and Engineered Barriers, Commission of European Communities, (1992).
16 Wicks, G.G. and Molecke, M.A., in Advances in Ceramics (G.B. Mellinger ed.), Vol. 39, p. 383, Am. Cer. Soc. (1994).
17 Williams, J.P., Wicks, G.G., Clark D.E., and Lodding, A.R., in Ceramic Transactions, Nuclear Waste Management IV, (G.G. Wicks, D.F. Bickford and L.R. Bunnell, eds.), Vol. 23, pp. 663-674 (1991).
18 Namboodri, C.G. Jr., S.L. Namboodri, Wicks, G.G., Lodding, A.R., Hench, L.L., Clark D.E., and Newton, R.G., in Ceramic Transactions, Nuclear Waste Management IV, (G.G. Wicks, D.F. Bickford and L.R. Bunnell, eds.), Vol. 23, pp. 653-662 (1991).
19 Schulz, R.L., Clark, D.E., Wicks, G.G., Lodding, A.R. and Van Iseghem, P., to be published in Advances in Ceramics (1998).
20 Wicks, G.G., McKibben, J.M., Plodinec, M.J. and Ramsey, W.G., in Disposal of Weapon Plutonium (E.R. Merz and C.E. Walter, NATO Series 1: Disarmament Technologies, Vol. 4, p.93 (1996).
21 Vernaz, E.Y., in "Glass as a Waste Form and Vitrification Technology", National Research Council Workshop, p. E26 (1996).
22 Clark, D.E., Pantano, C.G., and Hench, L.L., Corrosion of Glass, Books for Industry (1979).
23 Wicks, G.G., "Nuclear Waste Glasses: Corrosion Behavior and Field Tests", in Corrosion of Glass, Ceramics and Ceramic Superconductors (D.E. Clark and B.K Zoitos, eds.), p.218 (1992).
24 Hench, L.L., G.G., "Surface Modificaton of Bioactive Glasses and Ceramics", in Corrosion of Glass, Ceramics and Ceramic Superconductors (D.E. Clark and B.K Zoitos, eds.), p.298 (1992).
25 Wicks, G.G. and Wallace, R.M., in Scientific Basis for Nuclear Waste Management VI (D.C. Brookins ed.), Vol. 15, p. 23 (1983).

26 Pescatore, C. and Machiels, A.J., in Advances in Ceramics (G.G. Wicks and W.A. Ross ed.), Vol. 8, p. 508, Am. Cer. Soc. (1984).
27 Chamber, P.L., Kang, C.H., Dim, C.L. and Pigford, T.H., in Scientific Basis for Nuclear Waste Management XI (M.J. Apted and R.E. Westerman ed.), Vol. 112, p. 285 (1987).
28 Grambow, B., Lutze, W., Ewing, R.C. and Werme, L.O., in Scientific Basis for Nuclear Waste Management XI (M.J. Apted and R.E. Westerman ed.), Vol. 112, p. 531 (1987).
29 Plodinec, M.J., Jantzen, C.M. and Wicks, G.G., in Advances in Ceramics (G.G. Wicks and W.A. Ross ed.), Vol. 8, p. 491, Am. Cer. Soc. (1984).
30 Bourcier, W.L., in "Glass as a Waste Form and Vitrification Technology", National Research Council Workshop, p. E26 (1996).

PROGRESS OF CORROSION AND ELEMENT KINETICS OF HLW GLASSES IN GEOLOGICAL BURIAL; EVALUATION BY SIMS

Alexander R. Lodding
Chalmers University of Technology
S-41296 Gothenburg
Sweden

Pierre Van Iseghem
SCK•CEN
B-2400 Mol
Belgium

Lars O. Werme
Swedish Nuclear Fuel and Waste Management Company
Box 5864, S-10240 Stockholm
Sweden

ABSTRACT

The technique of in-depth concentration profiling by Secondary Ion Mass Spectrometry (SIMS) has developed into a versatile tool for evaluating the performance of nuclear waste form materials. Several long-term studies of glasses leached in interaction with geological environments have been evaluated with SIMS. Particularly extensive international burial tests of numerous waste forms leached under realistic repository conditions have been conducted in Sweden (granite environment), the US (salt & brine), and Belgium (Boom Clay). SIMS profiling quantitatively monitored the in-depth distributions of up to 40 elements, including very small concentrations of simulated nuclear waste, in residual corroded layers and precipitates on leached waste forms of about 25 different compositions. Variations in leachability in dependence of composition were charted, as well as typical elemental differences in kinetic response to leaching in the different geologies of long-term repository.

INTRODUCTION

The concentration profiling technique of Secondary Ion Mass Spectrometry, Dynamic SIMS [1-3], has been established in the study of nuclear waste glasses for about 15 years. The early 1980's saw a fast ripening in the consideration of glass as a preferred medium for immobilisation and long-term deposition of high-level radioactive material and other environmental waste [4-6]. Concerted efforts by numerous countries were directed to investigate interactions between glasses and different leaching media. Alkali borosilicate glasses with additions of different network forming, modifying and intermediate elements, within rather wide compositional ranges [7], were found to be acceptable from the points of view of production as suitable matrices for incorporation of radioactive waste. The main accent of research was to investigate the long-term durability of such glasses, and particularly the leachabilities of different consituent elements in different environments. Leaching studies in laboratories [8-9] implied that glasses of even moderately differing compositions could vary considerably in regard to their response to different leachants. The search for optimal waste glass composition became a dominant task, the main concern being the behaviour of such glasses in actually proposed geological environments as envisaged for long-term "burial" deposition of the waste form. It was gradually realised that laboratory testing under simulated geological conditions was not, by itself only, fully adequate for resolving the questions of waste glass behaviour in the actual repository. International consensus and collaboration in this field led to the establishment of long-term "in-situ" testing, for realistic apprisal of the effect of interactions between different waste forms, proposed package components, and environmental parameters of the burial sites.

To-date, such major projects, with test durations up to 5-12 years, have been conducted at Ballidon, UK (limestone environment) [10], Stripa, Sweden (granite environment) [11-13], Carlsbad, New Mexico (salt mine) [14-16], and at Mol, Belgium (clay environment) [17-19]. The setting-up and evaluation of these large scale ventures required an integrated technological-analytical-theoretical approach, involving exclusive techniques and specialised expertise. Consequently, considerable international collaboration was a precondition. Already in the laboratory stage of the glass leaching experiments, the evaluation of these projects required the integrated use of 5-10 analytical and microscopic techniques. A microanalytic-microstructural approach based on SIMS was developed for the study of corroded layers of glasses by the Chalmers University of Technology in collaboration with the University of Florida, the Savannah River Laboratory, and the

Swedish Nuclear Fuel and Waste Management Company [3, 20-21]. In addition to providing a sensitive quantitative analysis of the outermost glass layer, the technique yields detailed information regarding the variation in elemental composition as a function of depth. SIMS in-depth profiling has developed into a dedicated tool for characterising corrosion of glasses and ceramics [3,21-24], and has been taken into use in all hitherto initiated in-situ burial glass leaching projects. Table 1 gives relevant orientation on the scope of three such projects.

Table I. Simulated nuclear waste glasses in interaction with geological environment; three concluded long-term burial projects with employment of SIMS in post-leaching evaluation.

Location	Stripa, Sweden	WIPP, Carlsbad, NM	SCK•CEN, Mol, Belgium
Environment	Granite mine Ground water; Canister/overpack metals Bentonite backing	Salt mine Dry salt, brine; Canister/overpack metals Various backfills	Boom Clay Solid & slurry
Durations of exposure (months)	1, 3, 7, 12, 36, 48, 96	6, 12, 24, 60	24, 42, 60, 90
Temperatures of leaching (°C)	85, (10)	90, (10)	90, (15, 170)
Number of studied waste form compositions	6	16	10
Approx. number of multi-element SIMS profiles recorded	90	200	120

SIMS PROFILING OF ELEMENT CONCENTRATIONS IN GLASSES

The principle of the profiling technique involves sputtering the corroded glass surface while simultaneously recording the intensities of sputtered ions at the respective mass spectral peak of each element. The primary data supply the ion intensities, cyclically in parallel for all mass peaks of interest (for waste glass typically some 25 elements are cycled for each sputtered crater) as a function of sputtering time. The true depth co-ordinate for the leached layer, which may consist of successive sub-layers with different rates of sputter erosion, is obtained by taking into account the sputter rates, as determined by profilometry, for each sub-layer. Another computerised process converts the mass peak intensities to the respective element concentrations by use of pre-determined sensitivity factors. The resulting data are normalised to 100%, providing the elemental in-depth profiles in mole percent of the total sum of cations or anions. If it can be assumed that at least one of the glass constituent elements has remained practically undepleted in the leaching process (as frequently implied from leachate analysis for relatively "inert" elements in waste glasses, such as e.g. Al, Zr, Cr), the concentration data may be further processed to yield the actual amounts of leached-out or in-penetrated elements, i.e., the total loss or influx of any element in the interaction layer of leaching.

In nearly all modern glass leaching studies, the utilised SIMS equipment was of a commercial basic design. However, the task required some special modifications; in particular, a program had to be developed [3,20,25] to compensate for the build-up of surface charge during sputtering of the insulating specimen.

In the present stage of technique development, the elemental detection sensitivity of glass profiling is in 0.1-10 ppm range, and the depth resolution (depending on the as-corroded surface roughness and on the depth of sputter penetration) may vary between some 10 and 100 nm. The lateral resolution, in the order of 10 μm, allows the identification and selection of particular areas on specimen surface.

For more detailed presentation and discussion of SIMS applications to glasses, ceramics and biomineralizations, one may refer to recent review literature [3,25].

ELEMENTAL TENDENCIES IN REACTED NEAR-SURFACE LAYERS

Fig. 1 shows a set of typical in-depth variations in element concentrations in a leached waste glass, as determined by SIMS. One may readily distinguish successive layers and sublayers in the reacted glass, here numbered (in accordance with a convention largely based on SIMS profiling [7,21]) α (precitation layers) and β (depletion zones). The two outermost sub-layers are interpreted as consisting of re-precipitated products of reaction between the leached-out glass constituents and the outside fluid; in microscopic scrutiny, α_0 is implied to be composed of amorphous and microscopic salt deposits, while α_1 has a more homogeneous glassy aspect. The "major depletion layer", named β_0 (sometimes referred to as the "gel") represents the residual outer portion of the glass, selectively depleted in certain "mobile" elements, such as boron, the alkalis and alkaline earths, provided that they are not abundant in the leachant. The loss of such elements, and of elements of intermediate leachability, is at least partially compensated by an influx from the leachant, mainly hydrogen. Beneath β_0 lies the "gradient sub-layer", β_1, understood as containing the main reaction front of the glass in interaction with the leachant; and the "diffusion sub-layer", β_2, characterised mainly by an initial loss of mobile elements, compensated by corresponding substitution with hydrogen.

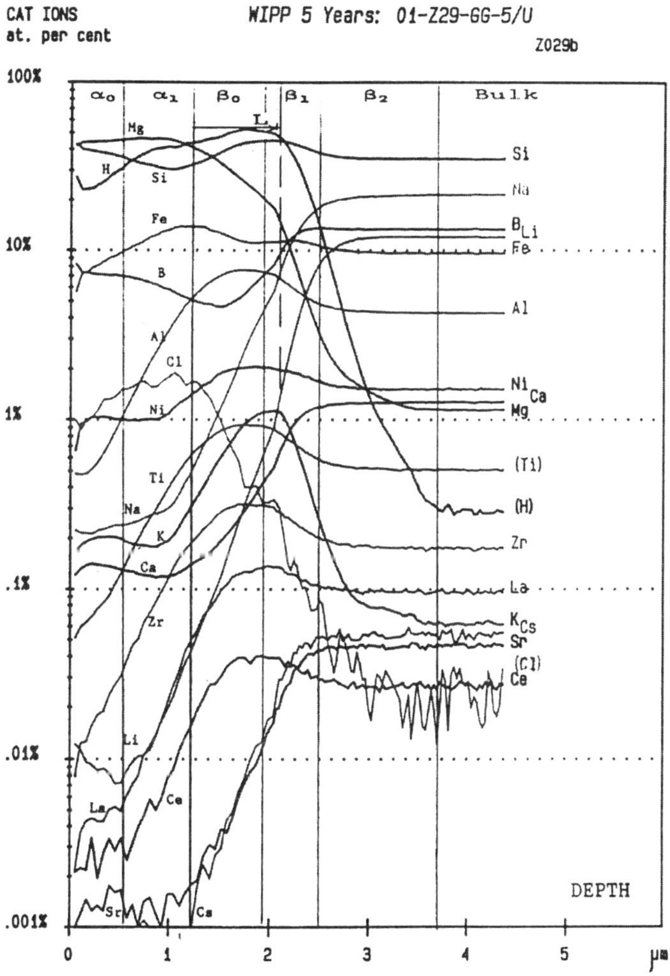

Fig. 1. In-depth concentration profiles of cations (mole%) at a glass-glass interface after five years' leaching in salt environment

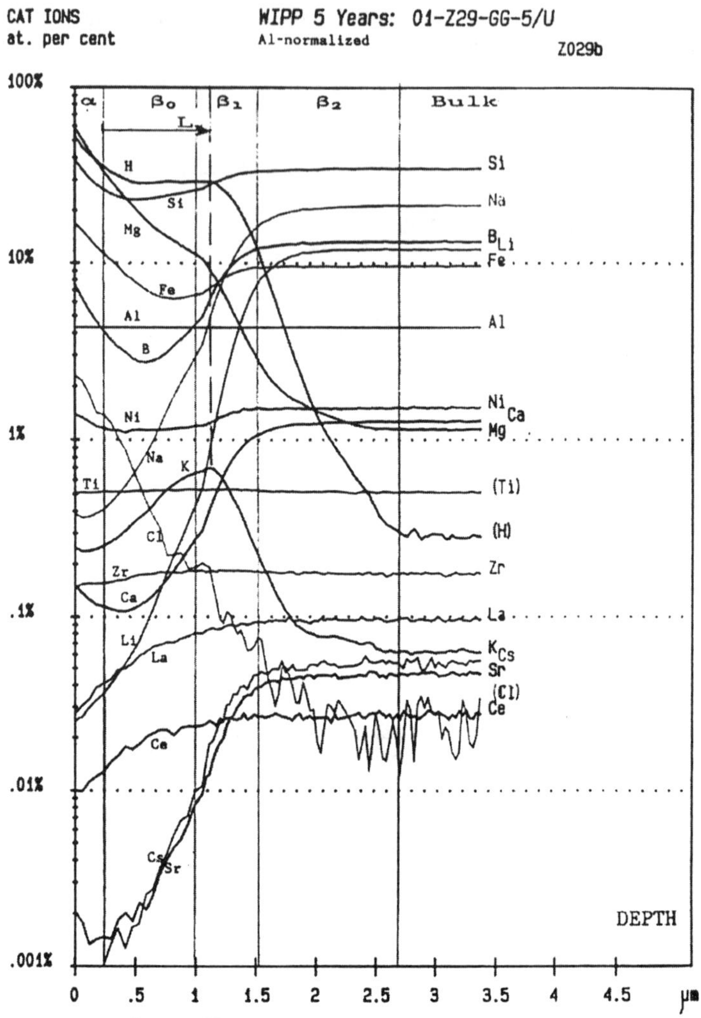

Fig. 2. Element concentration profiles, porosity corrected (by normalization to Al). α_0-sublayer omitted here

One may notice in Fig. 1 an apparent "enrichment" in the β zones of elements relatively inert to leaching, such as the main matrix component Si, Al, and some transition elements. However, this observed increase in some concentrations is related to the total sum of cations (except H), after the loss of nearly half of them into the leachant. Thus actually these "inert" elements remain nearly undepleted in the reacted glass (as confirmed e.g. by selective leachant analyses, conducted in parallel [9,26]). On the assumption that at least one of the "inert" elements remains preserved in the residual leached glass, one may normalise the profiles to the bulk concentration of such an element (here Al). The resulting "porosity-corrected" profiles (see Fig. 2) represent the volumetric concentration of each element in the β layers, which allows the quantitative determination (e.g. in $mmol/m^2$) of the loss or influx of each element. This at least in principle eliminates the need of separate leachant analysis, and conveniently establishes a very complete view of element kinetics in the leaching process. The above condition of at least one "inert element" of course poses a limitation to the applicability of such a procedure. If the glass is being leached by congruent dissolution, quantification requires complementary techniques such as weight loss determination [17], painstaking leachant analysis [26] and/or marker indication of the original pre-leaching surface [27]. However, it is found that the majority of "preferred" glass compositions, considered for practical use as waste forms, at least during the initial years of burial exhibit a selective rather than congruent mode of leaching, and thus preserve a residual network of "inert" elements. Fig. 2 [16] summarises the model of sub-layer formation as suggested by the results of elemental profiling by SIMS in glasses corroded by predominantly "selective-substitutional" (SS) [13,28] leaching mechanism.

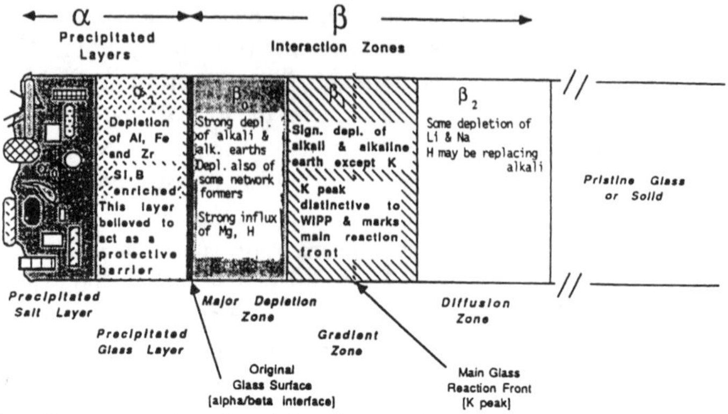

Fig. 3. Schematic representation of precipitated and depleted layers. Example of SS-type leaching of HLW glasses in salt environment [7]

Listed in Table 2 are typical elemental tendencies for three glasses recognised as durable waste forms in their respective burial environments. The depletion factors are defined as the residual fractions of the pre-leaching (bulk) concentrations (c_o) which are seen retained in the β_0 subzone after leaching. Thus DF=1 applies to "inert" elements, while low DF values mark "mobile" ones, highly depleted in the reacted glass. In the cases of elements which have entered the residual glass from the leachant (such as always hydrogen), their concentrations (related to the sum of bulk cations) are entered in brackets. Also listed for comparison are the depths of leaching attack for the three cases.

Table II. Depletion factors ($DF = c_{\beta 0}/c_o$) and interacted depths (L) of selected glasses after 5 years' leaching at about 90°C in geological burial environments (Brackets: *influx* in β_o-layer, in mole% of cations in bulk glass)

Glass sort Environment	Elemental depletion factors										
	Si	B	Al	Li	Na	Mg	Ca, Sr, Ba	La, Ce, Nd	Ti, Cr, Zr	Fe, Ni	H	L (μm)
ABS118 (JSS) Granite mine	0.8	0.05	0.95	.004	0.25	0.55	0.4	0.75	1.0	0.95	(20%)	2.3
SRL165 (US) Salt mine	0.35	0.3	1.0	.0003	0.01	(30%)	0.03	0.65	1.0	0.6	(25%)	2.2
SAN60 (Pamela) Boom Clay	1.0	0.003	0.9	.003	0.02	(4%)	0.4	0.6	0.9	0.8	(15%)	35 (ca)

RATE OF GROWTH OF REACTED LAYERS

As seen in figs. 1 and 2, the SIMS technique provides depth measurement of the residual glass layers. Again, a condition for assessing the whole reaction is that no material has been lost by congruent dissolution. Another complication, as sometimes noticed when the reacted depth exceeds a few tens of μm, may arise when the outermost layers are unavailable due to detachment ("flaking") from substrate. In such cases, the determination of total interacted depth may again require techniques complementary to SIMS [17,26,27].

Figs. 4 a,b,c show the progress of corrosion on some representative glasses, subjected to prolonged leaching in, respectively, the Stripa granite mine, the WIPP salt mine, and the Boom Clay at Mol. The depths L are counted from the outer edge of the β_0 sublayer to the reaction front in β_1 (see figs. 1-3). In Fig. 4c, for the case of burial in clay, results from laboratory simulation tests of intermediate durations have been included. As for the L assignment for the long-term Mol burial results, the SIMS profiles were affected by local flaking and by increasing prevalence of the congruent leaching mode. The reacted depths in clay at 90°C were therefore calculated from parallel weight-loss measurements [19].

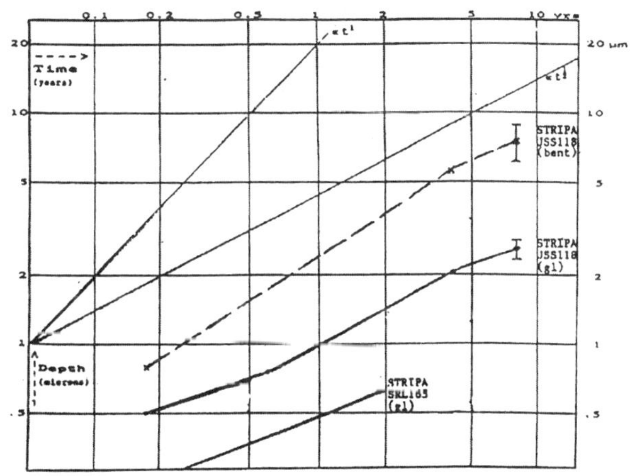

Fig. 4a. Leaching in granite environment. Reacted depth versus duration of burial

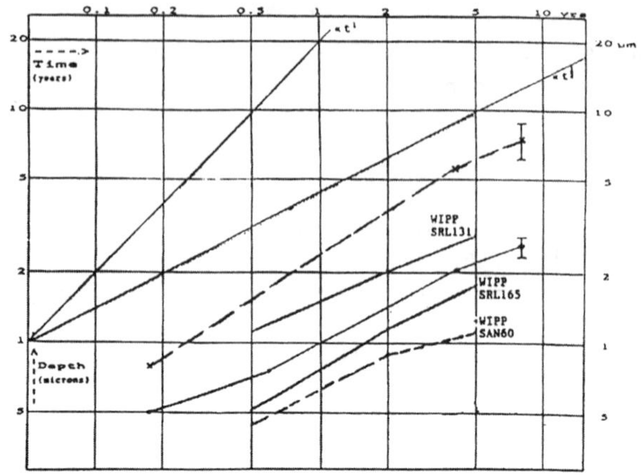

Fig. 4b. Leaching in salt and brine environment. Reacted depth versus duration of burial (details from fig. 4a included for comparison)

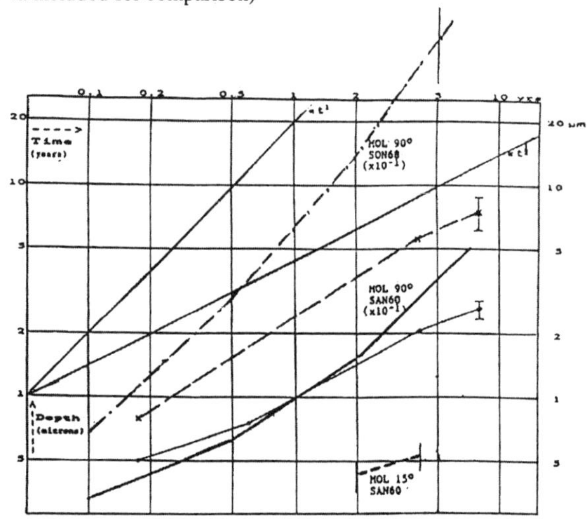

Fig. 4c. Leaching in Boom clay. Reacted depth versus duration of burial (details from fig. 4a included for comparison)

CHARACTERISTICS OF WASTE GLASS BEHAVIOUR IN DIFFERENT TYPES OF GEOLOGICAL BURIAL

a) Granite environment

In the Stripa mine, the Swedish ABS118 glass (similar in composition to the French Cogéma high level radioactive waste forms) was buried for 8 years, and other international HLW glasses for periods between 1 and 3 years. The behaviour of ABS 118, as shown in fig. 4a and in Table 2, may be considered as qualitatively typical also for the other 5 glasses tested in Stripa. One of them, the SRLY glass (165/28) is included in Fig. 4a for comparison. Between different glass compositions, the rates of corrosion were found to differ only moderately, viz. approximately within a factor 5 from the slowest to the fastest, as interpolated to 5 years' leaching at 85°C and to normal sample packing in the experimental assembly [12,13]. The slowest rate here corresponded to about 1 μm in 5 years.

As expected from laboratory experiments, the rates were found to vary appreciably with the SA/V (surface area to volume) ratio; at tight packing between neighbouring surfaces, the rate of attack was typically only 20% of that at surfaces where the flow of leachant was less restricted.

Glass surfaces in contact with bentonite clay (present in the assembly as a backfill alternative) were leached faster (by a factor 3 to 5 after four years) than normally stacked glass-glass interfaces; this might possibly be attributed to the sorption tendency of clay to silicic acid, inhibiting saturation at attacked glass surfaces [29].

The build-up of the precipitation layers I is in the order of 0.1 μm/year, and possibly beneficial [7] in promoting a selective-substitutional (SS) mode of long-term leaching, in which the growth of the reacted layer takes place with a root-of-time ($L \propto t^{1/2}$) characteristic. Although this is likely to give way gradually to an increasing share of pseudo-congruent mode (PC; [7,18]), with $L \propto t$ behaviour, one may conclude that 100 years of leaching of accepted HLW glass surfaces in granitic ground water environment, even if as elevated a temperature as 85°C is prevalent, will not affect more than some 30-50 μm in depth, corresponding to a glass leach-out in the order of 1 g/m²y.

b) Salt and brine

A considerable number of suggested and to some extent accepted waste forms were investigated in the international WIPP-MIIT project, involving up to 5 years burial in contact with salt or brine at 90°C. Under consideration were not only different compositions and storage variables, but also parameters such as different canister and overpack materials [7,30]. Some major points emerging from the SIMS study may be seen in Table 2 and in Fig. 4b.

The more than 15 international alkali-borosilicate glasses and related waste forms, studied at WIPP, ranged within wide limits with respect to concentrations of ingoing network-formers, modifiers, intermediates, and admixtures of simulated nuclear waste. Nevertheless, the depths of corrosion (L) after 5 years in the WIPP environment were found to differ only moderately between the various compositions, viz. roughly within one order of magnitude, from ca. 0.5 to 5 μm. This is similar to Stripa in order of magnitude. Also as in Stripa, the mutual differences in leaching performance could to some extent, but not entirely, be related to differences in elemental composition via the ratios of network-formers to modifiers [15].

As is illustrated in Fig. 4b by the recommended US composition SRL165, most of the waste forms in WIPP exhibited the selective leaching mode (SS), with an approximate root-of-time growth behaviour of L. The reacted layers exhibit (as in Stripa) a strong depletion of alkali and alkaline earth elements. The main network formers Si and B are here (see Table 2) also seen to be depleted, typically to some 30% of the bulk concentrations. This behaviour is in contrast to that in Stripa and in Boom clay, where the main network former, Si, behaves as one of the nearly "inert" elements, while the trivalent subsidiary network former, B, suffers nearly total depletion (see Table 2). In the salt and brine environment of WIPP, contrary to the other two repositories, boron is an abundant element of the leachant, which tends to preserve the subsidiary network in the borosilicate glasses. Another dominant element in WIPP geology is magnesium. To these compositional attributes of the leachant one may relate the observed distinctive features of leaching in WIPP: a) the formation of relatively thick (and possibly protective [7]) layers of surface precipitation with Mg and B as dominant cations; and b) the high influx of Mg as well as the relatively low loss of B in the reacted layers of the waste form [15]. The concentration of in-penetrated magnesium in the β layers is quantitatively comparable to the outflux of the original network formers, and appears to stabilise a residual network in the reacted glass [21]. As, moreover, the main tendencies were seen to be essentially similar for all waste forms leached in WIPP (even for glasses with compositions much deviating from the mean alkali borosilicate formula), and the predicted loss into the environment is throughout (even at an "unfavourable" assessment of ambient temperature) less than ca 2 $g/m^2.y$, the stable salt mine environment appears in several aspects attractive as a final repository for HLW glass.

c) Boom Clay

As may be seen in the last column of Table 2 as well as in the erosion-versus-time characteristics in Fig. 4c, the interaction at 90°C of the alkali borosilicate waste forms with the environment is considerably faster in the clay repository at Mol than it is in granite or in salt. For example, the extent of corrosion in the Pamela-type SAN60 glass in Boom Clay, about 35 μm in 5 years, may be compared (see fig. 3b) with the performance of the same glass in WIPP, with a reacted depth of only about 1 μm. One also notes an earlier onset of congruent dissolution ($L \propto t$) than was observed in the two other environments. Furthermore, the material consumption is found [18,19] to be even more avid for other glasses buried in Boom Clay. E.g., as seen in Fig. 4c, in the Cogéma glass SON68 the reacted layer after 5 years is by an order of magnitude deeper than in SAN60; and SON68 also exhibits at an earlier stage a predominantly congruent mode of dissolution. The difference in performance appears at least partially connected with the relatively high content of Al in the more resistant SAN60 glass.

The comparatively fast rates of dissolution of borosilicate glasses, as observed for interaction with Boom Clay, are almost certainly attributable to the sorption effects of the clay, affecting the saturation of silicic acid at the leached surface [20]. By contrast, and in line with the sorption argument, the non-silicate (Ti-Ca-Al-Zr based) Synroc waste form [31], after 3.5 years exposure to Boom Clay, was found by SIMS to be only slightly corroded (less than ca 1 μm in depth), and in distinction to the case of borosilicates the presence of clay enhanced the congruent solubility of transition element such as Ti, Zr and the rare earths [24].

In spite of the comparatively high leachability of alkali borosilicate waste glasses at 90°C, experience nevertheless suggests Boom Clay to be a satisfactory long term repository environment. Among the advantages one notes the homogeneity, plasticity and relatively low diffusive penetrability of clay. Further, the actual ambient temperature of storage is expected to be much lower than in the majority of the burial experiments, viz. soon to attain only 10-20°C. Laboratory experiments as well as burial studies [7,28,32] reveal very strong temperature dependence of the rates of leaching. Included in Fig. 4c are the results for the

SAN60 glass buried at about 15°C; SIMS profiling after 3 and 3.5 years has revealed [28] a reaction depth of corrosion in the order of 0.5 μm, which can be extrapolated to less than ca 20 μm in 100 years. This is of the same order of magnitude as in Stripa or WIPP at 90°C, and is considered as fully acceptable from the practical safety point of view of a long-term HLW repository.

PERSPECTIVES FOR FUTURE WORK

In interdisciplinary research on the production and properties of glasses and ceramics, the usefulness of SIMS is to-day widely acknowledged [3]. In the particular field of high level waste forms in interaction with different environments, the number of involved SIMS teams has hitherto been limited, but expanded applications of the technique are included in on-going and projected work [7,19,32].

Hitherto most applications have been directed to in-depth profiling of cation concentrations in corroded layers typically between some 10^{-1} - 10^{3} μm in depth. Complete microanalytical characterisation of the reactions involved in corrosion would require parallel information on anion distributions. Modified routines for in-depth profiling of elements such as carbon, nitrogen, oxygen, sulphur and the halogens have been satisfactorily tested [21,33]. For future work it is envisaged to include quantitative anion profiling in the established procedure.

Further, it is hoped to adapt the imaging facility of SIMS [2] to use on insulating materials. Imaging of element distributions in successive depth layers in glasses, represented as taper views of sputtered crater walls, has been achieved [33]. The advanced goal is the application of SIMS for detailed topographical imaging and stoichiometric identification of the various precipitates at the outermost surface (in the α layers) of leached glasses. This task is challenging due to problems of charging on insulating specimens; only scanty experience is hitherto available on ion imaging of glass surfaces. However, for related ceramics excellent SIMS images of corroded surfaces have been recorded [3], showing distributions of elements and phases with a sub-micron lateral resolution. In the study of waste forms in interaction with geological environments, the use of such a technique to define the composition of rate-relevant surface layers is of obvious interest.

REFERENCES

[1] A. Benninghoven, F.G. Rüdenauer, H.W. Werner: "Secondary Ion Mass Spectrometry, Basic Concepts, Instrumental Aspects, Applications and Trends"; Chem. Analysis Ser. 86, Wiley-Interscience, New York (1987).

[2] A. Lodding and U. Södervall: "Secondary Ion Mass Spectrometry, Dynamic SIMS"; Ch. 13 in "Surface Characterization, a User's Sourcebook" (D. Brune & R. Hellborg, Eds.), VCH Publ., Weinheim FRG (1997).

[3] A. Lodding: "Characterization of Corroded Ceramics by SIMS"; Ch. 4 in "Corrosion of Glass, Ceramics, and Ceramic Semiconductors" (D.E. Clark & B.K. Zoitos, Eds.), Noyes Publ., Park Ridge, NJ (1992).

[4] L.L. Hench, in "Ceramics and Glass Radioactive Waste Forms" (D.W. Readey & C.R. Cooley, Eds.), CONF-770102, 49 (1997).

[5] D.E. Clark, C.G. Pantano, L.L. Hench, "Corrosion of Glass"; Magazines for Industry Inc., N.Y. (1979).

[6] J.E. Mendel, "A Review of Leaching Test Methods and the Leachability of Various Solid Media Containing Radioactive Wastes, BNWL-1765, Pacif. Northwest Lab., Richland, WA (1973).

[7] G.G. Wicks, "Nuclear Waste Glasses; Corrosion and Field Tests"; Ch. 7 in "Corrosion of Glass, Ceramics and Ceramic Semiconductors" (D.E. Clark & B.K. Zoitos, Eds.), Noyes Publ., Park Ridge, NJ (1992).

[8] Materials Characterization Center, "MCC-1 Static Leach Test", PNL 3990, Pacific Northwest Laboratory, August 1980.

[9] D.E. Clark, A. Lodding, H. Odelius, L.O. Werme, Materials Sci. & Eng. 91, 241 (1978).

[10] R.G. Newton, Glass Tech. 29(1), 43 (1988).

[11] L.L. Hench, O. Werme, A. Lodding, in "Scientific Basis for Nuclear Waste Management V" (W. Lutze, Ed.), Elsevier, New York, 153 (1982).

[12] L.O. Werme, in "Testing of High Level Waste Forms under Repository conditions" (T. McMenamin, Ed.), CEC-EUR 12017 EN, 3 (1989).

[13] L.O. Werme, H. Odelius, A. Lodding, in "In Situ Testing of Radioactive Waste Forms and Engineered Barriers (T. McMenamin, Ed.), CEC-EUR 15629 EN (1993).

[14] G.G. Wicks, A. Lodding, P.B. Macedo, D.E. Clark, in "Scientific Basis for Nuclear Waste Management XV", Materials Res. Soc. 257, 65 (1992).

[15] A. Lodding, D.E. Clark, G.G. Wicks, in "In Situ Testing of Radioactive Waste Forms and Engineered Barriers" (T. McMenamin, Ed.), CEC-EUR 15629 (1993).

[16] G.G. Wicks, A. Lodding, M.A. Molecke, MRS Bulletin XVIII-9, 32 (1993).

[17] P. Van Iseghem, Mat. Res. Soc. 333, 133 (1994).

[18] A. Lodding & P. Van Iseghem, Mat. Res. Soc., 412, 229 (1996).

[19] P. Van Iseghem, Proc. CEA Summer Univ. on Glass (Méjannes, France, 1997), in press.

[20] A. Lodding, H. Odelius, D.E. Clark, L.O. Werme, Mikrochim. Acta, Suppl. 11, 145 (1985).

[21] A. Lodding, D.E. Clark, E.U. Engström, H. Odelius, M. Schuhmacher, B.K. Zoitos, G.G. Wicks: "SIMS Applications on Nuclear Waste Form", in "Ceramics Today - Tomorrow's Ceramics" (P. Vincenzini, Ed.), 3121-3129, Elsevier, Amsterdam (1991).

[22] V.M. Oversby & D.L. Phinney, J. Nucl. Mat. 190, 247 (1992).

[23] D.E. Clark, R.L. Schulz, A. Lodding, M. Friesel, G.G. Wicks, in "Environmental and Waste Issues in the Ceramic Industry", 409-426, ACerS, Westerville OH (1994).

[24] P. Van Iseghem, Wei Jiang, M. Blanchaert, K. Hart, A. Lodding, Mat. Res. Soc. 412, 305 (1996).

[25] A. Lodding, "SIMS of Mineralized Tissues: Present Trends and Potentials". Adv. Dent. Res. 11(4), 364-379 (1997).

[26] M. Brandys, M. Gong, R.E. Sasson, A. Barkatt, P.B. Macedo, Ceramic Trans. 9, 287, ACerS, Westerville, OH (1990).

[27] B.K. Zoitos, R.L. Schulz, D.E. Clark, A. Lodding, Ceramic Trans. 9, 297. ACerS, Westerville, OH (1990).

[28] A. Lodding, H. Odelius, P. Van Iseghem, in "In-Situ Testing of Radioactive Waste Forms and Engineered Barriers" (T. McMenamin, Ed.), CEC-EUR 15629 (1993).

[29] P. De Cannière, H. Moors, A. Dierckx, F. Gasiaux, M. Aertsens, M. Put, P. Van Iseghem, in “Radiochimica Acta” Vol. 82, 191-196 (1998), R. Oldenbourg Verlag, München (1998).

[30] D.E. Clark, R.L. Schulz, A. Lodding, A.W. Tipton, G.G. Wicks, in "In-Situ Testing of Radioactive Waste Forms and Engineered Barriers" (T. McMenamin, Ed.), CEC-EUR 15629 EN (1993).

[31] A.E. Ringwood, S.E. Kesson, K.D. Reeve, D.M. Levins, E.J. Ramm, in "Radioactive Waste Forms for the Future" (W. Lutze & R.C. Ewing, Eds.), 233-334, North Holland (1988).

[32] R.L. Schulz, D.E. Clark, G.G. Wicks, A. Lodding, R.G. Newton, K. Welcher, Cer. Trans. 61, 569, ACerS, Westerville, OH (1995).

[33] A. Lodding, E.U. Engström, B.K. Zoitos, D.E. Clark, G.G. Wicks, J. ACerS 75-10, 2702 (1992).

ACKNOWLEDGEMENTS

The authors wish to compliment Professor Larry L. Hench for his early insight, initiative and enthusiasm. His encouragement was an essential factor in the establishment of the SIMS technique in the study of nuclear waste glasses. We also gratefully acknowledge the valuable collaboration of Professor D.E. Clark and his team at the Univ. of Florida in important stages of the work. The contact with Dr. G.G. Wicks, Savannah River Laboratory, is a constant inspiration in our research.

FT-IRRS ANALYSIS OF SURFACE REACTIONS IN ALKALI SILICATE GLASSES

Greg Darby, David Clark and Joseph Simmons
Department of Materials Science and Engineering
University of Florida
Gainesville, FL 32611

ABSTRACT

Infrared specular reflectance spectra of a 33 mol% Na_2O/67 mol% SiO_2 glass were obtained at various degrees of surface polish. It was found that beyond a 600 grit polish, finer polishing could not keep up with the corrosion rate due to exposure to ambient humidity. The infrared specular reflectance spectra of three sodium silicate glasses were also investigated as a function of exposure time to 52% relative humidity and 81% relative humidity over a 24 hour period. It was found that under low humidity conditions, minor surface chemical/structural changes occured in a linear fashion. Under high humidity conditions, however, major changes ocurred to the surfaces in an exponential fashion. Surface carbonates were observed to have formed in these samples.

INTRODUCTION

Long term performance (durability) must be predicted through the use of short term laboratory and field tests. The focus of the paper will be on laboratory tests. These tests range from simple static leaching experiments to dynamic leach tests, where the solution is continuously replenished. Tests to predict long term glass performance continue to be devised, such as the pressurized unsaturated flow (PUF) test [1] and the drip test, discussed in this proceedings. In order to assess the validity of a particular test it is often necessary to run the experiment on simple, binary compositions so that the results may be better analyzed and correlated. Fourier Transform Infrared Reflectance Spectroscopy (FT-IRRS) has proven to be a useful tool for characterizing surface reactions on glasses. A drawback to studying binary compositions, however, is that these glasses are particularly susceptible to water corrosion, and protocols must be established so that valid results may be obtained. Sanders et al were aware of the effects of water vapor on their binary glasses and took precautions to minimize exposure [2]. The objective of this study is to 1) establish a protocol for polishing of glass samples to be analyzed by FT-IRRS, and 2) determine the effects of exposure time on sodium silicate binary glass samples different ambient humidity conditions prior to characterization, using FT-IRRS.

EXPERIMENTAL

Glass Preparation

Three glasses of the following compositions were prepared: 1) 20 mol% Na_2O-80 mol% SiO_2 (20N), 2) 25 mol% Na_2O-75 mol% SiO_2 (25N), and 3) 33 mol% Na_2O-67 mol% SiO_2 (33N). These compositions were prepared using reagent grade Na_2CO_3 and SiO_2. Each glass was melted in a platinum crucible at 1400°C in a Deltec high-temperature furnace and allowed to homogenize at this temperature for 24 hours. Each glass was then cast in a 1 inch diameter graphite mold, and the cast rod was annealed in a Thermolyne box furnace at 500°C for 1.5 hours. Following annealing, each glass rod was stored in a Blue-M drying oven.

Polishing Study

A 2mm thick disc was cut from the 33N glass rod using a Beuhler Isomet diamond wafering saw. This disc was then polished using SiC polishing paper and a polishing wheel. The following grit sequence was used when polishing: 240grit, 320grit, 400grit, 600grit, 17.5 micron Al_2O_3 powder and 1 micron Al_2O_3 powder. Following each stage of polishing, an FT-IR specular reflectance curve was obtained on the polished surface using an Impact 410 Nicolet FT-IR spectrometer over 32 scans at a resolution of 2 cm^{-1}. The minimum level of polishing required to obtain the optimum reflectance spectra was noted. The peak intensities of the Si-O-Si (located @ ~1070 cm^{-1}) bridging oxygen stretching vibration and the Si-O-Na (located @ ~970 cm^{-1}) non-bridging oxygen interaction were used as the standard for determining the optimum spectra [3]. Sanders et al showed experimentally that infrared reflections are proportional to the concentration of the vibrational species causing them [2]. In pure vitreous silica the Si-O-Si stretch occurs at approximately 1120 cm^{-1}. This stretching peak is lower in the 33N glass due to the presence of sodium.

Relative Humidity Experiments

One 2mm thick disc was cut and polished to 600 grit from each of the following cast rods: 20N, 25N and 33N. The reflectance spectra was measured for the as-polished surface and at the following time intervals, during which the samples were exposed to either 52% or 81% relative humidity conditions: 30 minutes, 1 hour, 3 hours, 6 hours and 24 hours. The experiments were run in an empty desiccator box for the low RH experiments and in a desiccator box including an open 60 ml teflon container filled with water for the high RH experiments. The desiccator box was allowed to equilibrate over a 24 hour period. These two conditions were intended to simulate the extremes in relative humidity in the laboratory on a given day. The relative humidity was monitored and recorded using a Fisher Scientific Digital Humidity Meter at the time of each measurement. All experiments were run at 30°C. Changes to the reflectance curves in these samples were noted..

RESULTS

Polishing Study

Figure 1 depicts the effects of polishing on the specular reflectance curves of the 33N glass. As the 33N glass was polished to finer levels, from 240 grit to 600 grit, it was observed that the peak intensities of the Si-O-Si stretch and the Si-O-Na increased. With the 33N composition, the Si-O-Si stretching peak occurs at approximately 1070 cm^{-1}, while the Si-O-Na interaction peak occurs at approximately 970 cm^{-1}. The increase in peak intensity with polishing may be attributed to the increase in surface reflectivity. This is consistent with a similar experiment conducted by Newton and Seddon on a simulated medieval glass [4]. With the 17.5 micron polish there appeared to be no significant increase in these peaks. In fact, with the 17.5 micron polish, the Si-O-Na peak appears to decrease somewhat. With the 1 micron polish, the Si-O-Na peak all but vanishes. The decrease in this non-bridging oxygen peak may be attributed to the susceptibility of the 33N composition to water corrosion, even at ambient humidity. The 17.5 micron and 1 micron polish required the use of Al_2O_3 powders, which was more time consuming a task than the 600 grit and lower polishes. Apparently the corrosion rate (growth of the alkali-depleted layer) was occuring at a faster rate than the polishing could remove in the finer polishing steps, resulting in the decrease in the Si-O-Na peak. Seddon and Newton observed increasing reflectivity up to 1 micron polishing, however, their results may be attributed to the higher durability of their simulated medieval glass to ambient humidity.

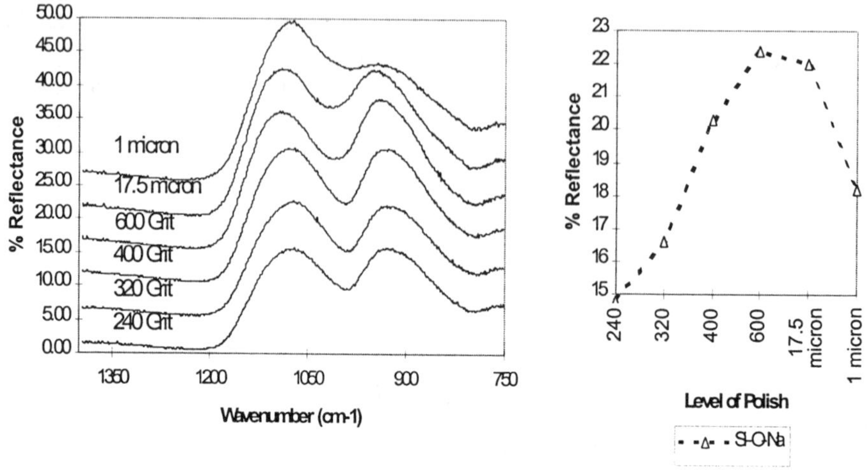

Figure 1. Effects of Polishing on the Reflectance Spectra of a 33N glass. The plot on the right shows the change in Si-O-Na peak intensity with increasing level of polishing.

Relative Humidity Experiments

Figure 2 shows the corrosion behavior of a 20N polished glass sample in average relative humidities of 52% (low) and 81% (high). Samples were exposed to each humidity condition for up to 24 hours. The plots on the left represent specular reflectance curves of the glass surface as a function of exposure time. Each spectrum is offset by 5% on the y-axis for ease of readability. The plots on the right depict changes in the Si-O-Si stretching vibration peak intensity and peak position with time. Zhang et al used peak position versus time plots to monitor changes in their binary glasses [5]. Notice that the y-axis scales are identical in these plots for the purposes of comparison. The solid and dashed lines were drawn in order to guide the eye. Comparing the plots of reflectance curves with exposure time, there does not appear to be significant changes over the 24 hour period. However, upon examining the change in peak intensity/position plots, there is an effect on the surface of these samples over the 24 hour period and this effect is dependent on the ambient humidity. Notice that the peak position shifts to higher wavenumbers in a linear fashion as time progresses. The slope of the line representing the change in peak position with time under the "high" RH condition is on the order of 4 times greater than that of the "low" RH condition. With the change in peak intensity with time curves, it is interesting to note the small and sudden increase in peak intensity intitially and then the gradual drop off as time progresses in both "low" and "high" humidity conditions. This small "bump" in peak intensity will be discussed later.

20N: 52% RH

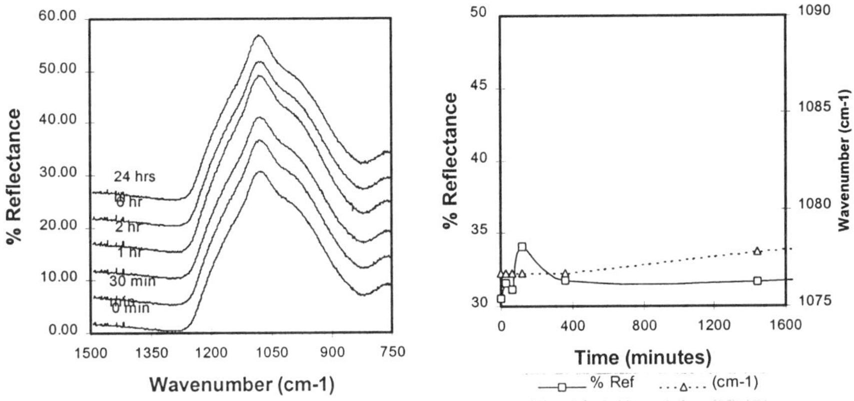

20N: 81% RH

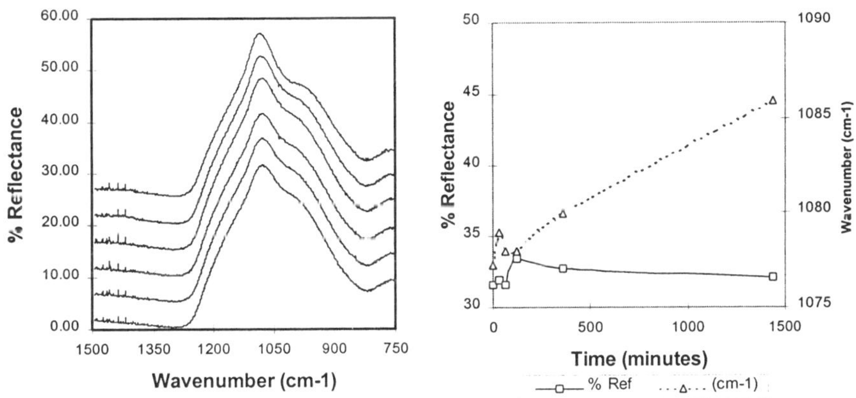

Figure 2. Corrosion behavior in a 20N glass at 52% relative humidity and at 81% relative humidity. Plots on the left show FT-IRRS curves and plots on the right show changes in Si-O-Si peak height and position versus time.

Figure 3 shows the corrosion behavior of a 25N polished glass sample in average relative humidities of 52% and 81%. The 25N glass showed similar behavior as the 20N glass under the "low" humidity condition (a small, linear increase in peak position and sudden increase in peak intensity followed by a gradual decrease with time). Under "high" humidity conditions, however, there was a noticeable change in behavior. Instead of a linear increase in peak position there appears to be more of an exponential increase with time. Also, referring to the reflectance spectra, there is the onset of a new peak at approximately 1450 cm^{-1} after about 2 hours of exposure to the "high" humidity condition. At this time a white crystalline precipitate can be seen to have formed on the surface of the glass. The peak at 1450 cm^{-1} has been attributed to the formation of surface carbonates, $Na_2CO_3 \bullet xH_2O$ [6]. Differential thermal analysis was run on crushed glass of the same composition that was corroded and a large endothermic peak appeared at approximately 850°C, which corresponds to the melting point of sodium carbonate (851°C).

25N: 52% RH

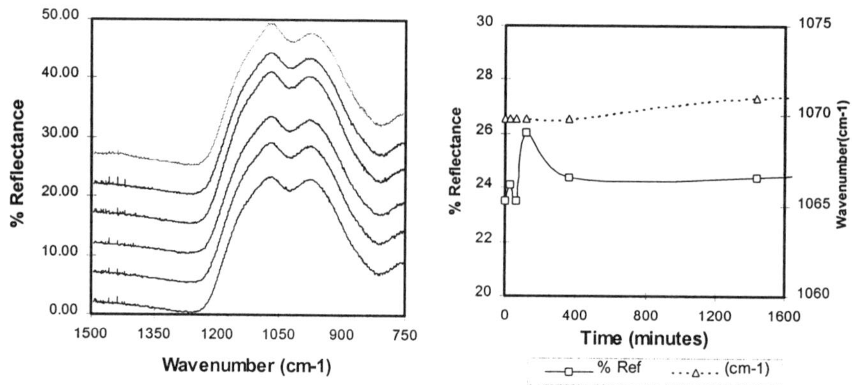

25N: 81%RH

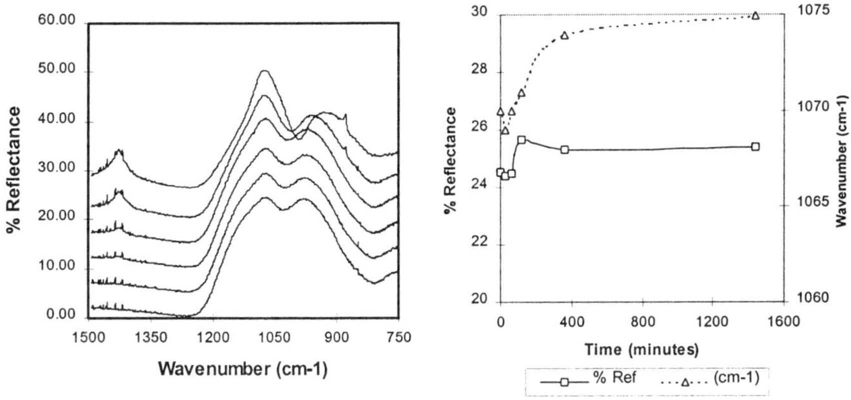

Figure 3. Corrosion behavior in a 25N glass at 52% relative humidity and at 81% relative humidity. Plots on the left show FT-IRRS curves and plots on the right show changes in Si-O-Si peak height and position versus time.

Figure 4 shows the corrosion behavior of a 33N polished glass sample in average relative humidities of 52% and 81%. Here, the carbonate peak becomes noticable after 24 hour exposure under the "low" humidity condition. There is a linear decrease in the peak position with time of the 33N composition under "low" humidity conditions. The small initial increase in peak intensity is present in this case as well. Under the "high" humidity condition, the formation of the 1450 cm^{-1} carbonate peak is immediate (30 min. exposure). The peak position of the Si-O-Si stretch changes exponentially with time to lower wavenumbers. The small increase in peak intensity at low times is barely noticeable with the drastic decrease in surface reflectance with time.

33N: 52% RH

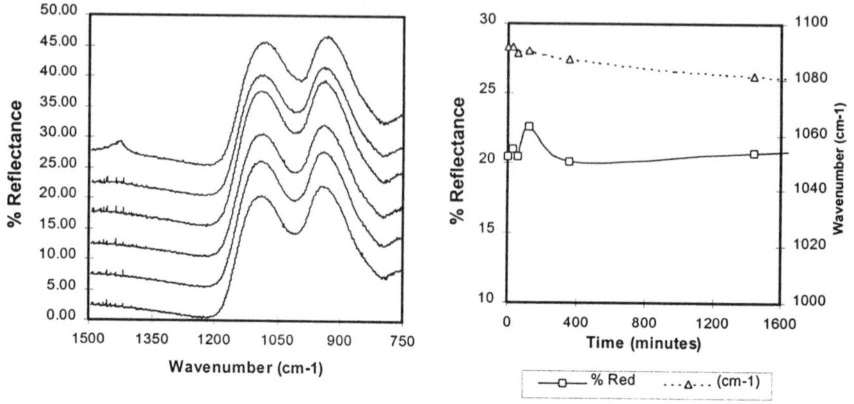

33N: 81% RH

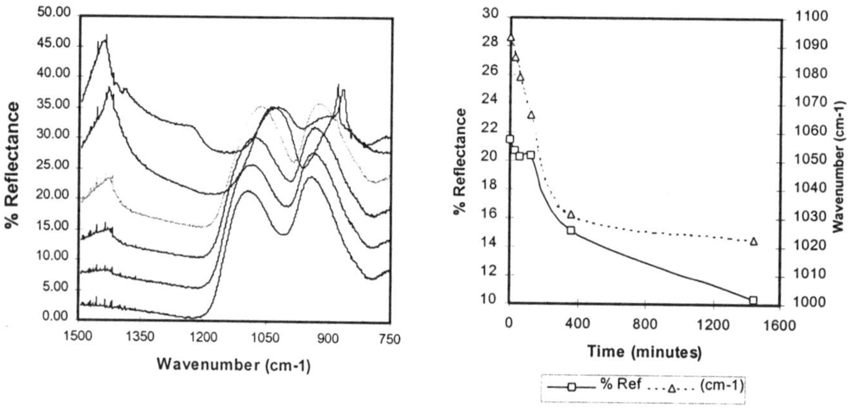

Figure 4. Corrosion behavior in a 33N glass at 52% relative humidity and at 81% relative humidity. Plots on the left show FT-IRRS curves and plots on the right show changes in Si-O-Si peak height and position versus time.

It is now necessary to comment on the sudden initial increase in peak intensity with time, then gradual decrease. Since the “low” RH and “high” RH experiments were run on different days, and this sudden increase appears in each composition under both humidity conditions, this “bump” in the peak intensity curves most likely has some significance. The authors believe the sudden increase in peak intensity to be attributed to the sudden reaction of the polished surface with the water in the atmosphere. Even at 30 minutes exposure, the alkali-depleted layer begins to form, leaving behind a silica-rich surface. This results in the corresponding increase in Si-O-Si peak as observed with the FT-IRRS spectra. Not too long after the surface reactions are taking place, however, the leached sodium ion reacts with the CO_2 in the atmosphere to form sodium carbonate precipitates. These precipitates can be seen, with the naked eye, to have formed on the glass surface. These precipitates hinder reflectivity of the glass surface, resulting in the gradual decrease in reflectivity with time. Further work must be done to confirm this. Figure 5 shows a schematic of the processes that may be occurring based on results obtained by FT-IRRS.

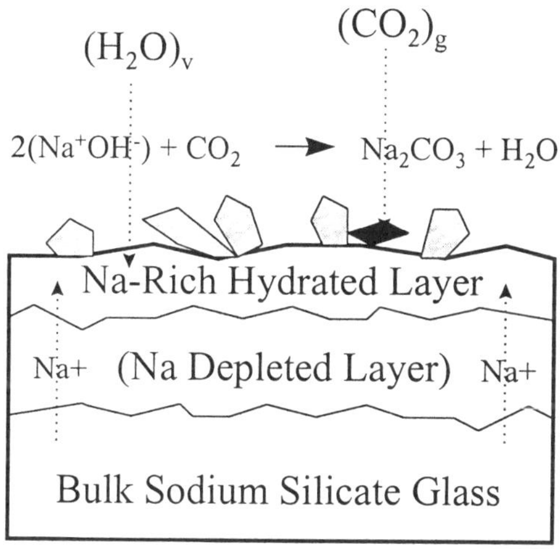

Figure 5. Schematic illustrating the various reaction zones in a sodium silicate glass exposed to ambient humidity.
Adapted from refs [7] and [8].

CONCLUSION

Polishing Study

It is important to establish polishing protocols when working with glasses which are susceptible to water corrosion. With the 33N glass compositions, polishing finer than 600 grit in ambient humidity does not appear to remove completely the corrosion layer which develops while the sample is exposed to air. In this case, polishing finer than 600 grit resulted in Si-O-Na peaks which were lower than would be expected, when analyzed using FT-IRRS.

Relative Humidity Experiments

During sample preparation or sample transportation of glasses which are susceptible to water corrosion, it is important to limit the time that the samples are exposed to the relative humidity. Under high humidity conditions, the structure and composition of the glass surface changes at very short times (on the order of 30 minutes or less). At these short times surface carbonates form, as observed using FT-IRRS. In Florida, 80% relative humidity conditions are not unusual. These "high" humidity conditions will effect the surface structure of sodium silicate glasses immediately. Simply transporting these glass samples to have some form of characterization performed may effect the outcome of that characterization and may lead to misleading results. Even under "dry" conditions, overnite storage of these glasses may alter the surface structure significantly.

ACKNOWLEDGEMENTS

The authors would like to acknowledge the Department of Energy for their financial support.

REFERENCES

1. B.P. McGrail, P.F. Marin and C.W. Lindenmeier, "Accelerated Testing of Waste Forms Using a Novel Pressurized Unsaturated Flow (PUF) Method," Mat. Res. Soc. Symp. Proc., Vol. 465, 1997.

2. D.M. Sanders, W.B. Person and L.L. Hench, J. Amer. Cer. Soc., 56, 247, 1974.

3. R.D. Husung and R.H. Doremus, "The Infrared Transmission Spectra of Four Silicate Glasses Before and After Exposure to Water," J. Mater. Res., Vol. 5, No. 10, 2209-2217, 1990.

4. R.G. Newton and A.B. Seddon, "The Durability of Silicate Glass in the Presence of a Saturated Leachant,", Corrosion Science, Vol. 33, No. 4, 617-626, 1992.

5. Z. Zhang, N. Soga and K. Hirao, "Water Corrosion Behavior of Densified Glass. I. Silicate Glass," Journal of Non-Crystalline Solids, 135, 55-61, 1991.

6. M. Prassas, J. Phalippou, L.L. Hench and J. Zarzycki, Journal of Non-Crystalline Solids, 48, 79, 1982.

7. D.E. Clark, R.L. Schulz, G.G. Wicks and A.R. Lodding, "Waste Glass Alteration Processes, Surface Layer Evolution and Rate Limiting Steps," Mat. Res. Soc. Symp. Proc., Vol. 333, 107-122, 1994.

8. X. Feng, "Surface Layer Effects on Waste Glass Corrosion," Mat. Res. Soc. Symp. Proc., Vol. 333, 55-68, 1994.

AGGRESSIVE CORROSION OF SODIUM SILICATE GLASS BY A DRIP TECHNIQUE

J. M. Moore, D. E. Clark, and J. H. Simmons
Department of Materials Science and Engineering
University of Florida
Gainesville, Florida

ABSTRACT

The corrosion of glass under drip conditions has been investigated in this study. Pre-heated deionized water was dripped at a constant rate on to the top face of a small sample of 20N glass, causing the gradual deterioration of the glass. The corrosion of the glass with time was monitored at fixed time intervals using Fourier Transform Infrared Reflection Spectroscopy (FT-IRRS). The analysis showed a progressive increase in the Si-O-Si stretch vibration (~1080 wavenumbers) and a progressive decrease in the Si-O stretch vibration peak (~950 wavenumbers), verifying the gradual leaching of the sodium out of the sample.

INTRODUCTION

There are many cases where glass is exposed to condensation and run-off in actual use. For example, glass nuclear waste storage containers may be exposed to the drip of ground water from the ceiling of the storage repository throughout their useful lifetime. Also, car windshield glass is exposed to drip and run-off of rain during a storm. Under these conditions, glass may corrode at a different rate than when fully immersed in water. Unfortunately, the majority of corrosion tests to date are static in nature and involve full immersion of glass in the leachant [1-3]. The objective of this study is to develop a quick and simple procedure for the testing and analysis of glass corrosion under the conditions of controlled drip and run-off.

The glass selected for study has a composition 20 mol% Na_2O-80 mol% SiO_2 (20N glass). It was selected for its chemical durability. The glass does not dissolve in air under ambient conditions, but shows sufficiently rapid degradation in water to permit testing in a reasonable time frame. The test was performed at an elevated temperature to further decrease the time of the experiment.

The primary analytic tool used to evaluate the corroded samples was FT-IRRS. Use of this technique for this purpose has been well documented in the literature [1-3].

EXPERIMENTAL PROCEDURE

Test Sample Preparations

Test samples were prepared from annealed cast cylinders of 20 mol% Na_2O-80 mol% SiO_2 glass poured from a glass melt maintained at 1400°C for 24 hours. 2 mm thick half disks with radii of 1.25 mm were cut from these cylinders and progressively polished down to 600 grit using silicon carbide polishing paper. Polished samples were stored in a Blue M oven maintained at 100°C while awaiting the drip test.

Drip Test

A schematic illustrating the drip test is provided in Figure 1. In the drip tests, the samples were placed in 60 ml Teflon leaching vessels and deionized water from a source reservoir was pre-heated to 90°C and allowed to drip onto one surface of the glass to induce corrosion. The water was delivered to the sample through action of a peristaltic cassette pump. The pump was capable of delivering water to the sample at rates ranging from 0 to 0.3 ml/min. A constant drip rate of 0.1 ml/min was selected for this study because it allowed the water sufficient dwell time on the surface of the sample, while still allowing for good water run-off.

The leaching vessels and Tygon tubing for conducting water were placed in a Blue M oven to achieve the 90°C test temperature. The water dripped off a Teflon mesh supported sample and was collected in the Teflon container. When the water level reached the removal tubing in the leach vessel, it was pumped out of the vessel and into a reservoir established to collect the deionized water/leachate. The level of the removal tubing was positioned sufficiently low to insure that the water level does not contact the sample, thereby maintaining a high sample surface area to volume of water ratio and creating higher pH in the solution that drips on the sample.

The collected water was monitored for pH, but the majority of analysis was done by FT-IRRS. FT-IRRS was used to examine the samples (both drip and bottom sides) prior to leaching and at periods of 0.5, 1.0, 3.0, and 6.0 hours throughout the drip tests. The FT-IRRS scans were collected 1400 to 400 wavenumbers, where major structural peaks occur.

The pathway of the leaching process in this experiment was governed by an ion-exchange reaction and a silica dissolution reaction [Figure 2]. This should result in the creation of a silica rich hydrated surface on the bulk glass. The thickness of the hydrated layer was governed by the competition between reactions 1 and 2. Reaction 1 favors a thick layer, while reaction 2 favors a thin layer.

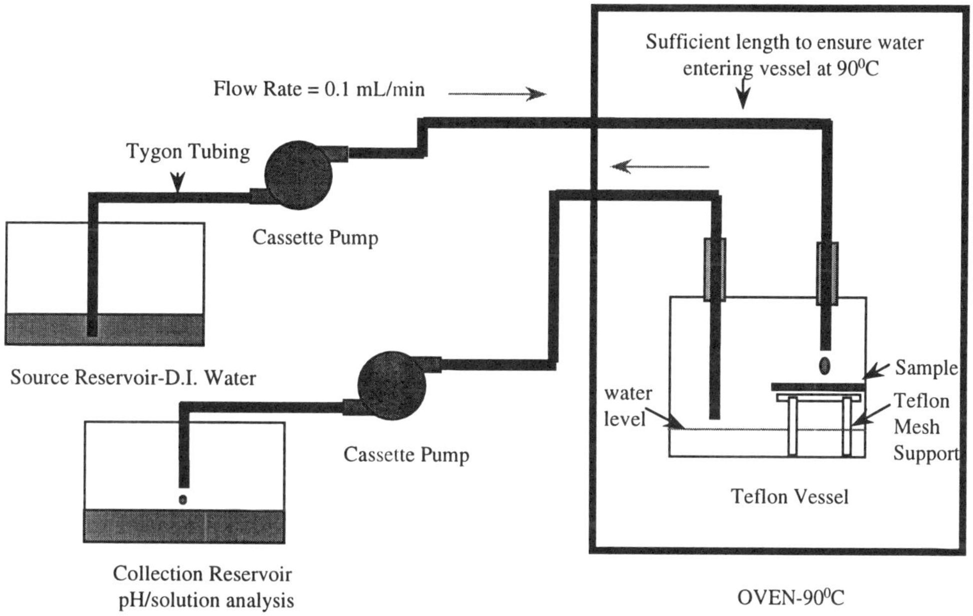

Figure 1. Flow-Thru Drip Leach Test

RESULTS AND DISCUSSION

The results of this analysis are shown for one sample in Figure 3. The analysis on the drip surfaces showed a progressive increase in the Si-O-Si stretch vibration peaks (~1080 wavenumbers) and a progressive decrease in Si-O stretch vibration peaks (~950 wavenumbers) with increasing time exposure to deionized water. These results suggested that sodium was gradually being leached from the surface of the samples (associated with a decrease in the Si-O stretch vibration), leaving a silica rich surface (associated with the Si-O-Si stretch vibration).

Comparisons of the FT-IR results of the current study to static leach tests performed on a similar specimen 20N glass [Figure 4] suggested that the mechanisms appear to be the same under static and drip conditions, though the kinetics may be different.

The interpretation of the results was also supported by pH analysis of the leachate. The initial pH of the deionized water in the source reservoir was **~7.0***. The leached solution collected after 24 hours of sample exposure had a measured pH of 9.75. This shift toward a more basic solution was expected as sodium ions from the sample continue to be added to the leachate resulting in the formation of NaOH.

*theoretically this should be 5.65

Ion Exchange

$$Na^{+}_{(g)} + H_3O^{+}_{(s)} \Rightarrow NaOH_{(s)} + 2H^{+}_{(g)} \quad (1)$$

Silica Dissolution

$$Si\text{-}O\text{-}Si + OH^{-} \Rightarrow Si\text{-}O^{-} + Si\text{-}OH \quad (2)$$

Figure 2. Equations Governing Glass Corrosion

SUMMARY

A quick and simple procedure for the testing and analysis of glass corrosion under the conditions of controlled drip and run-off has been developed. The procedure used FT-IR in the reflectance mode to evaluate the corrosion by deionized water of the drip surface of a small half disk 20 mol% Na_2O-80 mol% SiO_2 at several fixed time intervals. Results from the tests suggested that sodium was gradually being leached from the surface of the samples, leaving a silica rich surface. These results agreed with both static leach tests on similar samples and pH analysis of the leachate. The corrosion mechanisms appeared to be the same under static and drip conditions, though the kinetics may be different.

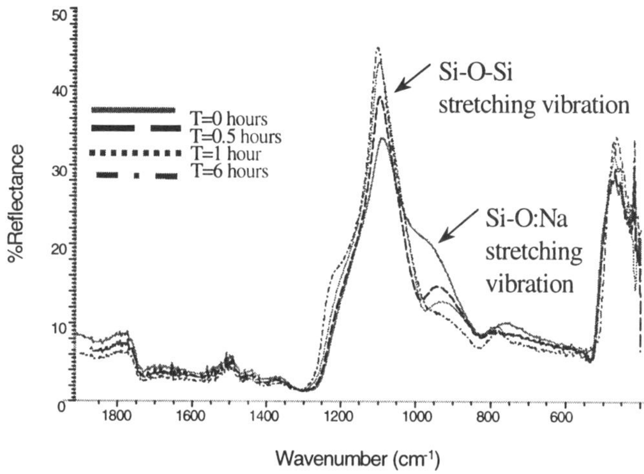

Figure 3. FT-IRRS Analysis on 20N Glass (Drip Surface)

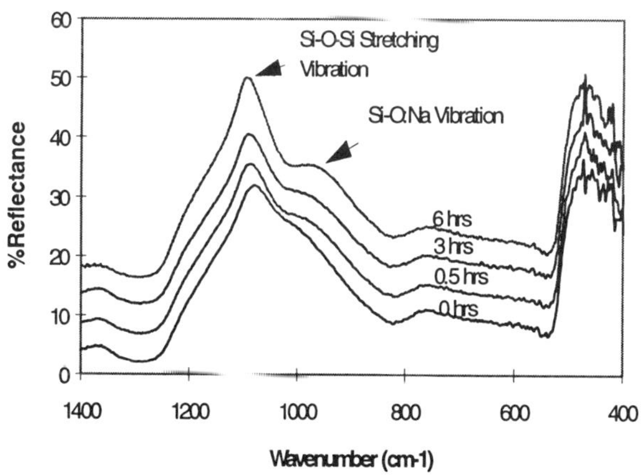

Figure 4. Static Leach Test Results for 20N Glass in Deionized Water at 20°C[5]

ACKNOWLEDGEMENTS
This was a DoE Sponsored Research Project
Contract Number # DE-FG07-96ER 45616

REFERENCES

1. D. M. Sanders, W.B. Person, and L.L. Hench, J. Amer.Cer. Soc., 56, 247, 1974.
2. D.E. Clark, C. G. Pantano, Jr. and L.L. Hench, Corrosion of Glass, (Magazines for Industry, New York, 1979), p.48.
3. D.E. Clark, C. G. Pantano, Jr. and L.L. Hench, Corrosion of Glass, (Magazines for Industry, New York, 1979), p.48.
4. LL. Hench and D.E. Clark, "Surface Properties and Performance Prediction of Alternative Waste Forms, U.S. NRC report 1981.
5. Figure Provided by Greg Darby, Dept. of Materials Science and Engineering, University of Florida, Gainesville, Florida.

Biomaterials and Bioreactivity

SCANNING THE CRACKS

Ellis Douek
Consultant Emeritus – Guy's Hospital
97 Harley Street
London W1N 1DF
Gillian Douek
24 Reynolds Close
London NW11 7EA

ABSTRACT

This paper was given at a symposium honouring the work of Larry Hench and his group. It describes in broad terms its relationship with medical and other groups in the application of Bioglass® to ossicular replacements and other hearing prostheses. It used the symbol of cracks in materials to discuss the manner in which Professor Hench related to physicians and other colleagues and identifies three groups in such a collaboration: 1) technical employees and the need for precision in defining the final product; 2) creative colleagues and the importance of conversation; 3) the commercial dimension which the authors touch on but cannot claim a complete understanding.

INTRODUCTION

In the 5th Century BC, Greece music was taught as a branch of mathematics. Throughout the thousand years of the Roman Empire learning rhetoric together with logic was considered the basis of methods to form the mind and prepare it for any future profession, be it medicine, military strategy or law.

In childhood, our own protests against geometrical theorems or Latin were put down with the necessity to "train the mind" by their exercises.

Even then it was clear that some were innately more capable of grasping the structures of language than the concepts of mathematics, that some could produce remarkable musical sounds whilst others could juxtapose form and colour producing beauty and art. The recognition that exchanging skills by helping each other with homework was perhaps our first attempts at co-operation between different specialities, different personalities and separate ambitions. Children who learn how and where to get help have a distinct advantage in later life.

This symposium is honouring Larry Hench and this paper refers to the study and understanding of the successful relationship of Professor Hench with people trained in different specialities. The cracks are an allusion to these separate identities but also to the nature of interfaces whether they be between different materials or personalities.

MINING THE CRACKS

The physicians have been very successful over the centuries even though there was little they could do in most cases. Their success depended on knowledge. If they could give a prognosis it was often sufficient for such less demanding times. If they knew the medicinal plants, in which season to collect them, at which time of day even, to enhance their pharmacological activity they could offer relief. The image of the aging, experienced and wise doctor was the cherished one in the past.

This is not the case any longer. Great changes have taken place over the last fifty years. Knowledge is there at the touch of a modem and patients present themselves with their Internet printout in their hand. Diagnosis by an experience physician is generally not accepted without backup from blood tests and MRI scans.

The creative physician, the one who needs to look forward towards advances in medicine, has to do this as part of a team in which he is one of a group of scientists representing separate skills.

The work discussed here relates to the surgery of deafness and in particular to the replacement of ossicles in the middle ear.

Originally, work with plastic materials such a polyethylene resulted in problems of rejection whilst techniques involving homografts from cadaver donor had to be abandoned with the onset of virus such as AIDS and particularly Creutzfeld Jacob Disease. A series of attempts using other inert materials and porous implants made of Teflon were not successful. Titanium has found its place but is not suitable everywhere and in particular not in the middle ear.

The collaboration with Hench and Merwin brought our surgical team in touch with Bioglass®. This was the introduction of bioactive as opposed to inert materials to otology.

Two examples illustrates how the study of the interface between the inorganic material and human tissue have affected our work.

The first is of a general nature. The greatest interest shown by the surgeons is in the nature of the bond between the inorganic material and the tissues, especially the soft tissues of the body. Inertness and its advantages was clear to everyone from the beginning. The use of inviting porous material such as titanium became clear as the osteoblasts grew into the spaces of the metal creating what amounted to a mechanical lock holding together the two. Bioactivity was a philosophical concept difficult to visualize until animal experiments and their histopathological findings became available, especially those demonstrated by the scanning electron microscope. These showed that the collagen fibres, the structural, strength-giving features of soft tissue were trapped by the loops of apatite that formed on the surface of Bioglass® when exposed to body fluid. This remarkable finding explains the solidity of the bond and explains the physical nature of bioactivity at least in this instance.

The second example is a surgical finding. An early example of Bioglass® middle ear prosthesis had been used to replace the incus despite the fact it had received a minor crack. This had occurred during reshaping by a drill because the rigidity of the resulting strut was felt to be entirely inadequate. The result was excellent and it was only four years later that the hearing failed suddenly. Re-operation showed that the crack had extended and continued to fracture until the whole structure had disintegrated. Soft tissue had grown into the crack and continued to grow in a manner not dissimilar to an abandoned house which begins to crack. Vegetation grows into the cracks dissolving the masonry and prying it apart. This finding could be viewed as a failure and added to unsuccessful experiment. As a surgical operation, it failed in the long-run but it also demonstrated the nature and effect of bioactivity. The soft tissue bonded beyond any doubt with an inorganic material, an event not seen before. In practical terms this finding went beyond scanning to mining the cracks for what it had to offer. It let the use of bulky prostheses, less likely to crack, which were marketed as Douek-MED™ (USBiomaterials).

If the cracks can be scanned for information and mined for the rich discoveries unclaimed by the specialities on either side, they must be spanned to make co-operation between them possible. This aspect of our work requires considerable psychological understanding if it is to succeed.

In the case of our research, it can stem in two ways; firstly, the physician can approach the scientist with his or her needs. In this instance the surgeon makes a general request to the scientific community, asking for a material strong and rigid enough to transmit acoustic vibrations and also which will bond to bone and to soft tissue such as the tympanic membrane.

Alternatively, the offer may come from the scientist to the physician. This has occurred often, for instance, in the offer of fibre optics and small cameras. These were made originally to look inside machines but were quickly inserted into the sinus, the stomach or the colon.

The development of a new prosthesis such as the Douek-MED™ (USBiomaterials) is typical of the contemporary process and is worth considering as an example of the situation in this type of research today.

TEAMWORK

The material Bioglass® was an offer to the surgeons after extensive animal experiments had been carried out to evaluate the behaviour of the material in vivo. This meant that biologists and histopathologists and their technicians had already participated in the project. The use of the material in ossicular replacements and later as a coating for cochlear implants also meant the introduction of audiologists, engineers and psycho-acoustic experts were introduced to the scheme in general or to parallel projects. The number of different specialists coming together from time to time fluctuated but proved extraordinarily enriching both in the immediate scientific sense and in the enjoyment of the process which is an essential though sometimes forgotten factor in the achievement of success.

If we are to examine the process objectively, it is best to divide the participating individuals into three groups:

The Technical Group

This group has to be able to visualise more or less completely the finished product. It is important to set a reasonable task with reasonable limits of time and cost and the product has to be inspected and to live up to expectations.

The work of technical colleagues has to be valued but more, such as a creative element, should not be expected or it will produce anxiety. Obviously, there are no firm lines and relationships can grow from experience but these principles should be kept in mind.

The Creative Group

The problems within this group are well known in scientific circles and rules, similar to those of etiquette have long been established regulating publication of material, presentation at meetings and a clear understanding of different ambitions.

Perhaps the most important element in these interpersonal relationships is controversial. Thedore Zeldin, in his series of lectures on BBC Radio 4, pointed out that creativity usually needs to be fuelled by more than polite chat. At the frontiers of knowledge, adventurous researchers have to be almost professional eavesdroppers, picking up ideas from the most obvious sources. The discovery of DNA was the outcome of conversations between Watson & Crick which went on ceaselessly for many years. Crick liked to speak to the scientists in whose work he was interested rather than read their papers in the journals. He asked naïve questions saying he needed to have the problems explained to him and claimed he gained new insight in this way. There is nowhere greater scope for this type of cross-fertilisation than in regular conversations between individuals with different specialities. There should be no rules other than a lack of inhibition and a promise to discuss any idea that comes to mind.

Difficulties arose within our creative group two of which are given as examples:

The first is a tension between the surgeons and the scientists. This occurred when the former made unreasonable demands such as requesting help from scientists in surgical interventions without forewarning. To the surgeons operations are scheduled to clinical need rather than according to plan. It is a rather messy pattern of work where the day or the week progresses according to what happens rather than following a timetable. This proved disturbing to the scientist who needed careful preparation. At the same time, the latter sometimes wished to test their techniques in a manner and at a time which were not clinically appropriate.

A second problem arose with creative individuals who were unable to let go. As a project neared completion, their fertile minds would have already moved on to other, better fields. Although exciting, these people proved very irritating to those anxious to get at least a prototype off the ground.

The Commercial Group

Today any scientifically developed product is required to have a commercial application and this means working closely with another group of people. In fact, the boundaries are again not clearly drawn as individuals on the commercial side have themselves often had a scientific training.

Initially, there seems few problems but different agendas emerge. The scientific side has to concentrate on the development of the product which is expected to satisfy a need. The commercial side has to identify the size of that need and to estimate the strength of the competitive products. The two groups have to do this together and it is in that situation that the different agendas become obvious.

Some of the funding usually comes from institutional donors and they too act as an audience that must be satisfied and which itself encourages commercial applications.

The commercial group also has to press for approval from state agencies and develop a difficult marketing network.

CONCLUSION

The challenges facing the teams of engineers, scientists, clinicians & businesses that interact to create new biomaterials and apply them to the clinical setting are substantial. The examples presented here have been used to demonstrate that it is possible to bring divergent groups together to gain technological advances for the benefit of man. Professor Hench's understanding and management of the dynamics of these relationships have bridged the cracks and allowed the successful development of Bioglass®.

BIBLIOGRAPHY

Bioceramics Volume 8: L. Hench, D. Greenspan, Eds., Pergamon Elsevier, Oxford, 1995

Work, Creativity and Social Justice. E. Jaques, Heineman, London, 1970

Conversation, T. Zeldin. The Harvill Press, London, 1998

APATITE FORMING MECHANISM ON AMORPHOUS PHASES IN BODY ENVIRONMENT

Tadashi Kokubo, Hyun-Min Kim, Hiroaki Takadama, Toshiki Miyazaki
Faculty of Engineering, Kyoto University, Kyoto 606-8501, Japan
Takashi Nakamura
Faculty of Medicine, Kyoto University, Kyoto 606-8507 Japan

ABSTRACT

Bioactive materials bond to bone through a bonelike apatite layer which is formed on their surfaces in living body. Amorphous phases play an important role in forming the apatite layer for most of the materials. It has been proposed that Si-OH and Ti-OH groups at the surfaces of the amorphous phases induce the apatite nucleation. In the present study, processes of apatite formations induced by the Si-OH and Ti-OH groups were investigated in a model system in a simulated body fluid (SBF) by using transmission electron microscopy and X-ray photoelectron spectroscopy. Apatite formation was found to progress in the sequence: 1) binding of calcium ions in SBF with the Si-OH or Ti-OH groups to form a calcium silicate or a calcium titanate, 2) late binding of phosphate ions in SBF with the calcium silicate or the calcium titanate to form apatite nuclei, and 3) spontaneous growth of the apatite nuclei by consuming calcium and phosphate ions in SBF. This indicates that Si-OH or Ti-OH groups on amorphous phases do not directly induce the apatite nucleation, but via formation of a calcium silicate or calcium titanate.

INTRODUCTION

Hench *et al.* first discovered that some glasses in the system Na_2O-CaO-SiO_2-P_2O_5 spontaneously bond to living bone, and designated them as Bioglass®.[1-3] Since then, various ceramics such as glass-ceramic Ceravital®,[4] sintered hydroxyapatite,[5] sintered β-tricalcium phosphate,[6] glass-ceramic Cerabone® A-W,[7, 8] glass-ceramic Bioverit®[9] have been found to bond to bone. Recently, even metals such as titanium, titanium alloys and tantalum were also shown to bond to bone, if they were priorly subjected to a specific chemical surface treatment.[10-12] Materials in this category are now called bioactive materials, and are occupying an indispensable position in the field of bone-repairing materials.

All the above bioactive materials bond to bone via forming a biologically active bonelike apatite layer on their surfaces in body.[2, 13-15] An essential requirement for an artificial material to bond to bone is therefore believed to be a formation of bonelike apatite layer on its surface in body environment.[2, 13] Except for the crystalline calcium phosphate family such as hydroxyapatite and β-tricalcium phosphate, bioactive materials are commonly composed of appreciable amounts of surface active amorphous phases, which are believed to play an important role in inducing the apatite formation. The amorphous phases are, for example, Na_2O-CaO-SiO_2-P_2O_5 glassy phase in case of Bioglass®-type glasses, calcium silicate-based glassy phase in case of glass-ceramic A-W, and an amorphous sodium titanate in case of bioactive titanium or titanium alloy prepared by chemical surface treatment. Hench *et al.* early emphasized the importance of silicate glassy phase for apatite formation on Bioglass®-type glass, and supported their position by molecular orbital calculations of the silicate structure.[16] Anderson *et al.* also reported that a silicate glassy phase is responsible for the apatite formation on bioactive silicate glasses.[17] Kokubo *et al.* showed on the basis of direct observation of apatite formation on oxide gels in a simulated body fluid that not only an amorphous silicate but also amorphous titanate and tantalate are capable of inducing apatite formation, and surface functional groups such as Si-OH, Ti-OH and Ta-OH groups formed on the amorphous phases are responsible for the apatite nucleation.[18-21] It is, however, not yet understood how the Si-OH, Ti-OH and Ta-OH groups induce the apatite nucleation.

In the present study, the process of apatite formation induced by these functional groups in body environment was investigated in model systems in a simulated body fluid by using transmission electron microscopy and X-ray photoelectron spectroscopy.

EXPERIMENTAL PROCEDURE

A collodion film which was supported on a titanium (Ti) grid (200 mesh) 3.0 mm in diameter was coated with carbon. It was faced to the parent glass (G) of glass-ceramic A-W, whose nominal composition is MgO 4.6, CaO 44.7, SiO_2 34.0, P_2O_5 16.2, CaF_2 0.5 wt%,[22] taking a distance of 0.5 mm, and soaked in 20 mL of acellular simulated body fluid (SBF) with pH (7.40) and ion concentrations nearly equal to those of human blood plasma at 36.5°C (see Table I). The collodion films soaked in SBF for various periods were observed under transmission electron microscope (TEM: JEM-2000FX III, JEOL Co., Japan) attached with an energy dispersive X-ray microanalyzer (EDX: VOYAGER III,

NORAN Inst. Inc., U.S.A.).

Commercially pure Ti (Kobe Steel Ltd., Japan) metal substrates 10×10×1 mm^3 in

Table I. Ion Concentrations of SBF and Human Blood Plasma

	Concentration / mM							
	Na^+	K^+	Mg^{2+}	Ca^{2+}	Cl^-	HCO_3^-	HPO_4^{2-}	SO_4^{2-}
SBF	142.0	5.0	1.5	2.5	148.8	4.2	1.0	0.5
Blood plasma	142.0	5.0	1.5	2.5	103.8	27.0	1.0	0.5

size were soaked in 5 mL of 5.0M-NaOH solution at 60°C for 24 h to form a sodium titanate gel layer on their surfaces, gently washed with distilled water, dried at 40°C for 24 h, and subsequently heat-treated at 600°C for 1 h in air atmosphere to form an amorphous sodium titanate layer.[10] They were soaked in 24 mL of SBF at 36.5°C. Surfaces of the Ti substrates soaked in SBF for various periods were analyzed by scanning electron microscopy (SEM: Model S2500CX, Hitachi Co., Japan), thin-film X-ray diffraction (XRD: Model 2651A1, Rigaku Co., Japan) and X-ray photoelectron spectroscopy (XPS: MT4300, ULVAC-PHI Co., Japan).

RESULTS AND DISCUSSION

Figure 1 shows TEM photographs of the collodion films as a function of soaking time in SBF. Black lines in the TEM image before soaking (0 h) are due to the Ti grid supporting the collodion film. TEM morphology changed slightly after soaking for 6 h, 12 h, 24 h, and an aggregate of needle-like particles was found finally after 96 h. This indicates that different kinds of products were formed on the collodion film after each of these soaking periods.

Figure 2 shows the electron diffraction pattern of the collodion film soaked in SBF for 96 h. An aggregate of needle-like particles shown in Fig. 1 were identified as crystalline apatite.

Figure 3 shows EDX spectra of the collodion films supported by Ti grid as a function of soaking time in SBF with glass G. The C and O, and Ti peaks detected before soaking

Fig. 1. TEM photographs of the collodion film supported by Ti grid as a function of soaking time in SBF.

(0 h) are due to the carbon and collodion film, and titanium grid, respectively. In addition to the C, O, and Ti peaks, Si was detected after 6 h. In addition to the C, O, Ti and Si, Ca was detected after 12 h. In addition to the C, O, Ti, Si and Ca, P, Mg and Cl were detected after 24 h. No additional element was detected after 48 h.

It is speculated from these results that the collodion film is attached with silicate ions rich in Si-OH, which were released from glass G, at the first stage. Only calcium ions in SBF bind with the Si-OH groups to form a kind of calcium silicate at the second stage. The P ions finally combine with the calcium silicate to form the apatite nuclei. Thus-formed apatite nuclei grow spontaneously by consuming calcium and phosphate ions in SBF. The aggregate found in the TEM image after 96 h in Fig. 1 is assumed to be an assembly of apatite nuclei grown in this manner.

Figure 4 shows SEM photographs of the surfaces of Ti metals subjected to the NaOH and heat treatments as a function of SBF. Porous network morphology of the surface of Ti metal before soaking (0 d) is due to the formation of amorphous sodium titanate on the metal by the NaOH and heat treatments. Particle-like assemblies of tiny needle-like crystals were formed on the sodium titanate after 2 d. They grew to cover whole surface by increasing soaking time to 3 and 7 d.

Figure 5 shows thin-film XRD patterns of the surface of Ti metal subjected to the NaOH and heat treatments as a function of soaking time in SBF. XRD patterns before soaking (0 d) indicates that the amorphous sodium titanate on the Ti substrate contains small amounts of crystalline sodium titanate ($Na_2Ti_5O_{11}$) and rutile (TiO_2). New XRD peaks, all of which were attributed to crystalline apatite, appeared after 2 d and their intensities increased with increasing soaking time to 3 and 7 d. This indicates that the particle-like assemblies shown in Fig. 4 are crystalline apatite.

Figure 6 shows Na_{1s}, O_{1s}, Ca_{2p} and P_{2p} XPS spectra of the surface of Ti metal subjected to the NaOH and heat treatments as a function of soaking time in SBF. The Na_{1s} spectrum at 1072.5 eV before soaking (0 d) is due to the Na in the amorphous sodium titanate. The Na_{1s} spectrum appreciably decreased its intensity with increasing soaking time, and disappeared completely after 2 d. The O_{1s} spectrum at 530.8 eV before soaking (0 d) was attributed to Ti-O bond. A small overlapping peak newly appeared at 532.5 eV on the shoulder of this spectrum after 30 min was attributed to Ti-OH bond. The O_{1s} spectrum shifted its half maximum position to higher binding energy of 531.7 eV after 2 d. The shifted binding energy position coincided well with that of the main O_{1s} spectrum of a sintered hydroxyapatite (HA), and thereby attributed to P-O bond. The Ca_{2p} spectrum was detected as a doublet peak with $Ca_{2p1/2}$ = 350.6 eV and $Ca_{2p3/2}$ = 347.1 eV after 30 min. These binding energy positions coincided well with those of a calcium titanate ($Ca_3Ti_2O_7$). The Ca_{2p} spectrum shifted its position to higher binding energy of $Ca_{2p1/2}$ = 352.0 eV and $Ca_{2p3/2}$ = 348.5 eV after 2 d. This change accompanied an abrupt increase in their intensities. The shifted binding energy positions coincided well with those of HA. The P_{2p} spectrum was first detected at 133.0 eV after 2 d, and gradually increased its intensity with increasing soaking time. The binding energy position was not shifted and coincided well with that of HA. Shift of binding energy position and abrupt increase or decrease in intensity were observed for all the XPS spectra after 2 d. This

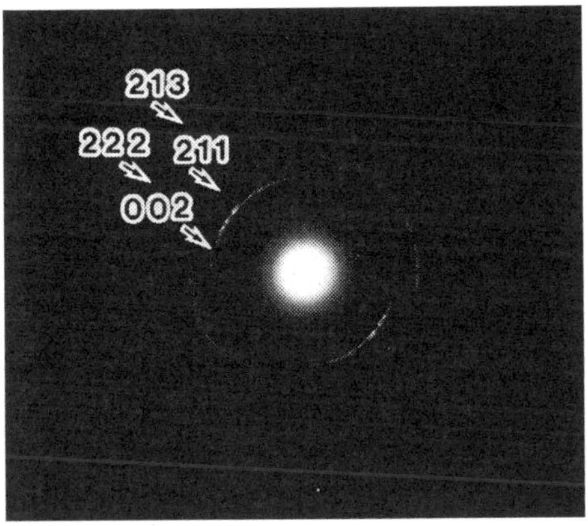

Fig. 2. Electron diffraction pattern of the collodion film soaked in SBF for 96 h.

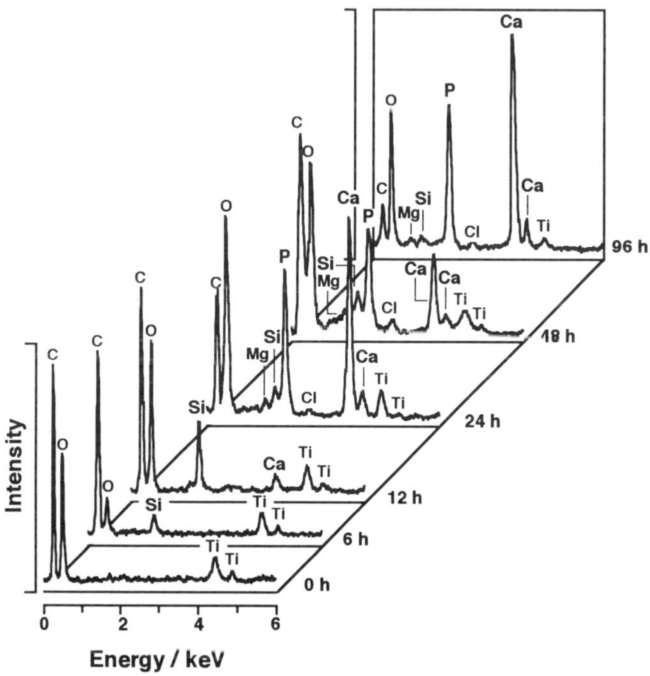

Fig. 3. EDX spectra of the collodion film supported by Ti grid as a function of soaking time in SBF.

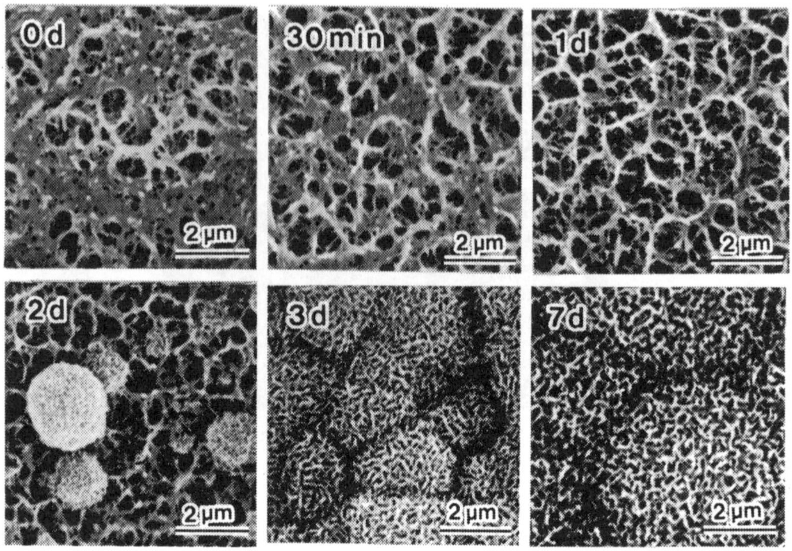

Fig. 4. SEM photographs of the surface of Ti metal subjected to the NaOH and heat treatments as a function of soaking time in SBF.

period corresponds to the soaking time when the apatite was first detected on the Ti substrate as shown Figs. 4 and 5.

The decrease in the Na peak from the first to the second stage is due to release of Na ions from the amorphous sodium titanate via ion exchange with H_3O ions in SBF.[10] Such ion exchange results in formation of Ti-OH group on the sodium titanate. Only the calcium ions bind with the Ti-OH groups to form a calcium titanate at the first stage. The phosphate ions bind appreciably later with the calcium titanate to form apatite nuclei at the second stage. Thus-formed apatite nuclei grow spontaneously by consuming calcium and phosphate ions in SBF.

Similar results were observed for tantalum metal prepared by a NaOH and heat treatments. An amorphous sodium tantalate exchanges its Na ions with H_3O ions in SBF to form Ta-OH groups. Only calcium ions bind to the Ta-OH groups to form a calcium tantalate. Phosphate ions then bind to the calcium tantalate to form apatite nuclei.

The apatite formation on the amorphous phases in body environment is therefore concluded to progress in a sequence as follows.

1) Formation of functional groups such as Si-OH, Ti-OH or Ta-OH on the amorphous phases.
2) Binding of calcium ions with the Si-OH, Ti-OH or Ta-OH groups to form a calcium silicate, a calcium titanate or a calcium tantalate.

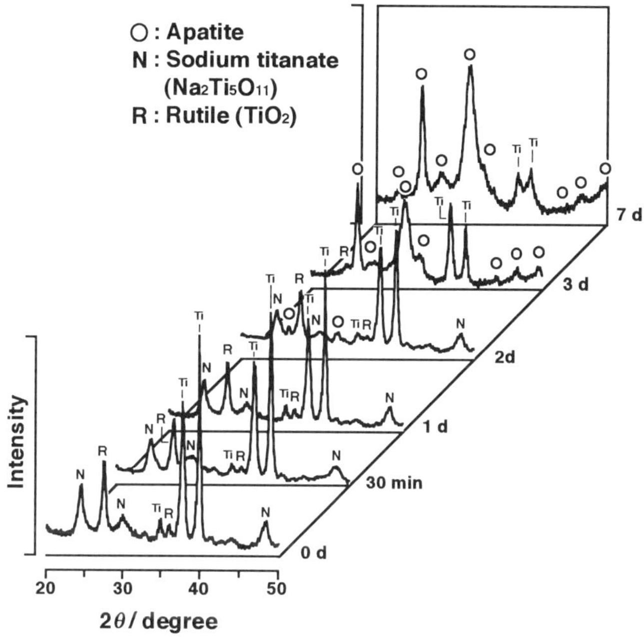

Fig. 5. Thin-film XRD patterns of the surface of Ti metal subjected to the NaOH and heat treatments as a function of soaking time in SBF.

3) Binding of phosphate ions with the calcium silicate, the calcium titanate or the calcium tantalate to form apatite nuclei.
4) Spontaneous growth of the apatite nuclei by consuming calcium and phosphate ions in body fluid.

This indicates that the Si-OH, Ti-OH or Ta-OH functional groups formed on the amorphous phases do not directly induce the apatite nucleation, but via formation of a calcium silicate, calcium titanate or a calcium tantalate.

CONCLUSIONS

In most bioactive materials, amorphous phases play an important role in forming a bonelike apatite on their surfaces in body environment. The apatite formation progress in the sequence of 1) formation of functional groups such as Si-OH, Ti-OH or Ta-OH on the amorphous phases, 2) binding of calcium ions with the Si-OH, Ti-OH or Ta-OH groups to form a calcium silicate, a calcium titanate or a calcium tantalate, 3) binding of phosphate ions with the calcium silicate, the calcium titanate or the calcium tantalate to form apatite

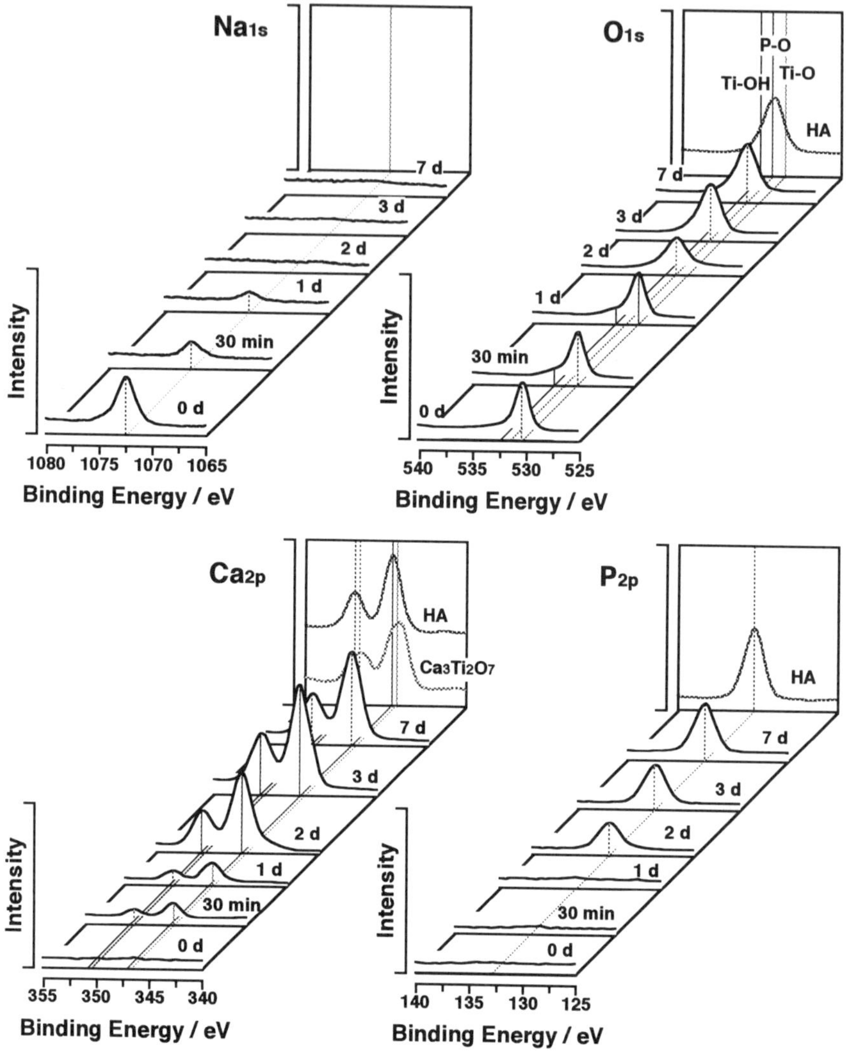

Fig. 6. Na_{1s}, O_{1s}, Ca_{2p} and P_{2p} XPS spectra of the surface of Ti metal subjected to the NaOH and heat treatments as a function of soaking time in SBF.

nuclei, and 4) spontaneous growth of the apatite nuclei by consuming the calcium and phosphate ions in body fluid. This indicates that the Si-OH, Ti-OH or Ta-OH functional groups formed on the amorphous phases do not directly induce the apatite nucleation, but via formation of a calcium silicate, calcium titanate or a calcium tantalate.

ACKNOWLEDGMENTS

This present work was supported by Grant-in-Aids for Scientific Research, the Ministry of Education, Science, Sports, and Culture, Japan.

REFERENCES

[1] L.L. Hench, R.J. Splinter, W.C. Allen and T.K. Greenlee, "Bonding Mechanisms at the Interface of Ceramic Prosthetic Materials," *J. Biomed. Mater. Res. Symp.*, **2**, 179-49 (1971).

[2] L.L. Hench, "Bioceramics: From Concept to Clinic," *J. Am. Ceram. Soc.*, **74**, 1487-510 (1991).

[3] L.L. Hench and Ö.H. Andersson, "Bioactive Glasses," pp. 41-62 in *An Introduction to Bioceramics.* Edited by L.L. Hench and J. Wilson, World Sci., Singapore, 1993.

[4] U.M. Gross, C. Müller-Mai and C. Voigt, "Ceravital® Bioactive Ceramics," pp. 105-24 in *An Introduction to Bioceramics.* Edited by L.L. Hench and J. Wilson, World Sci., Singapore, 1993.

[5] M. Jarcho, J.L. Kay, R.H. Gumaer and H.P. Drobeck, "Tissue, Cellular and Subcellular Events at Bone-Ceramic Hydroxyapatite Interface," *J. Bioeng.*, **1**, 79-92 (1977).

[6] B.V. Rejda, J.G.J. Peelen and K. de Groot, "Tricalcium Phosphate as a Bone Substitute," *J. Bioeng.*, **1**, 93-97 (1977).

[7] T. Kokubo, M. Shigematsu, Y. Nagashima, M. Tashiro, T. Nakamura, T. Yamamuro and S. Higashi, "Apatite- and Wollastonite-Containing Glass-Ceramic for Prosthetic Application," *Bull. Inst. Chem. Res., Kyoto Univ.*, **60**, 260-68 (1982).

[8] T. Kokubo, "A/W Glass-Ceramic: Processing and Properties," pp. 75-88 in *An Introduction to Bioceramics.* Edited by L.L. Hench and J. Wilson, World Sci., Singapore, 1993.

[9] W. Höland, K. Naumann, W. Vogel and J. Gummel, "Machinable Bioactive Glass Ceramic," *Wiss. Ztschr. Friedrich-Schiller-Univ. Jena, Math.-Naturwiss. Reihe.*, **32**, 571-80 (1983).

[10] H.-M. Kim, F. Miyaji, T. Kokubo and T. Nakamura, "Preparation of Bioactive Ti and Its Alloys via Simple Chemical Surface Treatment," *J. Biomed. Mater. Res.*, **32**, 409-417 (1996).

[11] W.-Q. Yan, T. Nakamura, M. Kobayashi, H.-M. Kim, F. Miyaji and T. Kokubo, "Bonding of Chemically Treated Titanium Implant to Bone," *J. Biomed. Mater. Res.*, **37**, 265-275 (1997).

[12] T. Miyazaki, H.-M. Kim, F. Miyaji, T. Kokubo and T. Nakamura, "Apatite Formation on Chemically Treated Tantalum Metal in Body Environment," pp. 317-320 in *Bioceramics Vol. 9.* Edited by T. Kokubo, T. Nakamura and F. Miyaji, Elsevier, Oxford, 1996.

[13] T. Kokubo, "Surface Chemistry of Bioactive Glass-Ceramics," *J. Non-Cryst. Solids*, **120**, 138-151 (1990).

[14] N. Neo, S. Kotani, T. Nakamura, T. Yamamuro, C. Ohtsuki, T. Kokubo and Y. Bando, "A Comparative Study of Ultrastructure of the Interface between Four Kinds of Surface-Active Ceramic and Bone," *J. Biomed. Mater. Res.*, **26**, 1419-32 (1992).

[15] M. Neo, T. Nakamura, C. Ohtsuki, T. Kokubo and T. Yamamuro, "Apatite Formation on Three Kinds of Bioactive Materials at an Early Stage *In Vivo*: A Comparative Study by Transmission Electron Microscopy," *J. Biomed. Mater. Res.*, **27**, 999-1006 (1993).

[16] J.H. West and L.L. Hench, "Reaction Kinetics of Bioactive Ceramics Part V: Molecular Orbiotal Modeling of Bioactive Glass Surface Reaction," pp. 75-86 in *Bioceramics Vol. 5.* Edited by T. Yamamuro, T. Kokubo and T. Nakamura, Kobunshi Kankokai, Kyoto, 1991.

[17] Ö.H. Anderson and K.H. Karlsson, "On the Bioactivity of Silicate Glasses," *J. Non-Cryst. Solids*, **121**, 145-151 (1991).

[18] P. Li, C. Ohtsuki, T. Kokubo, K. Nakanishi, N. Soga, T. Nakamura and T. Yamamuro, "Apatite Formation Induced on Silica Gel in a Simulated Body Fluid," *J. Am. Ceram. Soc.*, **75**, 2094-97 (1992).

[19] P. Li, C. Ohtsuki, T. Kokubo, K. Nakanishi, N. Soga, T. Nakamura, T. Yamamuro and K. de Groot, "A Role of Hydrated Silica, Titania and Alumina in Forming Biologically Active Bone-like Apatite on Implant," *J. Biomed. Mater. Res.*, **28**, 7-15 (1994).

[20] P. Li, I. Kangasniemi, K. de Groot and T. Kokubo, "Bonelike Hydroxyapatite Induction by a Gel-derived Titania Coating on a Titanium Substrate," *J. Am. Ceram. Soc.*, **77**, 1307-12 (1994).

[21] T. Miyazaki, H.-M. Kim, F. Miyaji, T. Kokubo and T. Nakamura, "Apatite-Forming Ability of Sol-Gel Derived Tantalate Gel," pp. 481-484 in *Bioceramics Vol. 11*. Edited by R.Z. LeGeros and J.P. LeGeros, World Sci., Singapore, 1998.

[22] M. Tanahashi, T. Kokubo, M. Minoda, T. Miyamoto, T. Nakamura and T. Yamamuro, "Apatite Coating on Organic Polymers by a Biomimetic Process," *J. Am. Ceram. Soc.*, **77**, 2805-2808 (1994).

IN-VIVO BIODEGRADATION OF GLASSES: IMAGES OF BIOACTIVITY

A.M. Gatti
E. Monari
Centre of Biomaterials
Dept. of Surgery
University of Modena
Modena, Italy

Larry L. Hench
Dept. of Materials
Imperial College, London, UK.

David C. Greenspan
USBiomaterials, Alachua
USA

ABSTRACT

The work takes into consideration the in-vivo degradative aspects of three different glasses, showing some aspects of their bioactivity. Sterilized granules of Bioglass, S53P4 and A-WG glass granules were implanted in rabbit dorsal muscles and in sheep's mandible. After 60 days of implantation, the materials with the living tissues were explanted, embedded in PPMA and sectioned for the histologic and scanning electron microscopic observations. Elemental microanalyses with an Energy Dispersive System were carried out to check the degradation of the materials through the topographic distribution of the elements. The three different glasses exhibited different states of degradation with different interactions with the soft and hard tissues for the same implatation tissues.

INTRODUCTION

Bioactive materials are the most interesting materials of 80's. The first bioactive glass was the so-called Bioglass® that was developed at the

University of Florida by Prof. Larry Hench. He first noted that some glasses, when put in contact with biological fluids, cells or tissues degrade and that the products of degradation can interact positively with the host tissue [1-6]. From in-vitro tests, simulating the body fluids, it was shown that the surface of the glass reacts developing a layer that is chemically similar to the inorganic component of bone.

Recently, other "active materials" have been developed for different biomedical uses, ranging from devices for drug delivery, to wound healing, to bone repair.

For these bioactive materials the definition of biocompatibility should be changed [7,8], since before their invention, the necessary condition for a biomaterial was its inertness. With the development of bioactive materials, there is a statement that "an appropriate host response in a specific application" must be validated. For this statement the European Community has elaborated three directives that define the protocols of testing of degradable materials before the in-vivo implantation [9-11].

At the moment there is no universal definition of bioactivity because the positive biological reactions produced by these materials have not been quantified. Since the material interaction with the host tissue depends on the structure of the material (crystalline, amorphous) and on the type of tissue (skin, muscles, bone) it is likely to be quite a while before there can be agreement on this definition.

The present work takes into consideration 3 different active glasses and shows their in-vivo behavior in soft and hard tissues.

MATERIALS AND METHODS

Three different glasses, Bioglass® developed at the University of Florida by Prof. Larry Hench, A-WG (apatite-wollastonite ceramic glass) developed by Prof. Kokubo at the University of Kyoto (Japan)[12], and S53P4 glass developed by Dr. Andersson at the University of Turku (Finland)[13] were implanted in rabbits and in sheep. The test in rabbits' dorsal muscular pockets, surgically created, had the aim to check the chemical degradation of the glasses. Previous experiments performed by our Centre in both the animal models have shown a correlation between the interaction of the glasses in soft (muscle) and hard tissues (bone). We have demonstrated that in the muscle the extracellular fluids degrade the glassy structure in a shorter time than that in bone, giving an idea of the type of the interface developed and of the times of reaction.

Table 1. Compositions of the glasses (in mole %)

Glasses	Granule size	SiO_2	CaO	Na_2O	P_2O_5	MgO	CaF_2
Bioglass	300-500μ	45	24,5	24,5	6		
A-WG	355-700μ	34,2	44,9		16,3	4.6	0,5
S53P4	300-500μ	53	20	23	4		

In anesthetized rabbits bilateral pockets were created in dorsal muscles, while in anesthetized sheep, after having exposed the mandibular bone, 4-mm holes were drilled with a microengine at low velocity. Ethylene oxide sterilized granules, packed in syringes, and were pushed in the cavities. The surgical sites were then covered with muscles and the surgical wound sutured. At the explantation, all the dorsal excised muscles and the jaw were fixed in 2.5% paraformaldheyde and dehydrated in ascending concentration series of alcohols. The x-ray microradiographs of the explanted samples were carried out in order to precisely locate the materials in the tissue for histologic preparation. After embedding in poly-methylmethacrylate, the samples were sectioned by a diamond saw (Accutom, by Struers, Denmark) in thick (200 micron) and thin (20 micron) slices respectively for the Scanning Electron Microscopy (SEM XL40 by Philips, The Netherlands) and for the histological observations (optical microscope Nikon, Japan). The thin sections were stained with fuchsin basic, suitable to show calcium deposition.

Elemental analyses and x-ray maps were carried out with an Energy Dispersive System (EDS, by EDAX, USA) in order to detect the topographic distribution of the elements before and after implantation. EDS analyses on the original glasses were carried out after the same preparation of the granules as above described; in order to show the homogeneous distribution of the components before degradation.

RESULTS

The time of implantation was sufficient in both cases to show the degradation of the materials and their interaction with the biological environment. Bioglass showed a higher rate of degradation than those of S53P4 and A-WG. The latter presented the slowest rate due to the crystalline structure (it is a glass ceramic) that makes it more stable. The degradation process changes the morphology of the materials and also their chemical compositions. By studying the morphology of the in-vivo created interface of S53P4 granules, it is possible to understand some steps of the degradation of all these materials.

Fig.1 shows a histological section of Bioglass granules implanted in muscles after staining with basic fuchsin. It is possible to observe that:

a) all the granules have fractures. Some of them are artifacts due to the diamond saw sectioning; others are developed in vivo. In fact some cells are present in the gap.
b) only the external part of the glass granules is stained: this means that the composition of the glass has changed and now there is the presence of Calcium.

The observation of a section of a S53P4 granule after 60-day implantation in rabbit's dorsal muscle under SEM in back-scattered mode (Fig. 2), shows three different zones: an undegraded core, an intermediate non-electron-dense layer and an outer "white" layer. The x-ray dot maps show that the core (c) is composed of the original elements and, at this stage, is not yet degraded. The intermediate layer is composed only by silicon, but probably it is a gel of hydrated silica (b). The "white" surface shell is electron dense and it is a calcium-phosphate (a). The glassy structure is totally changed and transformed. Previous works stated[14,15] that calcium and phosphorus in this zone comes from the core. Substantial amounts of phosphorus also come from the biological environment. In this aspect, this glass is like a sponge.

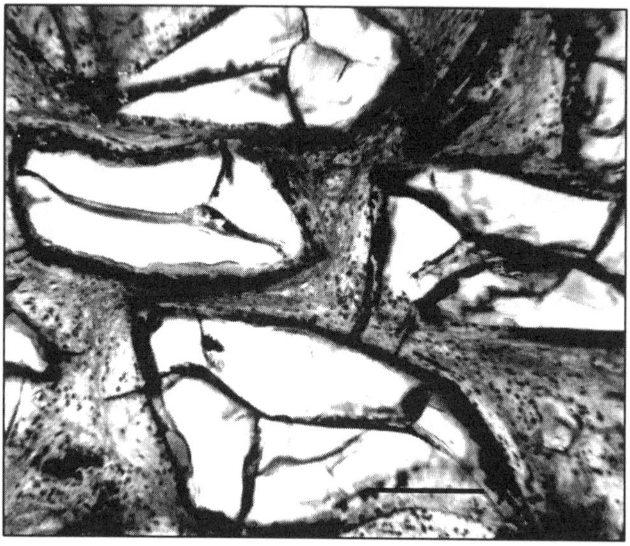

Fig. 1 Histological view under optical microscope of Bioglass granules implanted in rabbit's dorsal muscle and stained with basic fuchsin (bar= 100 μm).

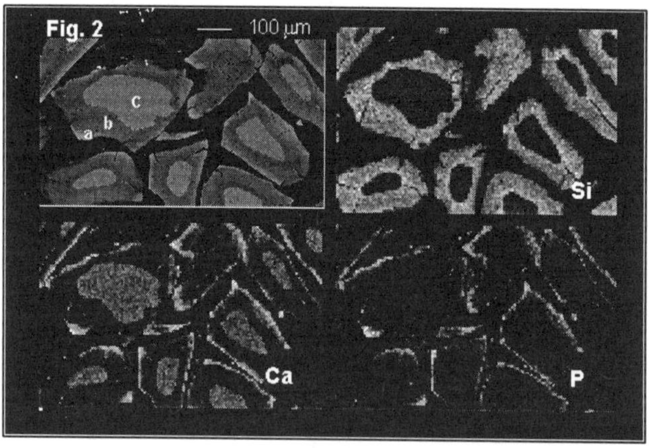

Fig. 2 SEM microphotograph of a S53P4 granule after a 60-day implantation in rabbit's dorsal muscle with the x-ray dot maps for Si, P, Ca and Na.

A-WG glass is less reactive; in fact the SEM images show a different morphology (Fig. 3). First, there is a physical degradation of the structure of the granule that changes its profile after a diffusion of the elements. So the thickness of the created interface is smaller.

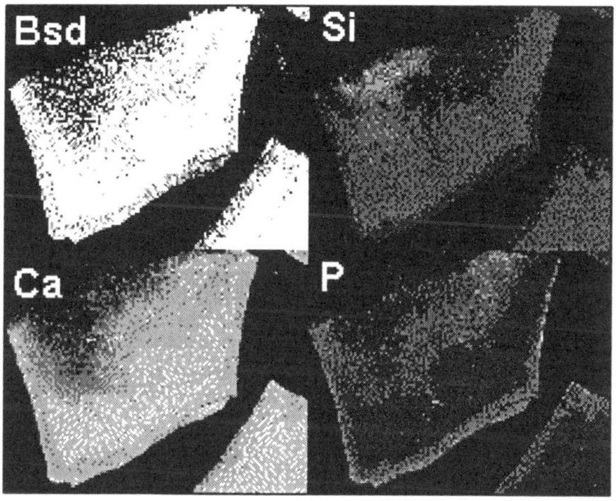

Fig. 3 SEM microphotograph of an A-WG granule implanted in dorsal muscle with the x-ray dot maps for Si, Ca, P, and Mg.

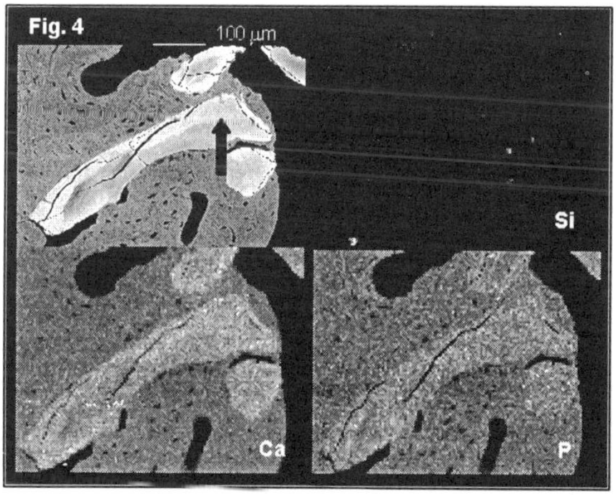

Fig. 4 SEM microphotograph (in back-scattered mode) of Bioglass granules surrounded by sheep's mandibular bone.

In the sheep jaw experiements, it is possible to see the "chemical bond" that these materials develop with the bone. Fig. 4 shows a small granule of Bioglass entrapped in bone and totally changed into calcium phosphate, as shown in the EDS spectra (Fig. 5b). It is possible to note that the silicon peak, present in the original composition of Bioglass (Fig. 5a) is now totally absent.

Also S53P4 glass (Fig. 6a.) and A-WG glass (Fig. 7) develop a bond with bone, although the total amount of new bone is greater for Bioglass and S53P4 than that for A-WG glass. The strength of the interfaces with bone is different in these materials as demonstrated by the lines of fracture occurred during the sectioning. In the case of A-WG this line is along the border of the glass, instead in S53P4 is inside the Ca-P rich layer.

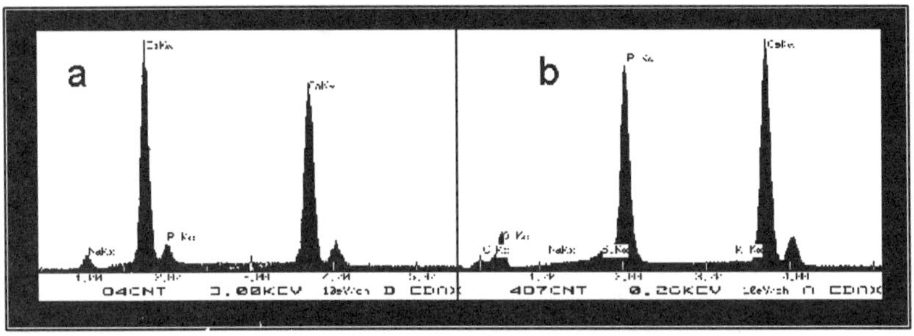

Fig. 5a, b EDS spectra of the original Bioglass and of the granule of Fig. 4. (black arrow)

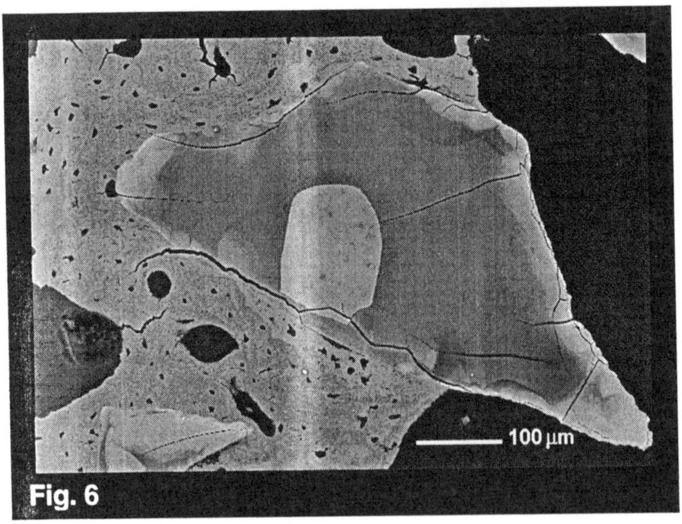

Fig. 6a, SEM backscattered imaged of 553P4 glass bonded to bone

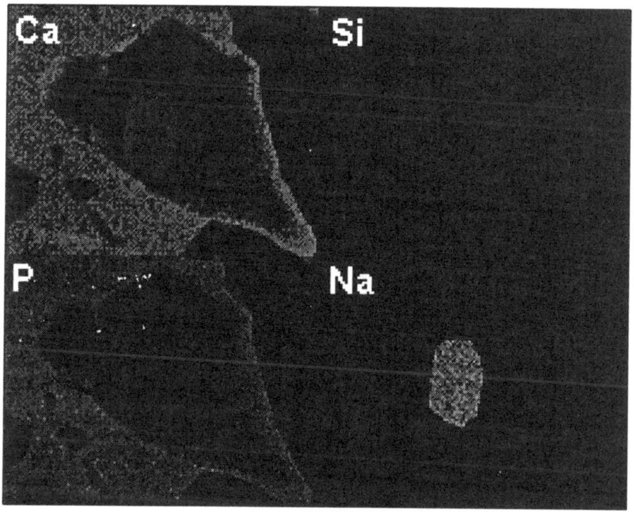

Fig. 6b. SEM microphotograph of a S53P4 granule surrounded by bone with the x-ray dot maps for Ca, Si, P and Na.

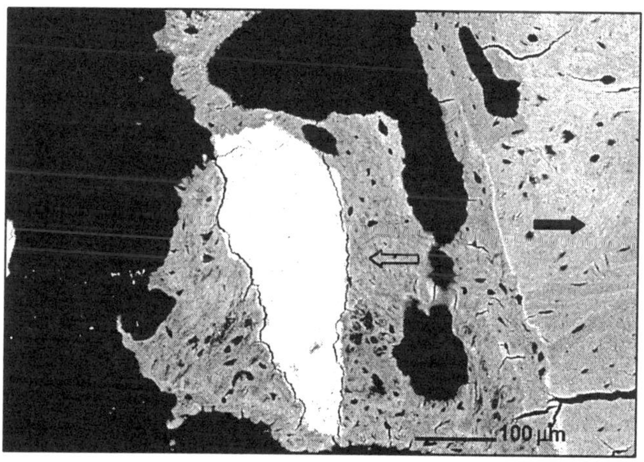

Fig. 7 SEM microphotograph of an A-WG granule surrounded by old (black arrow) and new grown bone (white arrow).

CONCLUSIONS

These bioactive glasses presented three different aspects of the in-vivo bioactivity. There is no doubt that their positive interaction can be modulated through the size of the granules. In fact, for equal volumes, small granules develop a greater surface area, and since they are exposed to the extracellular fluids, the total amount of released ions per time unit increases. These small granules can "stimulate" some biological favorable reactions, but it must be clear that a positive reaction can only be developed within a certain range of concentration of released ions. Under a threshold value or above the maximum concentration, the reactions will not to be positive. If the total amount of released ions per time exceeds the maximum, the cellular homeostasis is perturbed enough to possibly cause cell death. Under these conditions, a "bioactive" material can be toxic. Thus, for a material to be considered bioactive, one must demonstrate the beneficial effects *in-vivo*, and then within a particular environment. For example the same glasses that bond to bone could be toxic in the blood or induce negative reactions.

ACKNOWLEGMENTS

The Authors thank Ms S.Piaggi and R.Salvatori for their technical help. This work was supported by the National Council of Research of Italy (CNR-Progetto Finalizzato 1997-1999)

REFERENCES

[1]L.L. Hench, R.J. Splinter, W.C. Allen, T.K. Greenlee, "Bonding Mechanisms at the Interface of Ceramic Prosthetic Materials", *J. Biomed. Mater .Res.,* **2** 117-141 (1972).

[2]L.L. Hench, H.A. Paschall, W.C. Allen, G. Piotrowski, "Interfacial Behaviour of Ceramic Implants", *NBS Spec.Publ.*, **413** 19-35 (1974).

[3]L.L. Hench, C.G. Pantano, P.J. Buscemi, D.C. Greenspan, "Analysis of Bioglass Fixation of Hip Prosthesis" *J. Biomed Mater. Res.*, **11** (1977).

[4]L.L. Hench, D.E. Clark, "Physical Chemistry of Glass Surfaces," *J. Non-Cryst.Solids*, **28** 83-105, (1978).

[5]L.L. Hench, "Biomaterials", *Science*, **208** 826-831 (1980).

[6]L.L. Hench, E.C. Ethridge, *Biomaterials: An Interfacial Approach,* Academic Press, NewYork, (1978).

[7]"Clinical Applications of Biomaterials", pp. 11-16 in *Implant Materials in Biofunction: Advances in Biomaterials*, NIH Consensus Development Conference, De Putter et al., Eds. Elsevier Science, Amsterdam, Vol. VIII, (1988).

[8]II Consensus Conference on Definitions in Biomaterials, pp. 525-533 in *Biomaterial-Tissue Interface: Advances in Biomaterials,* Doherty-Williams-Williams, Ed. Elsevier Science, Amsterdam (1991).

[9]"Identification and Quantification of Degradation Products from Polymers ISO/EN 30993-13", *Biological Evaluation of Medical and Dental Materials and Devices*, **13** (1993).

[10]"Identification and Quantification of Degradation Products from Ceramic Materials ISO/EN 30993-14", *Biological Evaluation of Medical and dental Materials and Devices* **14**, (1993)

[11]"Identification and Quantification of Degradation Products from Metallic Materials ISO/EN 30993-15", *Biological Evaluation of Medical and Dental Materials and Devices* **15** (1993).

[12]T. Nakamura, T. Yamamuro, S. Higashi, T. Kokubo, S. Ito, "A new Glass-Ceramic for Bone Replacement: Evaluation of its Bonding to Bone Tissue" *J.of Biomedical Mater. Res.,* **19** 685-698 (1985).

[13]O.H. Andersson, K.H. Karlsson, "On the Bioactivity of Silicate Glass", *J Non-Cryst. Solids*, **149** 145-151 (1991).

[14]A.M. Gatti, G. Valdrè, A. Tombesi, "Importance of the Microanalysis in Understanding the Mechanism of Transformation in the Active Glassy Materials" *J. of Biomed. Mater. Res.,* **31** [4] 475-480 (1996).

[15]A.M. Gatti, L.L. Hench, E. Monari, F. Gonella, F. Caccavale, "Test of Bioactivity in Four Different Glasses" pp. 287-290 in *Bioceramics 10*, Sedel C. Rey, Editor, Elsevier Science Ltd, Paris, (1997).

OSTEOCERAMIC BONE REPLACEMENT AND GRAFTING

C. E. Olson, Urometrics, Inc., 445 Etna, Suite 56, St. Paul, MN

T. D. McGee, Materials Science and Engineering Department, Iowa State University, Ames, IA 50011

Dr. W. D. Hoefle, D.V.M., Orthopedic Surgeon, Veterinary Clinical Sciences, Iowa State University, Ames, IA 50011

Dr. R. Kudej, D.V.M., Ph.D., Allegheny General Hospital, 320 East North Ave., Pittsburgh, PA 15212

ABSTRACT

The Iowa State University Osteoceramic is a biologically-active composite of tricalcium phosphate and spinel. Hard tissue will bond to it. The osteoceramic has enduring strength. This paper reports its use as a replacement for a section of the femur of a dog. Seven years without mechanical support have been achieved. Successful bonding requires 1) a biologically active material with enduring strength, 2) tissue stress level compatible with physiological processes, 3) structure suitable for the imposed forces, and 4) provision for adequate blood supply. Radiographic and microscopic results are reported.

INTRODUCTION

We are pleased to contribute this paper to the Biomaterials Section of the Larry Hench symposium on the Surface-Active Properties of Materials. His contributions in Bioglass research and implants and his research on the tissue/implant surface interactions have had tremendous impact on the development of bioceramics during the last three decades.

This paper is a presentation of the developments of a crystalline counterpart to Bioglass that has similar bioactivity to Bioglass and hydroxyopatite, but that has enduring strength. Because of the historical nature of the symposium, this paper is also presented in historical context.

The Osteoceramic

The ceramic composite identified as an osteoceramic was invented in 1963, after the second author saw silicate ceramics intended for hip joint replacement at Haeger Potteries, Dundee, IL during a plant tour. The Haeger development was impregnated with a polymer to obtain similar elastic properties to bone and called Cerosium [1]. The polymer invoked undesirable tissue response.

To make a biologically active material, calcium phosphates seemed appropriate. But their solubility required a second phase for enduring strength. On physical chemical grounds, to prevent reaction with calcium phosphate, spinel ($MgAl_2O_4$) was chosen as a good second phase. Soft tissue experiments revealed the Ca:P ratio in calcium phosphate caused inflammatory response if it was too high (too basic) or too low (too acid). The range from 1 to 2 was acceptable, but the range from 1.5 to 1.67 was preferable. Tricalcium phosphate (TCP) $Ca_3(PO_4)_2$, has a ratio of 1.50 and hydroxyapatite (HA), $Ca_{10}(PO_4)_6(OH)_2$, has a ratio of 1.67. The HA in bone often has a ratio of about 1.62 and contains CO_3, F, Cl and OH anion substitutions. Some Mg and other cation substitutions occur in lesser amounts. Crystalline ceramics, glasses and cast glass-ceramics were produced. The original patent includes calcium phosphates with any inert material [2].

Commercial calcium phosphates are available in the ratio 1.5 to 1.67, with some at about 1.62. When fired at high temperatures water is lost and the anhydrous calcium phosphate crystallizes in various forms. Much of this research has been conducted with a "tricalcium phosphate" of ratio 1.62.

The structure of the Iowa State University Osteoceramic is like that of a sponge. The skeleton of the sponge is an interlocking network of microcrystalline spinel. The interconnected holes in the sponge are filled with calcium phosphate, which, despite its non-stoichiometry, is identified by x-ray diffraction as crystalline tricalcium phosphate. When polished to show relief between the soft calcium phosphate (CP) and the hard spinel (SP) phases, the structure is clearly visible (Fig. 1). Chemically removing the CP phase reveals the SP skeleton more clearly (Fig. 2). An energy dispersive x-ray microprobe evaluation shows that the CP and SP phases remain separate after firing at high temperatures (Fig. 3) [3]. The spinel skeleton has a finer grain size than the calcium phosphate in the holes in the spinel skeleton. As manufactured the maximum flaw size in the spinel is controlled by the maximum grain size of the calcium phosphate and this determines the strength. Chemical reactions that remove calcium phosphate at the surface do not change the flaw size in the spinel skeleton so the strength is unchanged. This gives the composite enduring strength. This is an improvement over glasses and monolithic phosphate systems that have increased flaw size and reduced strength when dissolution occurs. Physical properties are given in Table I.

Table I

Properties of the Composite	
Composition:	α-$Ca_3(PO_4)_2$ and $MgAl_2O_4$
Compressive strength	199 Mpa
Tensile strength	70 Mpa
Young's Modulus	114 Gpa
Reversible Thermal Expansion	$10.7 \times 10^{-6}/°C$
Bulk Density	3.09 gm/cc
True Density	3.37 gm/cc
Ca:P ratio	1.62

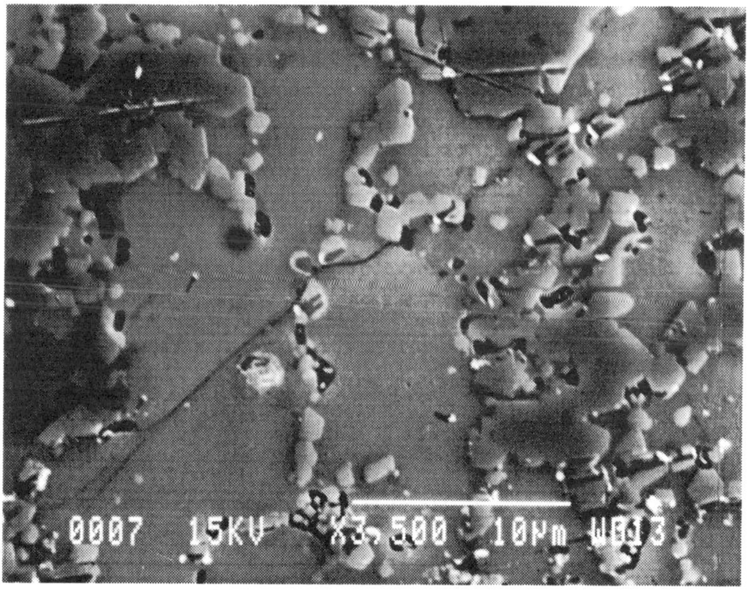

Figure 1 Polished section of osteoceramic 3500X

Figure 2 Same section as Fig. 1 with calcium phosphate phase removed by etching. 3500X

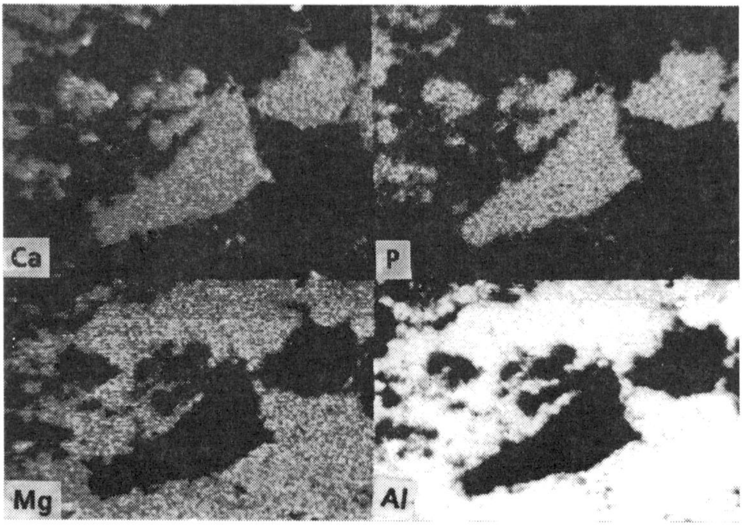

Figure 3 EDAX Map of osteoceramic 10,000X

The calcium phosphate phase in the fired composite was α-TCP at room temperature instead of the β-TCP expected at equilibrium. Thermal expansions of the spinel and the composite were linear with temperature. The TCP phase, when fired by itself, was α-TCP but the thermal expansion shows an inversion in the 100-150°C temperature range with hysteresis on heating and cooling. This was confirmed by scanning calorimetry. It appears, then, that the TCP phase is in a non-equilibrium form that should make it more soluble than the β form. Others have shown that HA forms on the surface of TCP so the tissue is in contact with HA [4]. But others have also found that TCP is much more soluble in the biological environment than TCP [5]. High solubility is perceived to be a benefit to early tissue response and to enhance mineralization.

CLINICAL EVALUATION

Tooth Roots

The first implants (1967) were made from an osteoceramic containing glass-forming components to lower the firing temperature, and produce a vitreous bond [6]. Later tooth roots were made with spinel and tricalcium phosphate without glassy additions. Both kinds contained interconnected pores of 200-500 μm diameter in the distal apex to allow bone ingrowth. The sub-gingiva portion was dense. They were implanted in the fresh alveolus of an extracted bimolar or canine tooth in the mandible or maxilla of a dog. In one case, after ten weeks, the mucosa was removed, exposing the top of the tooth root [7]. Attempts to extract the root were unsuccessful and the implant was left exposed. A year later the mucosa had covered the tooth. The tooth root was firmly fixed in place (Fig. 4).

Additional experiments compared the osteoceramic with three commercial implants, titanium alloy (Corevent), single crystal sapphire (Kyocera) and pyrolytic carbon (Pyrolite) [8]. Research orthopedic surgeons were recommending all three in 1981. When implanted according to the manufacturers procedure the tissue response was evaluated by measuring the amount of direct bone contact as determined by image analysis of light optical sections. The carbon implants did not stabilize and were lost. (Only one survived for histological analysis and it had no bony contact, only soft tissue contact.) The osteoceramic had superior tissue response (Fig. 5). Note that the inert implants (titanium alloy and sapphire) increased in tissue contact for about one year, then decreased. The growth of an interconnected fibrous capsule separating the implant from bone was responsible for the decrease. This is one of the mechanisms for failure for existing orthopedic implants. Loosening under stress as a progressively thicker capsule appears is sometimes found.

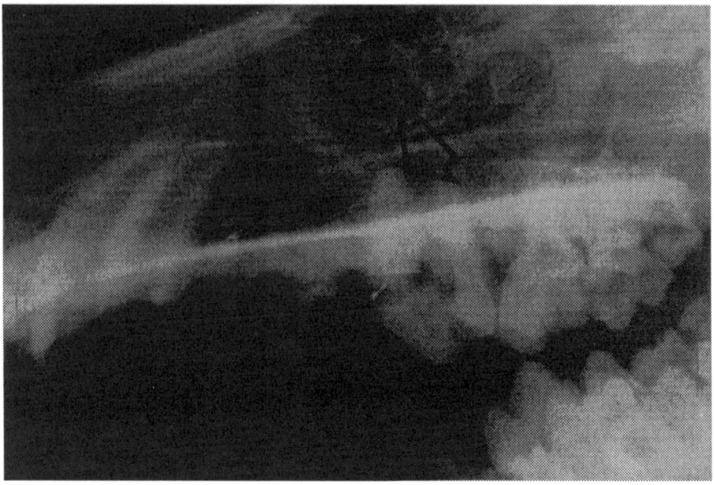

Figure 4 Radiograph of tooth root, 10 weeks

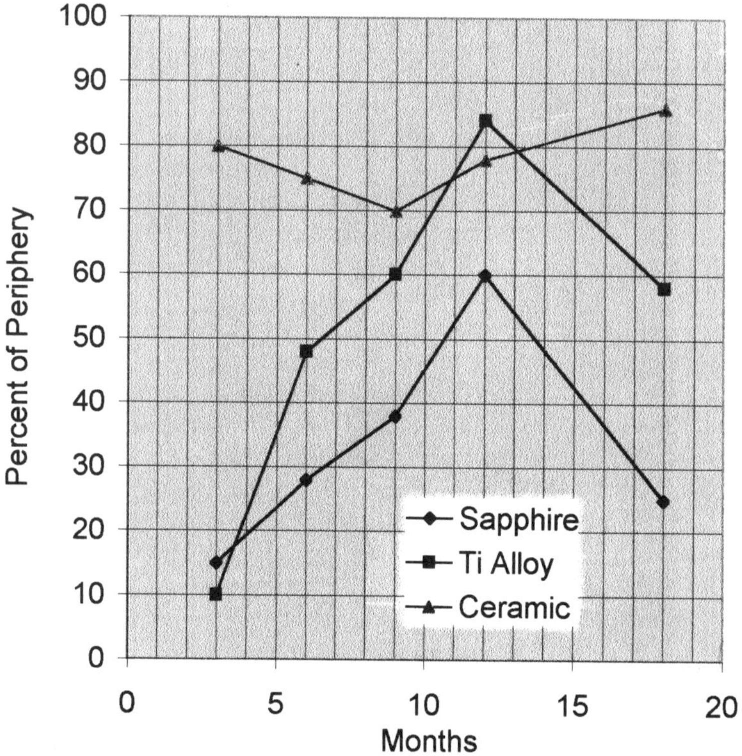

Figure 5 Average areas bony contact as a function of time

Porous Bone Replacements

In 1979 porous osteoceramic cylinders with an axial hole were implanted in the femur of dogs [9]. The CP phase was fluoroapatite. This tested the hypothesis that tissue would penetrate an entire implant for attachment, to bridge the proximal and distal bone by that penetration. Interconnected pores were large, 400-600 μm diameter. When implanted, they were stabilized with a stainless steel bone plate and screws. The implant was larger in diameter than the bone so the cortex of the bone opposite the plate was not aligned with the outside diameter of the implant. However, after 65 weeks, cortical bone had grown out to meet the implant (Fig. 6).

Figure 6 Fluoroapatite bone bridge after 65 weeks.

When the bone plate was removed the tissue had bonded to the implant (Fig. 7). The holes left by removal of the screws holding the bone plate are visible. The remodeled cortex has increased the diameter of the bone in contact with the implant. The limb was load bearing, at first, but in a few days the implant fractured through one of the holes used to clamp the implant to the bone plate (Fig. 8).

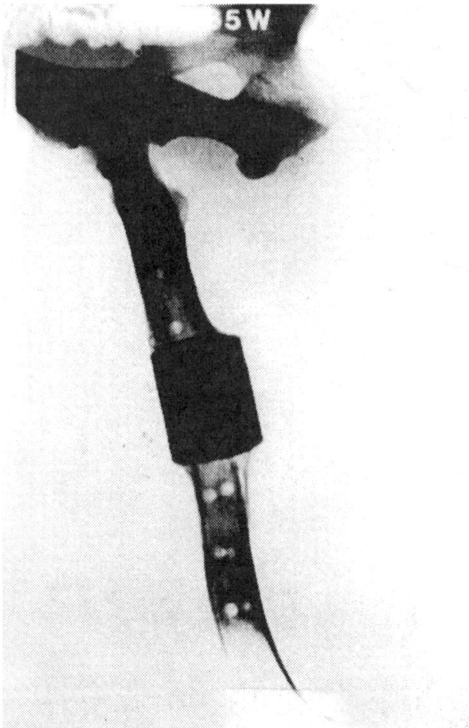

Figure 7 Fluoroapatite implant after removing the supporting bone plate.

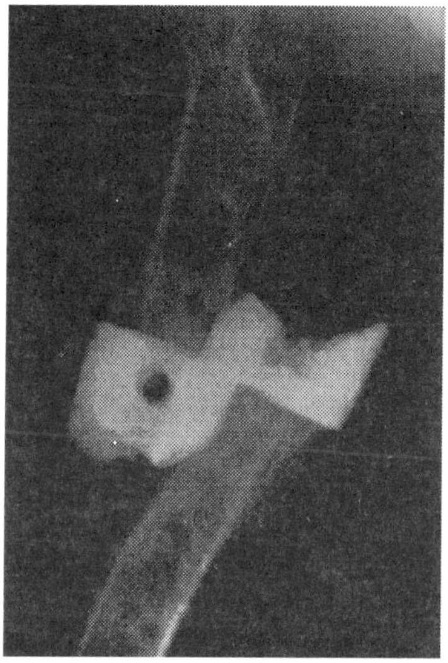

Figure 8 Fractured porous fluoroapatite implant a few days after removing the supporting bone plate.

The histology showed that cancellous bone had penetrated completely through the axial hole (Fig. 9). The stainless steel bone screws had been walled off with a fibrous capsule, but the osteoceramic was bonded to the bone by bone penetration into it. This bone only penetrated a few millimeters (Fig. 10). It could not compensate for the weakness inherent in the large pore size. The hypotheses that bone would penetrate from end to end failed. Therefore, the use of porous implants was discontinued.

Non-Porous Bone Replacement

The experiments with porous replacements and with non-porous tooth roots led to new design concepts: 1) Bone ingrowth areas must be large enough for Haversian systems and must have adequate blood supply. 2) Stress levels must be within the range normally encountered by the tissue, and 3) Provision for the forces such as tension, compression, and torsion in the anchoring bone must be provided. This led to a new bone replacement geometry Fig. 11). [10], [11].

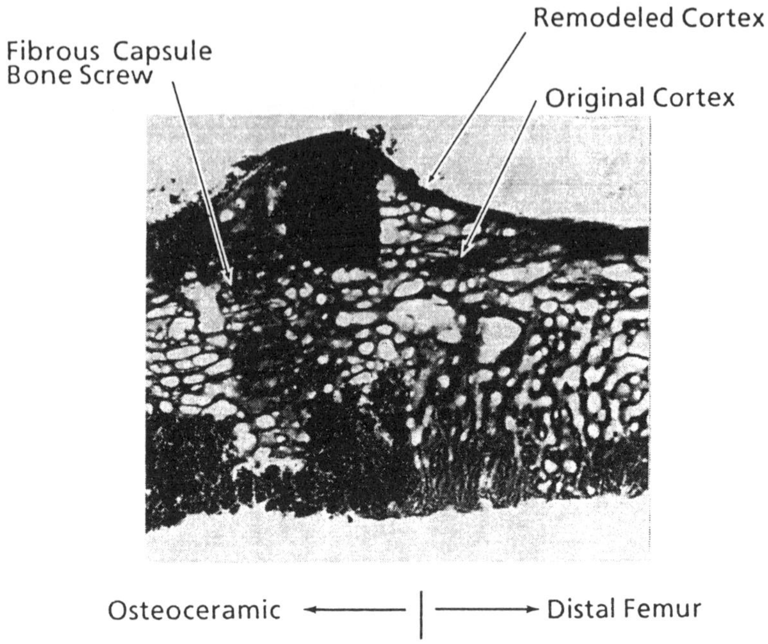

Figure 9 Microradiograph showing bone penetration into the fluoroapatite specimen. 2X.

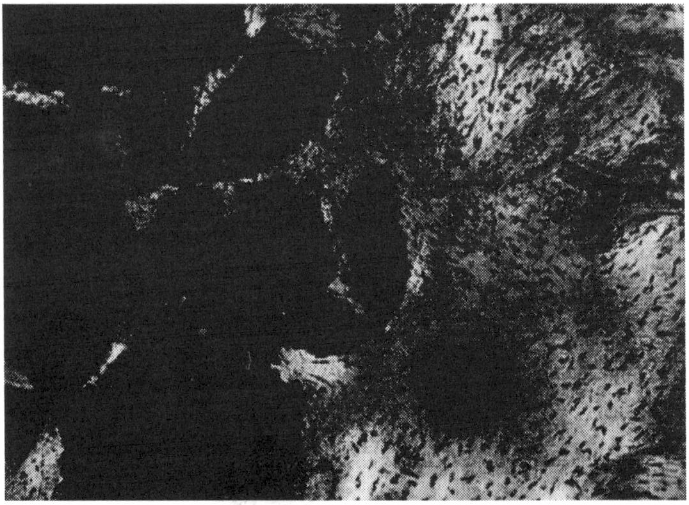

Figure 10 Polarized light thin section showing penetration of bone structure between the grains at the porous fluoroapatite surface. 40X.

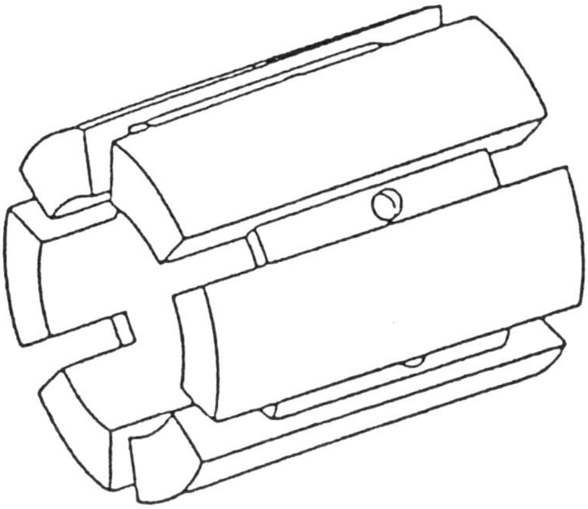

Figure 11 Geometry of the "bone-bridge" implant for replacement of a section of the diaphysis of a canine femur.

Bone replacements of this design had axial holes and radial holes for blood supply. Splined surfaces and dovetailed ends provided space for bone ingrowth to accommodate tension, compression, and torsion. These were implanted in the femurs of dogs (Fig. 12, 13).

After one year, the bone plate was removed. The implant was viable (Fig. 14). Three years later, the implant had remodeled (Fig. 15). This animal is active and behaves normally seven years after the removal of the bone plate.

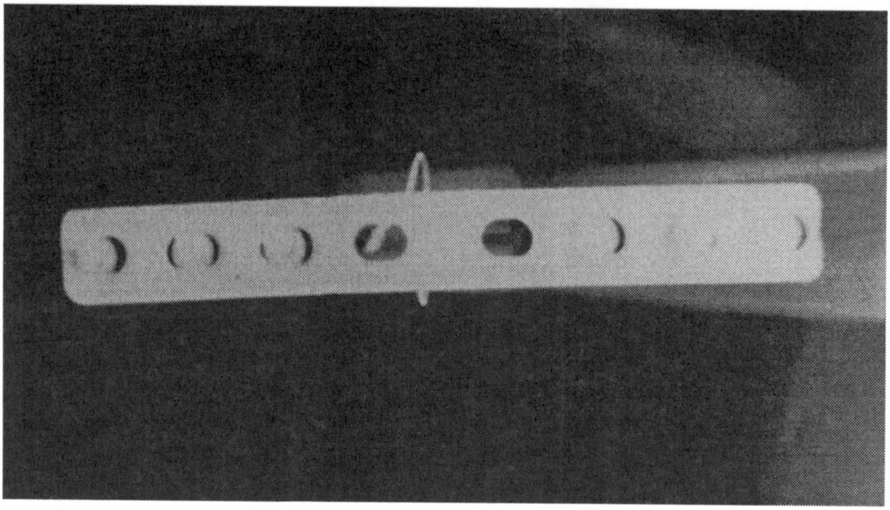

Figure 12 Radiograph taken immediately after surgery. 6/28/90

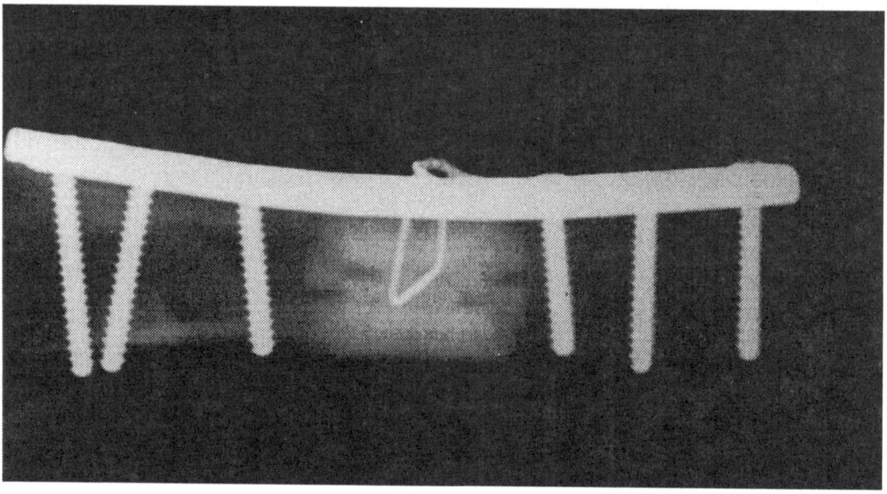

Figure 13 Radiograph after 55 weeks. 7/25/91

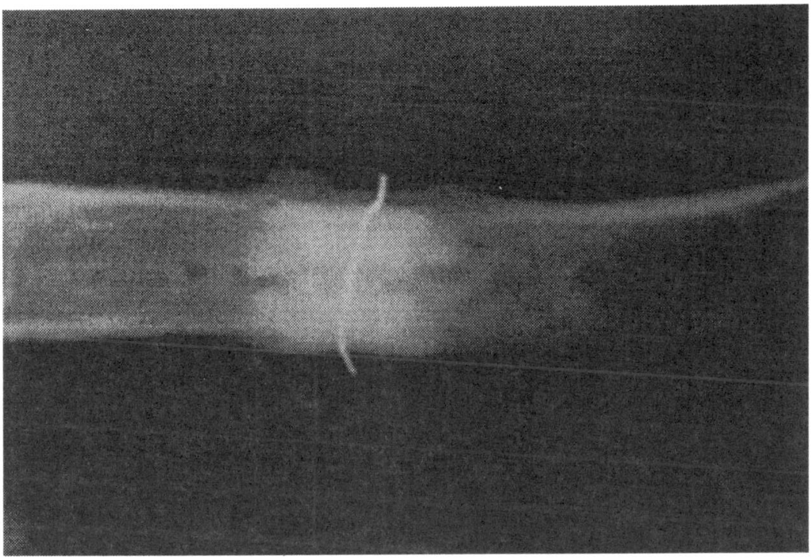

Figure 14 Radiograph removing the bone plate. 8/14/91

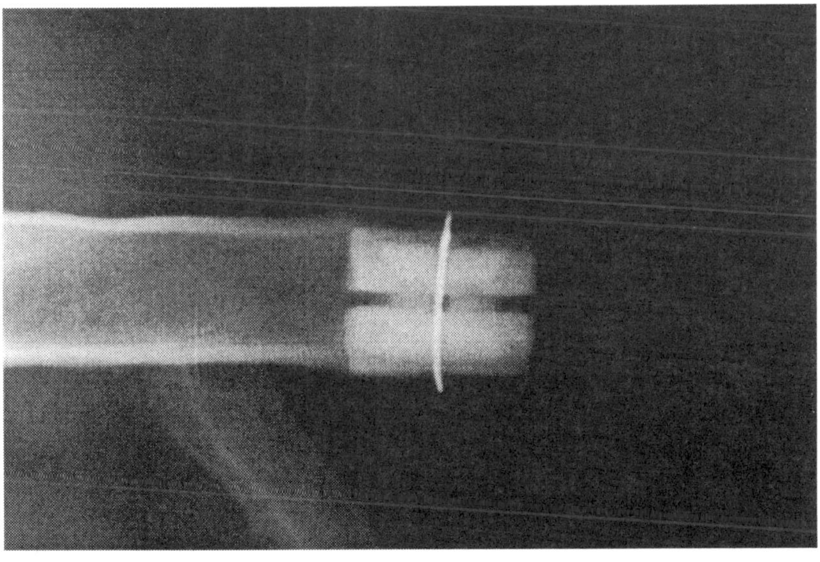

Figure 15 Radiograph two and one half years after removing the bone plate. 3/29/94

CONCLUSION

The requirements for successful bone replacement and grafting include:

1. An implant containing a calcium phosphate component to prevent the foreign body response.
2. An implant with enduring strength.
3. A geometry of design to accommodate the loads imposed.
4. A geometry that allows the normal physiological responses of the bone to develop supporting bone and to maintain that support, including stress level and blood supply.
5. Suitable surgical procedures to stabilize the implant until loads are transferred to the implant; and to remove the support when it is desirable to do so.

ACKNOWLEDGEMENTS

The authors are those most directly involved in the bone replacement experiments. During the years many others contributed to this research. Dr. Phillip Pearson was the first orthopedic surgeon to evaluate the osteoceramic in soft tissue and to perform the first tooth-root implants. Denginar Aksaci prepared the first fluoroapatite bone-replacement specimens, as a part of her PhD program in metallurgy. Ms. Ann Graves completed characterization of the osteoceramic. Catherine Tweden and Gabrielle G. Niederauer completed evaluation of tooth roots and made developments in histology as part of their PhD programs. Dr. John Kluge and Dr. Yosiya Niyo provided pathological analysis.

REFERENCES

1. L. Smith, "Ceramic-Plastic Material as a Bone Substitute" Arch. Surg., 87, 653-651, 1974.
2. T. D. McGee, "Artificial Bone or Tooth Prosthesis Material" US Pat. No. 3, 787, 900; 1974.
3. A. M. Graves, "Evaluation of a Bioactive Ceramic as an Endosseous Tooth Root Implant in Sheep" MSc Thesis, Iowa State University, 1988.
4. K. DeGroot "Biocompatibility of Clinical Implant Materials" 199-222. D. F. Williams, ed. Vol 1 CRC Press 1981.
5. R. Z. LeGeros, J. R. Parsons, G. Daculai, F. Driessens, D. Lee, S. T. Lier, S. Metsger, D. Peterson, and M. Walker, "Significance of the Porosity and Physical Chemistry of Calcium Phosphate Ceramics", Ann. N.Y. Acad. Sc. 523, 268-271, 1988.

6. T. Janikowski and T. D. McGee, "Artificial Teeth for Permanent Implantation" Proc. Ia. Acad. Sc. 76, 113-118, 1969.
7. T. D. McGee & J. L. Wood, "Calcium-Phosphate Magnesium-Aluminate Osteoceramics" J. Biomed. Mater. Res. Symp. No. 5 (Part 1) 137-144, 1974.
8. K. S. Tweden, "A Comparison of Four Endosseous Dental Implants" PhD Dissertation, Iowa State University, Ames, IA 1987.
9. D. Aksaci, T. D. McGee, W. D. Hoefle, A. Graves, and G. G. Niederauer, "Porous Fluoroapatite/Spinel Osteoceramics for Bone Bridges" Bioceramics, Materials and Applications, pp. 283-301, Am. Cer. Soc. 1991.
10. Curtis E. Olson, "The Effect of Geometry on the Tissue Response of Ceramic Bone Bridges in Canine Femurs" M.Sc. Thesis, Iowa State University, Ames, IA, 1992.
11. T. D. McGee and C. E. Olson, "General Requirements for a Successful Orthopedic Implant" pp. 69-82 in Encyclopedia Handbook of Biomaterials and Bioengineering Vol. 1 Part B (D. L. Wise et al., ed) Marcel Dekker, Inc., N.Y. 1995.

PATENT STATUS

Foreign and domestic patents are pending.

A CASE FOR SPECIFYING MATERIALS – THE PERSPECTIVE OF AN FDA ENGINEER

Gary Fischman, Ph.D.
US Food and Drug Administration
Rockville, MD 20850

INTRODUCTION

In recent years, the use of ceramics as biomedical materials has been given greater publicity. Even so, the development of ceramics for biomedical applications and their delivery to final product has been slow. Even though people are living longer and the United States is a major market for medical devices, Europe, as noted by Mecholsky[1] has put together a much more integrative effort toward the development of biomaterials (and thus "bio"ceramics) than has the U.S. His points ought be taken carefully, the change in demographics and the advancement of technology are requiring that more medical devices be placed into a larger number of people. Because of both technological advances and the market reality, the future of devices is that they will be implanted into an increasing number of people. As technology advances, devices will be implanted into patients that present with less compromising symptoms, they will be implanted into younger and younger patients and those devices will tend to work better. Thus we are seeing a need for materials that will be placed into devices that will enhance the longevity and/or performance of each device.

Ceramics, as a material class can fit into that need. In Fact, they already do – as a prime example, ceramics used in orthopaedic devices are helping change the field of hip arthroplasty. Oddly to the ceramist, smoothness is the significant property of these devices. The smoothness minimizes the wear of of the device and thus makes wear of the bearing surface a less significant cause of implant failure. Still, even in the orthopaedic industry, a fear of the brittleness and failure of ceramics prevails. A lack of knowledge of even the materials being used often extends from manufacturer to practitioner limiting the ultimate success possible.Within the Food and Drug Administration, concern over the reliability of new devices based on ceramic couples is significant. Even medical practitioners are trying amongst themselves to understand ceramics, but have a limited information source (see for example

Dambreville, et. al.[2] where Y-TZP is suggested as the complete zirconia ceramics family).

Ceramics are found in several applications in the human body. Aesthetically, ceramics are used in restorations such as dental crowns and bridges. Other ceramics from the biologically active materials in some alveolor ridge augmentation products to thrombogenically neutral pyrolytic carbons for heart valves prove the usefulness of ceramics in the biomedical industry. Still, the medical industry,as a whole, does not have the information or history for understanding ceramics in the same way that it does for metals and polymers. A good description of the clinical uses of ceramics can be found in texts such as that of Hench and Wilson.[3] However, looking at the number of ceramics that have made their way into the implant industry, there is a smaller number of ceramics in the market than one might expect. Certainly, not every material is a useful biomedical material. But new materials, and composites come to market slower than they might. It seems, from my interaction with the medical industry, that a major consideration is that the medical industry does not have a clear understanding of what ceramics are, and, therefore, how they can best be used.

I suggest that many medical companies do not design readily with ceramics for a variety of reasons. Among those reasons are the lack of information about the ceramics and the lack of consistency of properties amongst those same ceramics. It is easy to understand the concern when a lack of understanding can cause misunderstanding to proliferate. I believe that the companies developing biomedical devices and the agency that regulates these same devices would probably welcome ceramics more readily if they understood the materials better and believed that sufficient testing was done to trust these materials. This is important because the device designers and, clearly, the regulators do need to know the limitations of the materials as well as the advantages. We who know ceramics understand this difficulty in that the property of the material is a complex combination of starting material chemistry and phase composition as well as the processing history of the part itself.

In my tenure at the Food and Drug Administration, I have come across applications which suggest to me that the company sponsoring a given application does not clearly understand the ceramic(s) used in its device. In fact, it is not unreasonable to find a company actually describe the material as simply "ceramic" or describe crystalline as "glassy"(probably because it is shiny). Clearly this will not help an application. Companies also try to change their material for one of many reasons, often for marketability; in some cases, these changes would make any ceramist cringe – some have made me cringe. These issues slow down the progress of these materials to market and this makes the biomedical industry less interested in developing and using ceramics.

Ceramics will be used in a medical device if it is advantageous to use them. This is obvious, but advantageous means that the ceramic has to give a performance and/or a marketing advantage without major cost disadvantages. For success, the advantage of the ceramic must clearly outweigh its disadvantages.

What, then, can the ceramics industry do to try to improve the rate of development in this industry? I believe that the ceramics industry would like to develop this market. Some players within the ceramics industry already have, but others are still out there.

I believe we need to change the biomedical industry's perception of ceramics; there is no substitute for an informative decision. Two things that could help ceramics develop in the biomedical industry would be the development of clear standard specifications for the materials and appropriate use of the Master File system for describing the commercially confidential information.

STANDARDS DEVELOPMENT

The development of a standardized system of describing ceramics would go a long way to help. This would be similar, in some way, to the various standardized specification systems used for metallurgy, although I would strongly suggest a single universal standardization system for ceramics and glasses. At present, specifications amongst metals are significantly more common, even within the medical and surgical materials industry. Looking at ISO[4] and ASTM[5] standards, one quickly finds that the number of metals standards far outnumber ceramics, polymers and tissues.

Table I: Standards developed through medical based committees in ISO and ASTM.

	Recognized	Not Recognized	Totals
ceramic	8	1	9
metal	33	8	41
polymer	3	8	11
tissue	0	1	1
Totals	44	18	62

Table I gives the number of standards developed through medical based committees for both ASTM and ISO. The left-hand column refers to standards recognized through the FDA recognition process, the middle column refers to standards not recognized through the FDA recognition process and the right-hand column refers to totals from both columns.

Specifications are becoming more important in the regulation of devices. In 1997, Congress passed and the President signed the FDA Modernization Act which, in part, gives the FDA the authority to recognize standards and declarations of conformity to these standards as acceptable data for submissions.[6] Thus statements of conformity to recognized standards are an official means by which properties of a material will be accepted by the Food and Drug Administration. This, in my opinion, leads credence to the development of standard specifications for ceramics so that the base properties of the specified materials are known.

Within the specification, a baseline set of information should be given for a single material or a range of materials that a designer could use to develop product. Besides chemistry, phase composition and, if applicable, microstructural characteristics should be described. Also, basic materials properties including reliability information (such as Weibull modulus) should be available within such a specification. The materials specification should be a document that describes a material that could be used for most any device. Properties necessary for a particular device need not be incorporated into the specification, they could be developed separately.

It would also be helpful for the ceramics industry to devise a descriptive nomenclature. This nomenclature should describe chemistry, phase content and microstructure, so that a ceramic could easily be described and differentiated. Designers need to understand that alumina and zirconia are really names for families of materials and don't fully describe the materials themselves.

MASTER FILES

Often, material manufacturing companies are concerned about sensitive information. Companies are rightfully concerned that information that they have developed with respect to process, quality assurance, certain test results on the material as well as some properties should not become public knowledge and prefer not to share that information with the device manufacturer. Unfortunately, because the Center for Devices and Radiological Health, a center within the FDA, does not regulate materials, unless they are, in and of themselves, devices, there is little guidance on how to make and safeguard this information. The Center for Devices and Radiological Health, however, along lines similar to its sister Center for Drug Evaluation and Research does maintain documents called master files (MAFs).

An application to the Food and Drug, whether it be a premarket approval application (PMA), premarket notification (510(k)) or an investigatonal device exemption application(IDE) usually contains data and other information that the applicants has developed and regards as confidential commercial financial information (e.g., a trade secret). The information documented within the application is can be used by the FDA to determine the completeness of the application and is what is used for a final ruling on the application. A ceramic, in this case, is often used as a part or subassembly of the device for which the application is submitted and the ceramic manufacturer would rather not give the applying company certain sensitive information on the product. To ensure that the FDA can make a sound scientific judgement on the device application, a review of all the pertinent data is required. In such a case, the ceramic company can allow the FDA the ability to view pertinent information while witholding this same information from the applicant. This is done through the submission of a master file. The masterfile system helps to preserve the trade secrets of the ancillary medical device industry. In addition, the same masterfile may be used by several device manufacturers through permission of the masterfile owner as submitted with each individual application. The masterfile is considered as part of the submission by the FDA. The total application must still, however, have appropriate data to support the evaluation of that particular product and it may be necessary to have further testing done to the material based on the particular product use.

There are no specific requirements for a MAF. Data in the master file are reviewed within the context of the particular application (IDE, 510(k), or PMA) to determine if the information such as the material, process, component, etc. is sufficient to support the intended use of the medical device. The master file should be substantive in nature and should have information that may reasonably be regarded as trade secret or confidential commercial or financial information, if not, the FDA may not accept the MAF submission.

To actually submit a master file, the FDA requires two copies: one copy is immediately scanned but without optical character recognition and made available to review staff. The remaining hard copy is available for check out by staff, more specific guidance can be found from the FDA[7]. Remember that micrographs, photographs and other qualitative data will be scanned along with the rest of the document and may be more difficult to read once scanned. If the document needs to be reviewed, additional copies may be requested because of the inability to read the scanned document. To this end, electronic submission of the materials may be provided in addition to the two hard copies. It is reasonable to consider making CD copies of the MAF available to those evaluating a document upon request.

Some suggestions I can give on the submission of a master file are as follows:

1. Realize that a master file is a collection of information that you do not want divulged to the public, or to the company that will sponsor a device made, wholly or in part, with this product. The master file only need contain the information for which you would like to maintain this level of nondisclosure;

2. Remember, you want the person from the Food and Drug Administration's device evaluation team to read the master file. From this perspective, it is imperative to keep the master file

simply organized with information easily accessible, well indexed and cross indexed as necessary and well referenced. The clearer and more complete the information, the fewer questions that need to be asked about the information.

3. An application is based on submitted data. Therefore, if you have a master file, be sure that the information you want developed is filed within the master file lest the FDA request information from the sponsor that will require the sponsor to refer back to you.

4. As stated earlier, the current method by which documents are archived at the FDA is via scanning without optical character recognition. Addenda, in the form of supplemental sheets of any sort, are scanned and added to the beginning of the document and the optical file is repaginated. Amendments to the original file make that file more and more difficult to read.

5. I suggest that you consider developing your document as a soft document available on CD for an examiner. Although the FDA will accept electronic submission along with the two hard copies, I suggest that a well organized master file which states that the data is available on CD for the FDA upon request can be helpful.

These are simple suggestions that should help the person reviewing the master file in support of an application understand what you have and find where the information they need is located.

CONCLUSION

Finally, I cannot guarantee that these suggestions will solve all of the issues for ceramics going to market in the medical industry; we will only know if these suggestions are tried. I can say that these suggestions will make it easier for the designer and the regulatory agency to understand whether a material is being used properly and what device performance and analysis tests would be most useful in determining the appropriateness of the material for a particular device. It is up to the ceramic industry to develop the appropriate nomenclature and standards and to each individual company to develop its master file. Having this accomplished, I believe that the ceramics industry, as a whole, will be in a stronger position to sell to the biomedical industry.

[1] JJ Mecholsky, "An Office of Naval Research - European Office Materials Newsletter No. 10 (96-010) Report on Biomaterials Research in Europe" Office of Naval Research Europe.

[2] A. Dambreville, M Phillipe and A Ray, "Zirconia Ceramics "by night, all cats are grey," Maitrise Orthopaedique, http:// www.maitrise-orthop.com/corpusmaitri/orthopaedic/mo78_zircone/index.html.

[3] An Introduction to Bioceramics, ed. LL Hench and J Wilson, World Scientific Publishing Company, River Edge, NJ (1993), e.g., page 2

[4] International Organization for Stnadardisation, Geneva Switzerland

[5] American Society for Standardization and Materials, West Conshocken, PA

[6] The Food and Drug Administration Modernization Act of 1997, Public Law 105-115 Section 204.

[7] Guidance for Preparation of PMA Manufacturing Information, US Food and Drug Administration, CDRH, Office of Device Evaluation (available over the internet http://www.fda.gov/cdrh/ggpmain.html)

PROCESSING OF TITANIA CERAMICS AS SUBSTRATES FOR LIVER CELL CULTURE

K.-L. Eckert[1,2], B. Stieger[3], R. Gadow[1], P. J. Meier[3], E. Wintermantel[2]
[1]Institute for Manufacturing Technologies of Ceramic Components and Composites, University of Stuttgart, Allmandring 5b, D-70569 Stuttgart, Germany.
[2]Chair of Biocompatible Materials Science and Engineering, Department of Materials, ETH Zurich, Wagistrasse 23, CH-8952 Schlieren, Switzerland
[3]Division of Clinical Pharmacology and Toxicology, Department of Medicine, University Hospital Zurich, CH-8091 Zurich, Switzerland

ABSTRACT

Porous titania ceramics were prepared by thermoplastic shaping and tested as substrates for liver cell culture. A thermoplastic body was mixed from titania powders and wax, formed by pressing and sintered at 1350° C. The sintered ceramic material had an open porosity of 25 %. Primary rat hepatocytes were cultured on these substrates for 24 h and for 3 d. Cell distribution on the substrates was investigated by SEM. Cells cultured on titania ceramics had rounded morphology and were arranged in three-dimensional clusters. Hollow titania microspheres were developed as an alternative to disk-like substrates. The microspheres were shaped by spray drying. The wall thickness of the microspheres is high compared to the diameter.

INTRODUCTION

Titania ceramics have been demonstrated to be highly biocompatible and to have superior chemical stability in simulated physiological environments [1]. The potential advantages of titania ceramics compared to mostly used polymers are high chemical stability and high degree of hydrophilicity compared to standard polymeric substrates which require pretreatment for many types of cell cultures. Porous cell culture substrates can improve nutrition supply. Surface pores and the granular structure of titania ceramic materials lead to a degree of surface roughness which could influence the function of adhering cells [2]. Most work on inorganic materials as scaffolds for cell culture has been performed with regard to bone tissue, less attention has been paid to other cell types. In the present work, cell adhesion and cell distribution of primary rat hepatocytes on the scaffold surfaces were investigated. Standard light microscopal methods are hampered by the opacity of titania ceramics; therefore scanning electron microscopy (SEM) was used. The titania cell culture substrates were produced by thermoplastic shaping of a ceramic mass.

A second possibility to apply ceramic carriers in cell culture is as spherical microcarriers. Microspheres of different materials have been used as scaffolds for cell culture [3], as cell carriers in bioartificial liver assist devices [4] or as implantable cell carriers [5]. In addition to thermoplastic shaping for disk-type substrates, the feasibility of spray drying as a shaping method for titania microspheres was investigated in this work. A specific feature of the microsphere carriers to be developed is, that they are hollow and provide an opening. The opening serves to enhance metabolite exchange through the porous carrier material and allows blood vessel ingrowth in case of use as implantable cell carrier.

MATERIALS AND METHODS

Ceramic Sample Preparation

Twenty grams of titania powder with d_{50}=9.6 μm [6], 12 g of commercial titania powder (1171, Kronos, Germany), 1.6 g graphite powder (KS6, Lonza, Switzerland), and 3 g paraffin (MP 68° C, Fluka, Switzerland) were heated to 120° C and manually mixed into a thermoplastic body. Portions of the hot, plastic mixture were pressed between steel plates to a thickness of 1.2 mm and cut into discs of 10 mm diameter. Temporary binders were pyrolyzed by heating to 300° C within 48 h. To produce hollow microspheres, a slurry with 50 % titania powder content (Kronos 1171) and 0.25 % deflocculator (Dolapix CE 64, Zschimmer&Schwarz, Germany) was prepared and spray-dried in a laboratory spray-dryer (Buechi, Switzerland) at an inlet temperature of 185° C. All samples were fired at 1320° C for 20 min.

Cell Culturing:

Rat hepatocytes were isolated with a two-step collagenase perfusion method and cultured as described by U. A. Boelsterli et al. [7]. Titania sample discs were put into 35 mm Falcon Primaria culture dishes and seeded with $3x10^6$ cells in 3 ml of supplemented Williams Medium E onto the discs without further treatment. For control culture, 35 mm Falcon Primaria dishes were coated with 200 μg Collagen R (Serva, Heidelberg, Germany, collagen type I) in 1 ml solution for 1 h at 37° C prior to cell seeding. The samples were incubated for

24 h and for 3 days with daily medium replacement. Cells were fixed for 1 h at 4° C with 3 % glutaraldehyde in PBS and critically point dried with CO_2.

Morphological Characterization

Open porosity and pore size distribution were analysed by mercury pressure porosimetry (Porotec, Germany). Morphology of the titania surfaces and of the cells was analyzed using scanning electron microscopy (SEM, Hitachi S-2500C). The specimens were coated with platinum in a sputter coater prior to SEM analysis.

RESULTS

Ceramics

The surface of the fired titania discs showed a rough topography, due to the granular structure of the material and due to the surface porosity (fig. 1). The surfaces of the individual grains were structured by crystal growth steps (fig. 2). The bulk material exhibited an open porosity of 25 % and 2.2 μm mean pore diameter (fig. 3). Spray-dried granules were of spherical shape, most of them being hollow (fig. 4). The internal volume of hollow granules was, however, small compared to the total granule volume (fig. 5).

Cell Culture Testing

On titania ceramics, most of the cells adhering to the titania surface were arranged in three-dimensional aggregates (fig. 6). After 24 h of incubation, a minor fraction of the cells was flattened. After 3 d, the total cell number had decreased probably as a result of the shear stress applied during the daily replacement of the culture medium (fig. 7). Consequently, the number of spheroidal cells decreased with time, whereas the number of spread cells increased. However, these differences were small. The overall appearance of the cultures remained constant during the culture period.

Hepatocyte morphology after culture on collagen-coated dishes displayed significant differences compared to titania ceramics. After 24 h the cells were spread and spherical residues were found on the flat cells (fig. 8). Increased flattening was observed after 3 d of culturing (fig. 9).

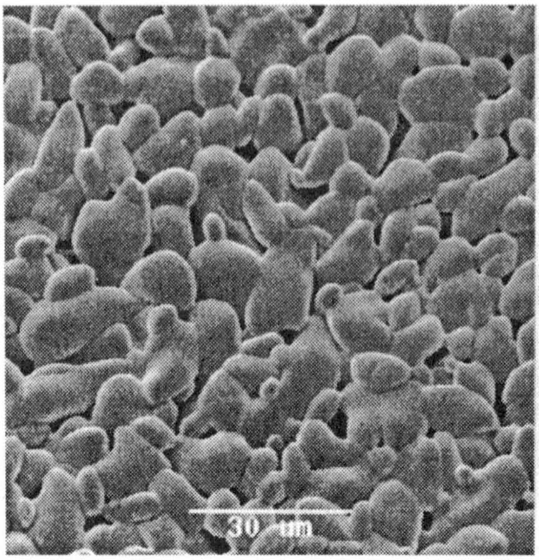

Fig. 1: Survey of titania ceramic surface. The topography is defined by the granular structure of the material.

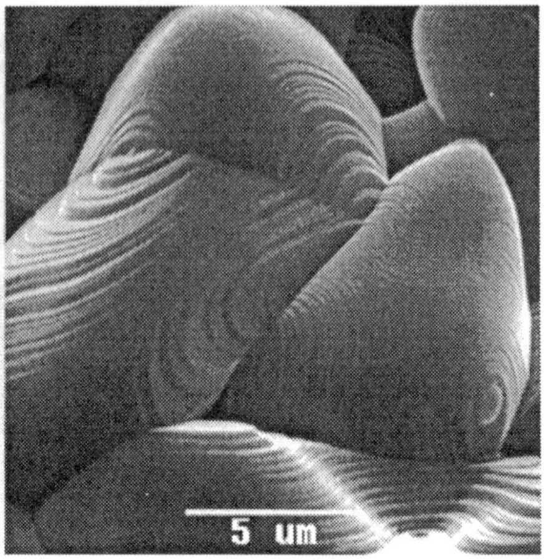

Fig. 2: The surface of individual grains is structured by growth steps.

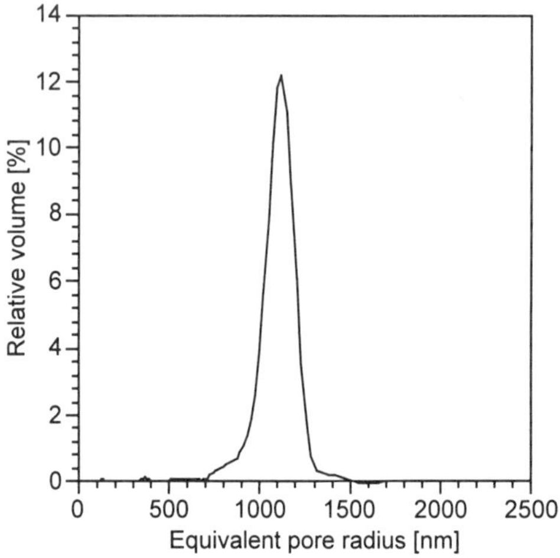

Fig. 3: Pore size distribution of the ceramic material. The mean pore diameter is 2.2 μm

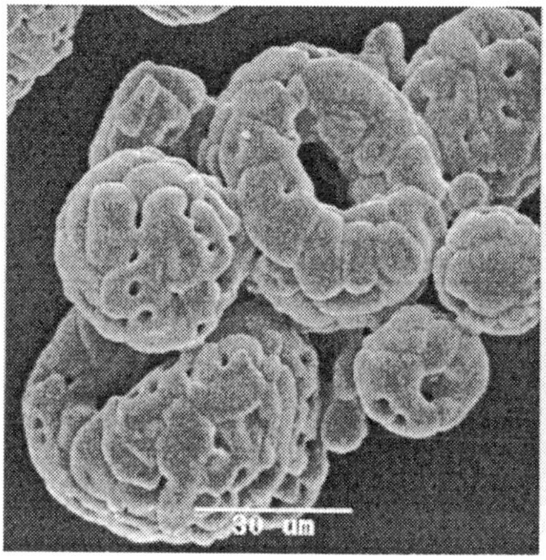

Fig. 4: Spray-dried titania granules after sintering. Most of the granules are hollow and have an opening which would make the internal surface accessible to cells.

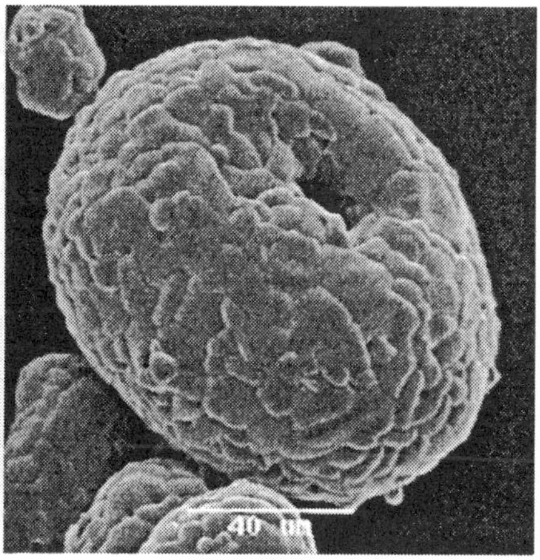

Fig. 5: Spray-dried and sintered granule. The internal volume of the granules is small compared to the total volume.

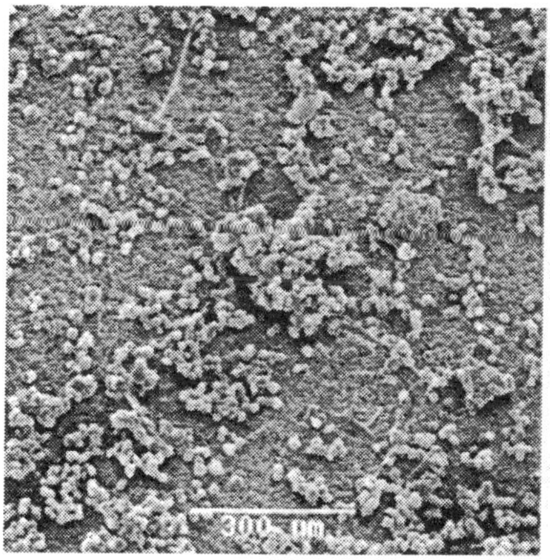

Fig. 6: Cells cultured for 24 h on titania ceramic surface. The cells form three-dimensional networks.

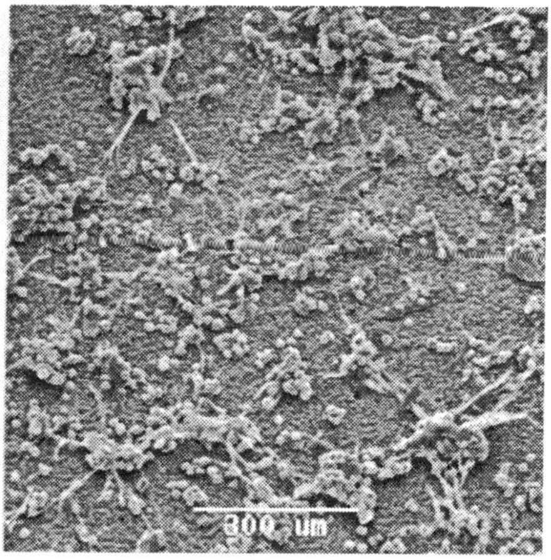

Fig: 7: Cells cultured for 72 h on titania ceramic surface. The total cell number is decreased, probably due to shear stress during culture medium exchange.

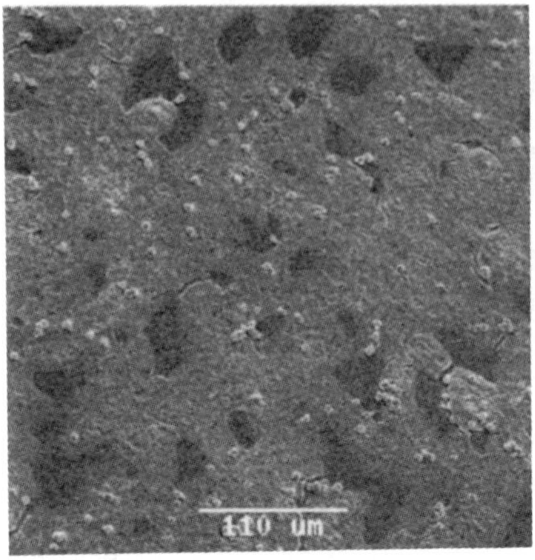

Fig. 8: Hepatocytes cultured for 24 h on collagen-coated cell culture dish. The cells have flattened on the polymer surface.

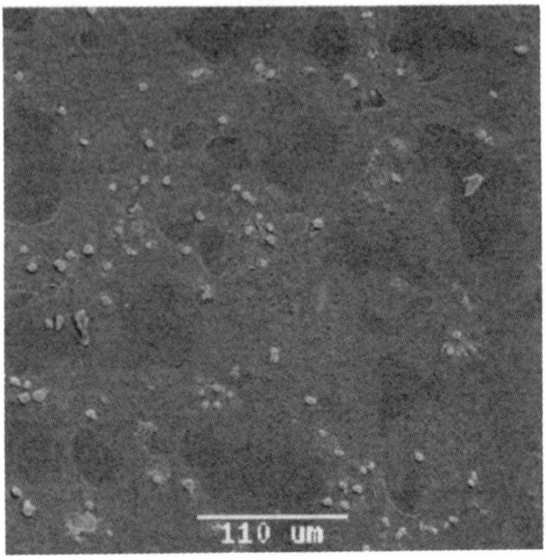

Fig. 9: Hepatocytes cultured for 72 h on collagen-coated cell culture dish. Cell flattening is increased after the extended culture period.

DISCUSSION

Hot molding of ceramic discs by using steel plates was an effective method for molding ceramic samples for biomedical testing. The ceramic body composition and the process parameters chosen yielded a material with an interconnected pore structure as well as a regular and reproducible surface morphology.

Hepatocytes cultured on these titania ceramic discs were well attached on the surface without applying cell-adhesive coatings like collagen. Large differences in cell morphology could be observed compared to standard culture conditions using collagen-coated culture dishes. On titania ceramics the cells did not spread, but appeared as sphereoids yielding three-dimensional clusters whereas on collagen-coated cell culture polystyrene the cells flattened out rapidly. Similar findings have been reported for adult rat cardiomyocytes, where cell shapes remained in vivo-like on polycrystalline titania ceramics. Latter effect could even be improved with sol-gel derived coatings on the titania surfaces [8]. Therefore, the preservation of a rounded cell shape of hepatocytes cultured on titania ceramics can be attributed to the effect of the substrate material. In this work SEM-analysis was performed to investigate the cell adhesion and the development of cell shape and cell distribution over the culture duration. Although the observed intercellular contacts and aggregation combined with the low degree of cell flattening are seen as favorable indicators of cell functionality

[9, 10], measurements of liver-specific enzyme activity and metabolic processes will be necessary in order to definitely work out the effects of porous titania ceramics as scaffold material for hepatocyte culture.

Spray drying as shaping method for hollow spheres can be can be applied to produce titania ceramic carriers for cell culture systems and cell transplantation. However, further investigations are necessary to optimize the spraying process with regard to requirements of biomedical use, such as control of porosity and structure, achievement of a defined surface topography and minimizing the wall thickness of the hollow microspheres.

ACKNOWLEDGEMENTS

The authors would like to thank Dr. G. Kahr, Institute of Geotechnical Engineering, ETH Zurich, for mercury porosimeter measurements. This work was supported by the Swiss National Science Foundation, grant 3138-051172.97/1.

REFERENCES

[1] B. Fartash, H. Liao, J. Li, N. Fouda and L. Hermansson, "Long-Term Evaluation of Titania-Based Ceramics Compared with Commercially Pure Titanium *in Vivo*," *Journal of Materials Science : Materials in Medicine*, **6** 451-454 (1995).

[2] A. Curtis and C. Wilkinson, "Topographical Control of Cells," *Biomaterials*, **18** 1573-1583 (1997).

[3] Q.-Q. Qiu, P. Ducheyne, H. Gao and P. S. Ayyaswamy, "Formation and Differentiation of Three-Dimensional Rat Marrow Stromal Cell Culture on Microcarriers in a Rotating-Wall Vessel", *Tissue Engineering*, **4** 989-1001 (1999)

[4] W. S. Arnaout, A. D. Moscioni, R. L. Barbour, A. A. Demetriou, "Development of Bioartificial Liver: Bilirubin Conjugation in Gunn Rats", *J. Surg. Res.*, **48** 379-382 (1990)

[5] K.-L. Eckert, P. Groscurth, R. Gadow and E. Wintermantel, "Implantable Ceramic Cell Carriers for Liver Cell Transplantation"; in *IEKC6 "Advanced Ceramics and Composites"*. Expert Verlag, Stuttgart, (1998), in press.

[6] J. Blum, K.-L. Eckert, A. Schroeder, M. Petitmermet, S.-W. Ha and E. Wintermantel, "In Vitro Testing of Porous Titania Dioxide Ceramics"; pp. 89-92 in *Bioceramics*, vol. 9. Edited by T. Kokubo, T. Nakamura and F. Miaji. Elsevier Science, Oxford, 1996.

[7] U. A. Boelsterli, B. Zimmerli, and P. J. Meier, *Am J. Physiol.*, **268** G797-G805 (1995) .

[8] J. Elbel, L. Polonchuk, K.-L. Eckert, H. Eppenberger and E. Wintermantel, "Sol-Gel Derived Titania Coatings for Cell-Culture Substrates"; pp. 305-

308 in *Bioceramics*, vol. 11. Edited by R. Z. LeGeros and J. P. LeGeros. World Scientific Publishing Co. Pte. Ltd., Singapore, 1998.

[9] P. V. Moghe, F. Berthiaume, R. M. Ezzell, M. Toner, R. G. Tompkins and M. L. Yarmush, "Culture Matrix Configuration and Composition in the Maintainance of Hepatocyte Polarity and Function", *Biomaterials*, **17** 373-485 (1996)

[10] B. A. Naughton, B. Sibanda, J. P. Weintraub, J. San Roman and Vava Kamali, "A Stereotypic Transplantable Liver Tissue Culture System", pp. 65-91 in *Frontiers in Bioprocessing*, vol. 3. Edited by P. Todd. Humana Press., Totawa N.J., 1995.

THE DIFFERENTIAL RESPONSE OF CULTURED MAMMALIAN CELLS TO COMMERCIAL SILICATE GLASSES

Amy B. Jedlicka and Alexis G. Clare
New York State College of Ceramics
2 Pine Street
Alfred NY 14802

ABSTRACT

A simultaneous study of chemical durability, film formation and surface chemistry carried out in identical conditions to cell culture studies was correlated to cell behavior on seven silicate glass substrates. "Normal" cell behavior was noted on durable glass without boron and on glass with boron that formed a leached layer. Extensive cell coverage observed on bioactive glasses correlated to calcium phosphate film formation. Early cell behavior was dominated by initial interfacial chemistry whereas later cell behavior was regulated by the dynamic nature of the glass surface chemistry.

INTRODUCTION

The target of this work was to observe and explain the fundamental nature of the *in vitro* response to several commercial glasses. The sequence of glasses chosen is shown in figure 1. Six of the seven glasses are commercial glasses with varied applications. Sodium trisilicate, while not a commercial glass was investigated to interrogate the role of adding a modifying ion to a network forming (silica) glass. As shown on the diagram the sequence of glasses studied range from a network former (SiO_2) to network former + one modifier (Na_2O-SiO_2) to commercial soda-lime silicate glass (network former with two modifiers and small additives of alumina and magnesia). At this point the sequence branched, and the left branch consisted of two silicate-based known bioactive glasses 45S5 Bioglass® and 45S5.4F, a fluorine containing Bioglass® (samples were donated by U.S. Biomaterials, Alachua Florida). The right hand branch consisted of two fiberglass compositions for which biocompatibility data are available due to extensive studies on exposure of fiberglass to the lung. These glasses are extremely durable; the main difference between the two compositions is that FS contains higher silica and lower boron than FB. The fiber glass samples were donated by Johns Manville

Corporation of Colorado.

The philosophy of the experiments described here was threefold: first, to take a fairly systematic variation of glass composition for which there were individual data in the literature and study them together; second, to carry out durability experiments in conditions as nearly identical as possible to the cell culture experiments; and finally, to use a universal cell type for exposure to the materials to fundamentally understand cell behavior as a function of surface chemistry. One could argue that a fundamental understanding of the chemical interactions of the surface might begin with the protein attachments but cells are much easier to observe and their behavior can be categorized both qualitatively and quantitatively. A full description of the experiment will be published later.

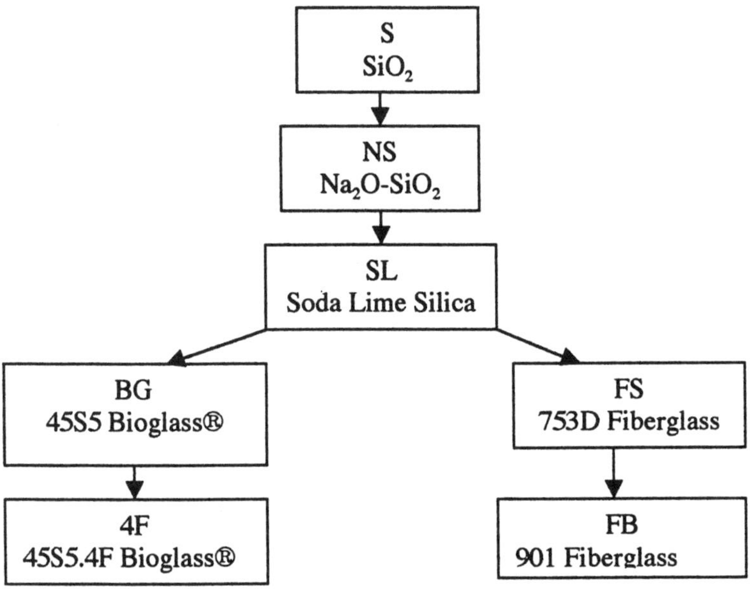

Figure 1 The Experimental Samples

EXPERIMENTAL

Chemical durability experiments were carried out in as identical situations as possible to the cell culture experiments. In order to correlate results to previously published data, three increasingly complex liquids were used: distilled water, Dulbecco's phosphate buffered saline (PBS) solution 1X[δ], and Ham's F-12 1X cell culture media with L-Glutamine[ς] (with added Nu-Serum[ϒ] and penicillin/streptomycin solution[ϖ]).

The glasses were soaked in these solutions, at a material surface area to solution volume (SA/V) ratio of 0.05 cm^{-1}, for specific time intervals related to the times spent in the cell culture study – generally 1, 4, 12, 24, 48, 96 and 168 hours, with times added as necessary. The glasses were placed in the respective static solutions in 125 ml Nalgene[Z] bottles. The bottles were then placed in a humid CO_2 incubator[Ψ], and the temperature was kept constant at 37 °C. Controls, containing no sample, were also exposed to the same conditions to account for behavior due to parameters other than the glass dissolution processes. The solutions were analyzed using weight loss, inductively coupled plasma spectroscopy and pH analysis. pH titration with NaOH was also performed on some powders. Diffuse Reflectance Infrared Fourier Transform Spectroscopy (DRIFTS) was used to examine the surface chemistry of samples after exposure to similar conditions.

Chinese hamster ovary (CHO) cells[ρ], from a certified cell line of the American Type Culture Collection, were chosen for analysis. Aseptic technique was followed for all procedures performed during cell culturing. The CHO cells were cultured in a CO_2 incubator in tissue culture grade flasks[ℜ] with Ham's F-12 medium, 9.9 volume percent Nu-Serum and 0.98 volume percent penicillin/streptomycin (pen/strep). The cells were subcultured at 3-4 day intervals (at ~80% confluence).

The CHO cells were subcultured approximately 20 times before exposure to the glass samples. This permitted the cells to stabilize after frozen storage, and ensured a normal, acclimatized, and hardy population. For the *in vitro* analysis, cell concentrations were measured after subculturing. Cell counting was performed using a hemocytometer[℘] counting number of cells per 0.1 microliter. The cell concentrations were adjusted to 1.25×10^5 cells/ml, and the cells were then plated onto the glass substrates. The CHO cells were allowed to attach to, and grow on, the glass substrates for times of 1, 4, 12, 24, 48, and 96 hours. Longer exposure times

[δ] Dulbecco's Phosphate Buffered Saline Solution, 1X, Cellgro, Mediatech, Herndon, VA
[ς] Ham's F-12 1X Cell culture media with L-Glutamine, Cellgro, Mediatech, Herndon, VA
[ϒ] Collaborative Biomedical Products, Becton Dickinson Labware, Bedford, MA
[ϖ] Penicillin/Streptomycin Solution, Cellgro, Mediatech, Herndon, VA
[Z] Nalge Company, Rochester, NY
[Ψ] Model 2325, VWRbrand, VWR Scientific, Bridgeport, NJ
[ρ] ATCC CCL-61, CHO-K1, American Type Culture Collection, Rockville, MD
[ℜ] Falcon, Becton Dickinson, Franklin Lakes, NJ
[℘] Brightline Hemocytometer, Improved Neubauer, Hausser Scientific Company, Hoarsham, PA

were not feasible due to deterioration of the media, requirements of the cells for replenished nutrients (which would not have correlated to the chemical durability study), and the appearance of contaminants in the stagnant cultures. In the absence of abnormal cell behavior, the glass samples were treated with a two percent gluteraldehyde solution[Σ] to 'fix' the cells in their natural state. The samples, in gluteraldehyde, were placed in a refrigerator for 24 hours, then removed, rinsed with distilled water, and allowed to dry.

The CHO cells adhered to the various substrates were analyzed using optical[θ] and scanning electron[®] microscopy. Three parameters – rate of cell reproduction, cell adhesion (or degree of cell spreading) and cell morphology were used to quantify the biological response to the substrates. Rate of reproduction was measured by counting the number of cells adhered to the substrate for specific magnifications. Degree of spreading was determined by measuring the percent of the glass surface covered by the CHO cells, analyzed using Image Pro Analysis software[M], and dividing this calculated area by the number of cells attached to the surface. The CHO cell morphology was a more qualitative measure of biological response, obtained from optical and scanning electron microscopy.

RESULTS AND DISCUSSION

Chemical Durability

The results of solution analysis can be extended beyond mere ionic concentrations to generate a clearer picture of dissolution processes. The parameter α, introduced by Schmidt, indicates dissolution behavior [1]. The equation for α is:

$$\alpha = \frac{\dfrac{\text{ppm SiO}_2}{\text{mwSiO}_2}}{\dfrac{\text{ppmR}^+}{2\text{mwR}^+} + \dfrac{\text{ppmR}^{2+}}{\text{mwR}^{2+}}} \quad \frac{1 - P(\text{SiO}_2)}{P(\text{SiO}_2)} \qquad 1$$

where $P(SiO_2)$ is the mole fraction of SiO_2 in the glass and mw is molecular weight. When α tends toward a value of zero, the alkali and alkaline earth ions are in abundance in solution, with little Si, indicating that leaching is occurring. If α is close to one, total dissolution is occurring, as shown by equivalent amounts of silica being lost along with the alkali. [2, 3, 4] Intermediate values indicate that both

[Σ] Fisher Chemical, Fisher Scientific, Fair Lawn, NJ
[θ] Polyvar-MET, Reichert-Jung Optische Werke AG, Austria
[®] SEM, Amray 1810, AMRAY, Inc., Bedford, MA

dissolution mechanisms are occurring concurrently. The α parameter also provides information regarding the formation of a silica-rich gel layer. When α tends toward zero, with leaching the dominant dissolution mechanism, a silica rich surface layer is formed. This layer is quantified by the "film formation parameter", ε, where $\varepsilon = (\text{ppm } SiO_2) \cdot ((1-\alpha)/\alpha)$. The dependence of this parameter on α results in an indeterminate value when $\alpha = 0$. This restricts its quantitative value, however comparison of ε trends can still be drawn. Due to the calculation of α depending on bulk solution data and not specific leached amounts, total network dissolution may begin to occur while the calculated α is still less than one. The epsilon parameter provides a more accurate measure of the switching of the rate controlling mechanism. A maximum in the plot of ε vs. time represents the time at which total dissolution becomes rate controlling.

These two parameters, α and ε were calculated for each set of glasses soaked in the three solutions. One result for the Bioglasses in media is shown in figure 2. The limitations of these equations have clearly been reached with α going below zero and beyond one. These dissolution parameters are generally used for studying corrosion of binary alkali silicates and ternary alkali-alkaline-earth silicates in water. [2, 3, 4] In this study, problems were encountered applying these equations to more complex glass systems. Also, due to the inherent high sodium concentration in PBS and media, Na concentrations could not be accurately measured for those systems, causing the values of α and ε to be subject to error.

In an effort to provide a broader and more applicable representation of dissolution behavior, a new dissolution parameter, α', is presented here. The premise is similar to that for the α calculation – the amount of SiO_2 lost is calculated as a function of the other components of the glass, and the relative percentage of silica in the glass. This new α' parameter accounts for the possibility of other network formers incorporated into the glass (as is the case for Bioglass and fiberglass) and does not depend solely on the sensitive measure of species concentration that confines α to simple fluids. The equation for α' is:

$$\alpha' = \frac{\text{ICP calculated weight of NWF in solution}}{\text{weight loss x wt\% NWF}} \qquad 2$$

where NWF = network former. The dependence of this α' parameter on weight loss permits use of this equation in fluids where salt concentrations cannot be accurately measured. Essentially, this α' parameter represents the weight of network formers lost, as calculated by ICP measurement, with respect to the weight lost by the glass – assuming network dissolution. This α' parameter varies in a similar manner as the α parameter. As α' tends to zero, leaching is indicated, weight loss calculated for NWF (assuming congruent dissolution) is greater than the weight loss measured for NWF by ICP, and as α' tends toward one, network dissolution is occurring. This α' parameter may also be used to provide information on reprecipitation, as is the case for the Bioglass aqueous interaction. An α' value greater than one represents a greater calculated weight loss, due to the network former dissolution, than was actually measured – an indication of reprecipitation.

The α' parameter for glass dissolution in media is shown in Table I. S, SL, FS and FB are fairly durable in this fluid; a blank indicates that exposure times were too short to get a measurable weight loss. Durable glasses need a higher surface area form. NS exhibits leaching then dissolution and reprecipitation. This reprecipitation in media is not evidenced through other analysis techniques. The two Bioglass_ compositions exhibit initially high α' values, and exceed one after one hour soak time, indicating that the calcium-phosphate rich film forms faster on these glasses in media than in other solutions studied.

Table I Alpha prime values calculated for glasses soaked in media.

	1	**4**	**12**	**24**	**48**	**96**	**168**
S α'							
NS α'	0.161	0.0179	0.108	0.480	1.055	1.068	0.459
SL α'							0.112
BG α'	0.974	1.027	1.032	1.172	1.107	1.391	1.438
4F α'	0.880	1.274	1.085	1.136	1.141	1.356	1.278
FS α'					0.107	0.0134	0.262
FB α'							1.204

This α' parameter combines solution and material characterization and therefore provides a more accurate measure of dissolution than the solution-dependent α parameter. This new α' parameter can be used to describe dissolution processes of complex glasses in salt-containing solutions, whereas the α parameter does not permit this type of complex analysis. Further characterization of the material surface, in this study provided by DRIFT spectroscopy, reveals more detailed information regarding the chemical changes occurring at the glass surface.

DRIFTS and Cell Culture

One of the most interesting results from the DRIFTS was that in most of the liquids S, SL, FS and FB showed similar spectra, with leaching followed by some silica-rich layer formation. The Bioglass compositions were consistent with data in the literature, forming a hydroxyapatite film, with α' indicating earlier dissolution for the 4F composition. However, when observations were made with cells present, the cells exhibited quite different behavior on glasses of indistinguishable durability. Figure 3 shows cell coverage as a function of time for each of the seven glasses.

Two separate cell processes – cell spreading and cell reproduction – are incorporated into this percent coverage variable. Several general observations can be made from this percent coverage data. First, assuming a cell cycle duration greater than 12 hours (as observed for a polystyrene control), increased percent coverage in the early stages is attributed to cell attachment and spreading. (All plated cell concentrations were 1.25×10^5 cells/ml, with sample SA/V ratios of ~0.1 cm^{-1}.) Relative trends of cell attachment and spreading between the varying compositions can be determined. At 4 and 12 hours, the two Bioglass® compositions were covered with CHO cells to a much greater degree than any of the other substrates in this study. This may be attributed to the large amount of adsorbed protein (observed in the DRIFT spectra) and the high Ca concentration of the glass, promoting cell adhesion and spreading. Also, at 12 hours, the difference between the BG glass and the 4F glass is significant – the percent coverage increases dramatically on the BG glass between 4 and 12 hours, while remaining unchanged on the 4F glass. This may be due to the relative degree of dissolution between these two glasses. As determined in the chemical durability experiment, the Si concentration in the 4F media solution is greater than that for the BG sample, and, from band assignments in DRIFT spectra, the surface of 4F is changing more rapidly than the BG surface. This decreased cell coverage with surface change is also observed for the NS glass - where the initial cell response is favorable, but the dynamic nature of the surface impedes further attachment and growth

The other four glasses behave relatively similar to each other in terms of percent cell coverage. The FB glass is the lowest in percent cell coverage, and this may be the cellular response to increased boron content. At longer times, accurate measurements of the area of cell coverage for each substrate were not possible. Only the cell behavior on the four durable glasses was determinable. At 96 hours, there is clear distinction between the percent cell coverage of S, SL, and FS, and that of FB. However, the magnitude of the difference is less at this time than at lower exposure periods. It appears that the initial inhibitory reaction of the cells to the FB composition is altered at longer times, attributed to the dissolution behavior of this glass. The increased boron content in this composition may cause initial cell response to be unfavorable, however, with increased glass dissolution, favorable surface sites become available for cell attachment.

Due to the corroded nature of the NS substrate at long soak times, it became difficult to distinguish between surface features of the glass and adsorbed cells. At 48 hours, the cell coverage on this glass is low, indicating disturbed cell processes. While NS initially does not cause cell death, clearly at later times the excess alkali leached into the solutionand the associated rise in pH proves toxic to the cells.

The apparent sensitivity of the cells to boron in the glass was supported by the pH titration experiments. The pH titration indicated that there were about three times the number of acidic sites on FS glass than on S or SL (FB was not determined because leaching, as demonstrated by the α' parameter, resulted in potential competing ions in the titrate). In addition FS and FB (when FB was in the early dissolution stage) showed abnormal cell morphologies. A "normal" cell cycle (for example, that seen on polystyrene) would consist of attachment, spreading, elongation and division. However the FS showed cells with a very spiked morphology similar to the type of morphology seen in dendritic cell growth though not as extreme. Figure 4 shows the cells exposed to the FS glass as a function of time. Similar morphology is displayed by the FB glass at short times but this is modified later when leaching presents a surface more similar to that of S and SL (as borne out by DRIFTS and α'). It is proposed that the pointed behavior results from cells putting out pseudopods to try to avoid acidic sites instead of the very equatorial spreading that is observed on the densely covered Bioglass® compositions.

CONCLUSIONS

Selected results from a comprehensive study of the dissolution, surface film formation and cell response to commercial silicate glasses have been presented. It appears that early cell attachment is governed by initial surface chemistry of the glass whereas long term stability of the cells depends on how dynamic the glass surface is with respect to the surrounding medium. Durable glasses elicit "normal" cell behavior when boron is not included. Boron tends to cause anomalies in cell spreading behavior. Cell behavior on Bioglass is related to the film formation indicated by DRIFTS and the α' parameter. A high percentage coverage results from the increased attachment and spreading of the cells on the biocompatible glasses.

REFERENCES

1. Y.A. Schmidt, "Structure of Glass," *Conf. Struct. Glass, 2nd, 1953,* p. 319.

2. E.C. Etheridge, "Mechanisms and Kinetics of Binary Alkali Silicate Glass Corrosion"; Ph.D. Thesis, The University of Florida, Gainesville, Florida, 1977

3. D.E. Clark, C.G. Pantano, and L.L. Hench, Corrosion of Glass, Magazines for Industry, New York, 1979.

4. D.M. Sanders and L.L. Hench, "Mechanisms of Glass Corrosion," *J. Am. Ceram. Soc.*, **56** [7] 373-377 (1973).

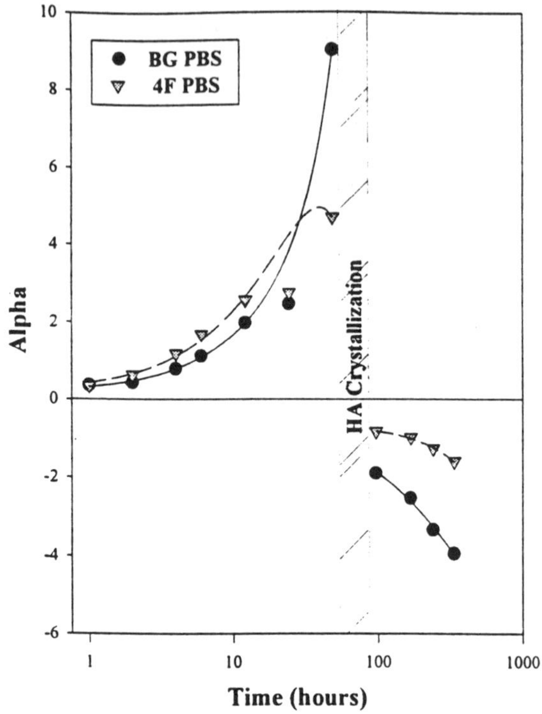

Figure 2 Alpha Dissolution Parameter for BG and 4F soaked in PBS

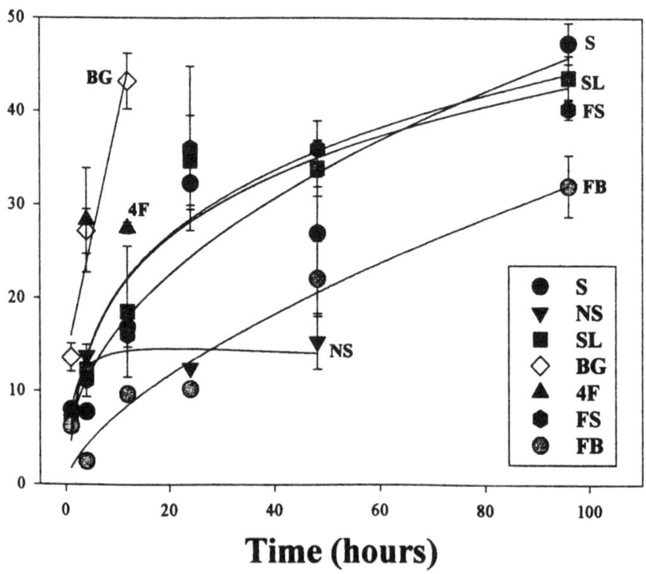

Figure 3 Percent CHO Cell Coverage on Glass Substrates as Function of Time

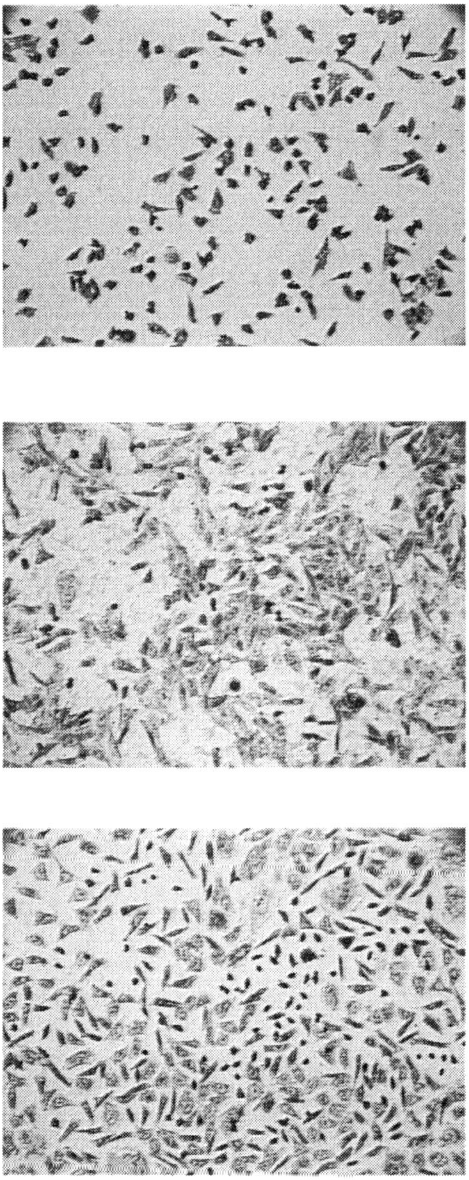

c

Figure 4 CHO cells attached to FS glass at 100x, (a) 12, (b) 24 and (c) 96 hours.

SPECIFIC INTERACTIONS OF GLASS FIBERS WITH EXTRACELLULAR MATRIX PROTEINS

Alan H. Goldstein, Fierer Chair of Molecular Cell Biology
Division of Biology and Biomedical Materials Engineering Science Program
Alfred University, Alfred, NY 14802

Joel S. Salesky
College of Ceramic Engineering and Materials Science
Alfred University, Alfred, NY 14802

ABSTRACT

When inhaled glass fibers reach the alveolar tissue, primary interfacial phenomena would be expected to occur between these fibers and the extracellular matrix (ECM) of the alveolar cells. We have developed a microscale *in vitro* system to identify and characterize ECM proteins adsorbed to the surface of these fibers. Preliminary studies have focused on three fiber types: MMVF 10 (rapidly dissolving borosilicate), MMVF 34 (high alumina HT), and RCF 1 (~50:50 alumina:silica). Adherent ECM proteins stripped from the fibers represented a subset of ECM proteins. While most of the adsorbed proteins were common to all fiber types, some differences in protein profiles were observed. In several cases the amount of protein adsorbed to the fiber appeared significantly enhanced over the amount of the same protein in the ECM. These data suggest that the fiber surfaces had high affinity for some ECM proteins. Numerous adsorbed proteins were in the low molecular weight range which includes growth factors, hormones, and signal molecules involved in both tissue inflammatory and healing responses.

INTRODUCTION

The underlying hypothesis for this research is that biological responses to long glass fibers by alveolar lung cells must initially be mediated by the extracellular matrix (ECM). We have focused on 'long' fibers (~1μm in diameter x ~20μm in length). These fibers are considered too large to be cleared by resident alveolar macrophages, so-called 'dust' cells (1-4). We also ignore situations where the alveolar cells are penetrated by the glass fibers resulting in further departure from physiological homeostasis. Thus, we have focused on the bioelectrochemistry, biochemistry, and ultimately the molecular physiology of the dissolution of long fibers at the interphase region of alveolar cells.

It has been shown that the adverse physiological responses associated with the presence of asbestos fibers in the lung are the result of the insoluble nature of these materials (2, 3). As a result, there is great interest in understanding the environment in which fibers dissolve in the lung. In addition, a high dissolution rate may create an array of physiological responses, some of which may impact the dissolution. This is because localized regions of high ionic strength as well as specific ion effects may trigger ECM- and/or plasmalemma-mediated responses by alveolar cells (5). These effects could occur via modification of localized electrochemistry at or near the cell surface (e.g. modification of the Debye layer near the plasmalemma or the Donnan potential in the ECM). Alternatively, specific ion effects could directly trigger physiological responses in alveolar cells either by acting as a signal transducing agent (e.g. free Ca^{++} activation of membrane receptors) or indirectly by forming precipitates (e.g. hydroxyapatite) that are capable of inducing cellular responses (5). Therefore, the development of soluble glass fibers requires the simultaneous development of a database of relevant physiological responses which, in turn, requires the development of efficacious *in vitro* and *in vivo* systems for: 1. The study of interfacial phenomena involving the fibers, the ECM and alveolar cells. 2. The quantitation of dissolution rates and secondary chemical/biochemical reactions associated with dissolution in the ECM. 3. Characterization of cellular responses to dissolution phenomena *in vivo*.

MATERIALS AND METHODS

Fiberglass samples were provided to Alfred University through industry members of the NSF Industry-University Center for Biosurfaces (NSF IUCB). The relevant properties of fibers used in this study are shown in Table I.

Table I. Relevant properties of fibers used in this study.

Fiber	SDS-PAGE Label	Pathology*	Dissolution in vivo*	Dissolution in vitro*	Al_2O_3 wt %#	CaO+ MgO wt %#
MMVF 34 High alumina HT	HT	No	Fast	Medium	23%	25%
MMVF 10 Borosilicate	10	No	Fast	Fast	5%	12%
RCF 1 ~50:50 Al:Si	1A	Yes	Slow	Slow	48%	1%

*Personal communication, Walter Eastes, Owens-Corning, Granville, OH
#MMVF10 and RCF1 as described in ref. 2, MMVF34 as described in ref. 3

In vitro dissolution means dissolution in *in vitro* systems such as those described in 4 and references cited therein. *In vivo* dissolution means dissolution in whole animal studies (cf. 2 and 3).

Matrigel® (Beton Dickinson, ref. 6) is a solubilized basement membrane preparation extracted from the Englebreth-Holm-Swarm (EHS) mouse sarcoma, a tumor rich in extracellular matrix proteins. Its major component is laminin, followed by collagen IV, heparan sulfate, proteoglycans, entactin and nidogen. It also contains TGF-beta, fibroblast growth factor, tissue plasminogen activator and other naturally occurring growth factors. Matrigel® supports attachment, differentiation and growth of a wide range of mammalian cells types including neurons, oligodendrocytes, and hepatocytes. It is the basis for several well established assay systems involving angiogenesis and tumor cell invasion. ECM proteins adsorbed to fibers were identified using the following protocol.

1. Thaw Matrigel®* from –20C to 'sol' phase on ice ($\rho \sim 0.86$). Pipette 100 µL of Matrigel® onto precooled Perti plate on ice. Add 5 µL of a 1 mg/mL fiber slurry. The aqueous slurry also contained 20% w:v dipalmitoyl phosphatidylcholine (DPPC), the major alveolar surfactant. Mix thoroughly and overlay onto 200 µL of 20% sucrose solution ($\rho \sim 1.16$) in a thin-walled polymerase chain reaction (PCR) tube. Contact with room T sucrose solution causes Matrigel® to undergo sol → gel transition.
2. Incubate @ 37C for 1 hr then place on ice for 15 min in order for Matrigel® to undergo gel→ sol transition.
3. Microfuge for 5 minutes at 13,000 x g to pellet glass fibers through 20% sucrose pad.
4. Cut the tip of the PCR tube containing the fiber pellet into a 1.7 mL microfuge tube and add 50 µL of reducing Laemmli (6) sample buffer (standard Laemmli sample buffer plus 5% β-mercaptoethanol as a reducing agent). Vortex for 30 sec. and boil for 3 min.
5. Total sample volume was measured and gel load was normalized to represent equal masses of glass fibers. Given that the fibers contained approximately the same distribution of sizes, this would normalize the sample for fiber surface area (see Table I for fiber types). This step is designed to take into account the variability in volume that occurs when the tip of the PCR tube is cut and placed into the microfuge tube.
6. Load sample volumes representing equivalent masses of fibers onto SDS-PAGE (various gel configurations; 4-12%, 10-20%, or 12%). Protein electrophoresis carried out as per Laemmli (7).
7. Visualize proteins via silver stain (Gel Code®, Pierce Chemical Co., ref. 8) followed by quantitative and qualitative analysis using BioRad's Multianalyst video imaging system and Q1 software.

RESULTS AND DISCUSSION

We have successfully developed an *in vitro* system for characterization of interactions between glass fibers and protein components of the extracellular matrix. This microscale assay has provided important information about both specificity and quantity of protein binding by different glass fiber samples. Figure 1. shows a characteristic result of our assay system.

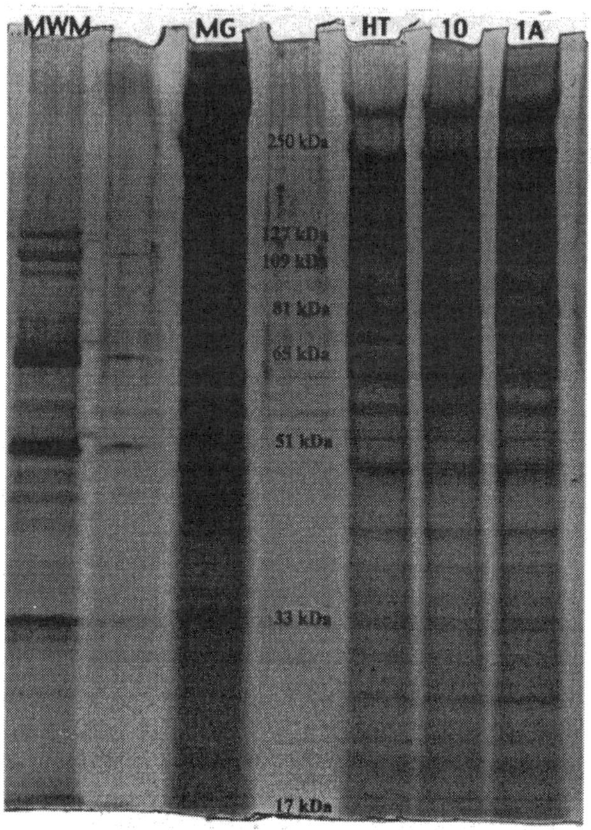

Figure 1. 12% SDS-PAGE of proteins adsorbed to glass fibers. Lanes designated as follows; MWM, molecular weight markers (note that Gel Code® stains proteins in colors from black to orange so that all bands are not equally visible here); MG, Matrigel® at 1.3 μg; HT, 10 and 1A as per Table I. It is probable that the densely staining bands above and below the 250 kDa position are laminin A chain (400,000 Da) and unresolved laminin B1 and B2 chains (215,000 Da and 205,000 Da respectively). While this gel ends at ~17 kDa, the fibers show a similar affinity for the entire ECM protein profile down to ~5 kDa (Fig. 2). Fibers alone showed no stained bands (not shown)

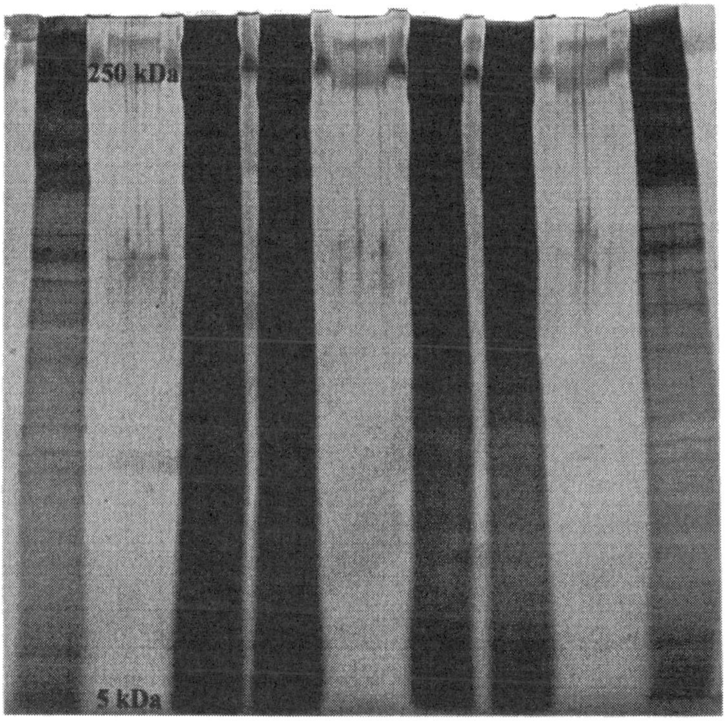

Figure 2. Various dilutions of Matrigel® to show the numbers and distribution of ECM proteins. Note the large numbers of low molecular mass proteins.

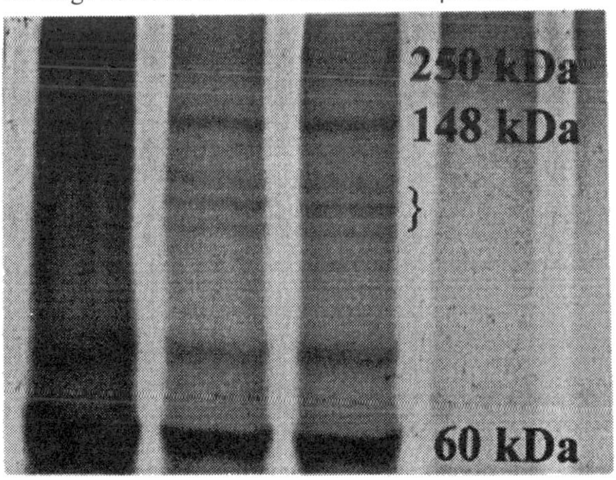

Figure 3. Using nonreducing conditions with no boiling, relatively high molecular mass protein complexes (bracketed) appear to adsorb to fibers 10 and 1A but are absent in HT (gel lanes are HT, 10, and 1A left to right as in Fig. 1)

Data generated to date may be summarized as follows. The chemical interactions between glass fiber surfaces and specific protein components of the ECM are of sufficient strength such that these ECM proteins separate with the fibers when the fibers are driven through the sucrose density pad at 13,000 x g. The general protein profile is similar for the three types of fibers tested, however there are both qualitative and quantitative differences in these profiles. Some major proteins in the adsorption profile are not major proteins in the total profile of ECM proteins (Figure 1) so that some surface interactions between the fibers and the ECM are specific and not always simply proportional to the amount of a given protein in the ECM. Coating the fibers with the major alveolar surfactant, dipalmitoyl phosphatidylcholine (DPPC), has no visible effect on the protein binding profiles of the various fibers (not shown). In addition, no visible differences were noted if the fibers were 'spun out' of the ECM directly from the gel phase, i.e. if one eliminated the 15 min @ 4C component of Step 2 of the experimental protocol above (not shown). This indicates that the forces binding the proteins to the fiber surface were stronger than the forces maintaining these components in their normal supramolecular gel state within the ECM.

Elimination of reducing agent and 3 min boiling from Step 3 of the experimental protocol (simple SDS denaturation and solubilization of adherent proteins) resulted in the appearance of novel high molecular mass components visualized after SDS-PAGE (Figure 3). These data indicate that, at least in some cases, fibers in the ECM are interacting with supramolercular complexes rather than individual proteins. A large number of low molecular weight proteins including polypeptide-sized molecules were found in all fiber profiles (Fig. 1). As discussed in the legend for Fig. 1, the protein profile below 17 kDa was similar to the Matrigel® itself, indicating either a very high affinity of the fibers for low molecular mass proteins and/or that the enhanced mobility of these low molecular weight proteins and polypeptides may allow them to compete more effectively for binding sites on the fiber surface. Low molecular weight proteins and peptides include a number of important signal molecules that mediate cellular responses (5).

We are currently in the process of extracting quantitative data from our gels. However, qualitative analysis allows us to draw important conclusions that suggest immediate future experiments. The sensitivity of the Gel Code® silver stain system is well characterized (8). Known protein standards (Lane MWM in Fig. 1) may be visualized at an average minimum concentration of 0.1 to 0.01 μg (recall that, due to color staining, apparent optical density will vary dramatically in these images). For reference, the Matrigel® standard in Fig. 1 contains 1.3 μg of Matrigel®. Representative gels have been analyzed via video imaging with the BioRad Fluor-S MultiImager®. We have obtained values in the range of 0.9 to 1.4 μg adsorbed protein/ cm^2 fiber surface, or 0.76 to 1.14 μg adsorbed protein/μg fiber. Detailed quantitative analyses will be reported in a future publication, however these numbers are in close agreement with the average value of 1μg/cm^2 reported by Horbett for adsorption behavior of proteins at solid-liquid interfaces (9). Therefore, these data appear to support our original hypothesis that the fiber surface actively adsorbs ECM proteins. In addition the ECM contains large amounts of complex carbohydrates including hyaluronan and glycosaminoglycans so that additional macromolecules may be adsorbed to the fiber surface that are not visualized in our experimental system. Given this biophysical picture, it is reasonable to assume that at least some fiber dissolution processes will occur across an interphase region influenced by the presence of biocolloids.

CONCLUSION

Future directions will include use of this system to do functional proteomics. Replication and quantitation of appropriate molecular mass standards will be required to develop a quantitative picture of the protein profile for each fiber. We will use the BioRad MultImager/Q-1 software® video gel analysis system to quantitate both the numbers and molecular masses of adherent proteins. In addition, we intend to conduct direct ICP-AA analyses of representative protein profiles for evidence of protein binding of mineral components of the glass fibers. We have considered that major differences between *in vitro* and *in vivo* dissolution rates may result from biocolloid-mediated phenomena occurring at the fiber-ECM interphase (adsorption, chelation, etc.). Such phenomena could dramatically influence dissolution rates, local concentrations of ions, and other variables of state with respect to the response of cells in the region of the fiber. Finally, we intend to develop a cell culture system to study the effects (if any) of matrix-embedded fibers on cell growth and development. Tumor invasion and angiogenesis assays are conducted by coating microwell tissue culture plates with a thin (0.5 mm) layer of Matrigel® with the cells plated on top. Likewise, we should be able to coat these wells with Matrigel® + or - fibers and plate fibroblast or alveolar cells on top. Cell growth may then be observed. Fiber densities would be equivalent to ~1 fiber per alveolar sac as calculated based on some mean number of alveolar epithelia per sac.

ACKNOWLEDGMENTS

This work was supported by the NSF Industry-University Center for Biosurfaces. We thank Walter Eastes and Russ Potter (Owens-Corning Science & Technology Center, Granville, OH), and Alexis Clare and Bill LaCourse (S.U.N.Y. College of Ceramic Engineering and Materials Science at Alfred University) for useful discussions and essential input.

REFERENCES

[1]W. Eastes and J.G. Hadley, "A Mathematical Model of Fiber Carcinogenicity and Fibrosis in Inhalation and Intraperitoneal Experiments in Rats", *Inhalation Toxicology*, **8** 323-343 (1996)

[2]T. W. Hesterberg, W. C. Miiller, E. E. McConnell, J. Chevalier, J. G. Hadley, D. M.. Bernstein, P. Thevenaz, and R. Anderson, "Chronic Inhalation toxicity of Size-Separated Glass Fibers in Fischer 344 Rats", *Fundamental and Applied Toxicology*, **20** 464-476 (1993).

[3]T. W. Hesterberg, G. Chase, C. Axten, W.C. Miiller, R.P. Musselman, O. Kamstrup, J. Hadley, C. Morscheidt, D.M. Bernstein, and P. Thevenaz, "Biopersistence of Synthetic Vitreous Fibers and Amosite Asbestos in the Rat Lung Following Inhalation", *Toxicology and Applied Pharmacology* , **151** 262-275 (1998).

[4]R.M. Potter and S.M. Mattson, "Glass Fibers in Simulated Lung Fluid: Dissolution Behaviour and Analytical Requirements", *Ann. Occup. Hyg.*, **38**[6] 857-877 (1994).

[5]H. Lodish, D. Baltimore, A. Berk, S.L. Zipursky, P. Matsudaira, and J. Darnell, "The Extracellular Matrix: Primary Components and Functions"; pp. 1123-1149 in *Molecular Cell Biology*, 3rd ed., H. Lodish, et. al. W.H. Freeman and Company, New York, 1995.

[6]Beton Dickinson Product Specification Sheet 40234-19, 5/97.

[7]U.K. Laemmli "Cleavage of Structural Proteins during the Assembly of the Head of Bacteriophage T4", *Nature*, **227** 680-685 (1970)

[8]D.W. Sammons, L.D. Adams, T.J. Vidmar, C.A. Hatfield, D.H. Jones, P.J. Chuba, and S.W. Crooks "Applicability of Color Silver Stain GELCODE® System to Protein Mapping with Two-Dimensional Electrophoresis"; pp.112-127 in *Two-Dimensional Gel Electrophoresis of Proteins*. Edited by J.E. Celos and R. Bravo. Academic Press, New York, 1984.

[9]T.A. Horbett "Chapter 3, Some Background Concepts"; pp.136 in *Biomaterials Science: An IntroductionTo Materials In Medicine*. Edited by B.D. Ratner, A.S. Hoffman, F.J. Schoen, and J.E. Lemons. Academic Press, New York, 1996.

Porous Materials

SOL-GEL DERIVED BIOGLASS®: A NEW RESORBABLE BIOMATERIAL

D. C. Greenspan, USBiomaterials Corp., Alachua, Florida 32615,
J. P. Zhong, USBiomaterials Corp., Alachua, Florida 32615,
D. L. Wheeler, University of Florida, Department of Orthopedics, Gainesville, Florida 32611

ABSTRACT

Melt derived bioactive glass (Bioglass®) has been used clinically since 1985. The ability of this material to form a bond with bone and soft tissue has been related to a series of surface reactions that occur when the material is subjected to an aqueous environment. The formation of a high-specific surface area silica layer and subsequent precipitation of hydroxy-carbonate apatite are the critical steps that lead to this bonding layer. Recently, the sol-gel process has been used to create a series of bioactive glasses with an initial surface area greater than 125 m^2/g. These materials have demonstrated the ability to enhance bone regeneration and are absorbed in the body. Long term tests are underway to validate the utility of these glasses as bone graft substitutes.

INTRODUCTION

Professor Larry Hench first published his results on a new bioactive glass material (now known as Bioglass®) that could form a bond with living tissue in 1972(1). This first paper demonstrated, in a rat long bone model, that the bioactive glass reacted after implantation, and that this reaction caused a mechanically strong bond between the glass and the host bone. Following this discovery, Professor Hench, and others, published many papers which demonstrated that a series of surface reactions occur on the surface of these bioactive glasses, and that these reactions are responsible for the unique ability of these materials to bond with living tissue(2-8). As has been pointed out by Hench(2,9-10), the initial sodium-hydrogen ion exchange at the glass surface, followed by dissolution and repolymerization of the silica at the glass surface are the critical steps in determining the ability of these bioactive glasses to bond with bone. It is this early reaction which allows the formation of the biologically active hydroxy-carbonate apatite (HCA) layer on the surface of the glass that is responsible for the bone bonding ability of Bioglass®.

Recognizing the critical nature of the high surface area silica rich surface layer on the Bioglass®, Li and Hench(11) used a sol-gel process to create a porous matrix containing silica, calcium and phosphorous. These materials were partially stabilized, had a high specific surface area, and at least 40% volume porosity. When reacted in buffer solutions, this material showed the formation of a HCA layer very similar in nature to that formed on the melt derived Bioglass®. Numerous studies have demonstrated the bioactive nature of these sol-gel derived glasses(10-14), and their use is now being investigated as a bone graft material for dental and orthopedic use(15).

This paper reports on the in-vitro and in-vivo response of the sol-gel bioactive glasses, and compares their behavior to the original bioactive glass developed by Hench (known as 45S5 Bioglass®).

MATERIALS AND METHODS

Glass Preparation

The preparation of melt derived Bioglass® has been described in detail previously(1,7). Briefly, reagent grade SiO_2, $CaCO_3$, Na_2CO_3, and P_2O_5, were mixed thoroughly, melted in a Pt crucible at 1350°C for 8 hours, quenched in distilled water, dried with acetone, and milled in an alumina ball mill. The particle size used in the in-vitro studies was 710-300μm, and in the in-vivo study was 710-90μm.

For the sol-gel materials, tetraethylorthosilicate(TEOS), triethylene phosphate(TEP) and calcium nitrate were used. The TEOS was hydrolyzed for 30 minutes, using 8 moles of H_2O/mol of TEOS. HCl was used as the catalyst. Following hydrolysis, triethylene phosphate (TEP) was added, and finally calcium nitrate was added to the solution. The mix was stirred for 1 hour, then transferred to a polypropylene mold for aging. The aging was carried out at 60°C for 55 hours. After aging, the pore liquor was removed and the gels transferred to a drying vessel in a 95% relative humidity environment. The vessel was loosely covered and the gels dried at 180°C for 3 days. Following drying, the gels were partially stabilized at 700°C for an additional 3 days. The stabilized gels were then ground in an alumina ball mill to a final particle size of 710-300μm for both in-vitro and in-vivo testing.

In-Vitro Reactivity

To test the in-vitro HCA layer formation and resorbability of the glasses, samples were reacted in an ionic solution, developed by Kokubo(16). This solution approximates the ionic concentration of blood plasma, and is useful as a screening test to determine the reactivity of a biomaterials surface. The 0.2g of materials were placed in 200 ml test solution for periods of time up to 2 weeks in a dynamic shaker bath at 37°C and 175rpm. Due to the high level of reactivity of the sol-gel materials which had been previously reported(17,18), the solutions were replaced at 6 hours, 24 hours, then every 2 days until completion of the tests to avoid ionic saturation, and more closely approximate the living environment. After reaction at various times, the pH of the solution was measured and the ionic concentrations were determined by ICP.

The formation of the reactive surface layers on the materials was determined by FTIR, as has been described previously(11). Basically, after reaction in the SBF for the various time points, the samples were removed from the solution, filtered onto 8.0μm filter paper, and rinsed three times with acetone to stop any reactions. The samples were then placed in a diffuse reflection stage in a Maga-550 FTIR, and a reflection spectrum was obtained.

In-Vivo Testing

The materials were implanted in the femoral condyles of New Zealand white rabbits weighing between 2-3kg. Following anesthesia, 6mm holes were drilled bi-laterally from the lateral cortical surface through the medial cortex, using a low-speed saline-cooled drill. The bone marrow chamber was evacuated by repeated saline washes. Following ablation, the test materials were introduced into the right condylar defect and the left limb served as the unfilled control. Twelve additional animals provided data for normal bone. Animals were euthanized with a sodium pentaphenobarbitol solution (60mg/kg), either 4, 8, or 12 weeks post-operatively. The distal femurs were harvested, sectioned and prepared for histologic analysis.

Specimens were processed for undecalcified histology using standard dehydration, infiltration, and embedding techniques. Thin sections (10-20μm) were cut using an Exakt bone saw and microgrinder system. The specimens were stained with a modified VanGeison stain. Sections were analyzed using a Leica DMRB microscope and an Optimus image analysis system.

RESULTS

In-Vitro

Fig. 1 shows the change in pH as a function of time for the three materials tested. The pH increase in the 45S5 melt derived Bioglass® was greater than that of either of the sol-gel materials. The change in pH was more rapid for the 45S5 composition, and continued for a longer period of time than for the sol-gel derived material.

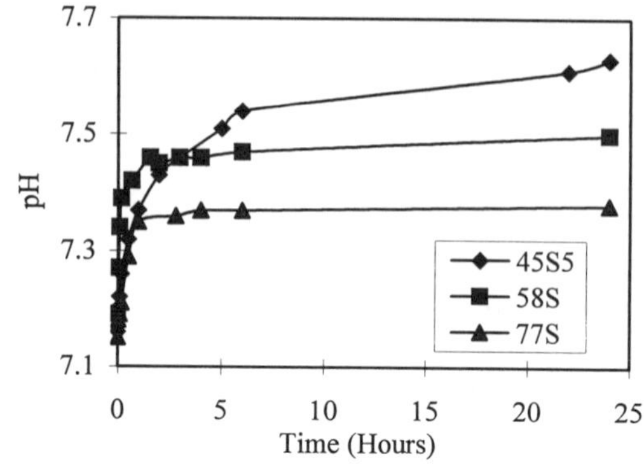

Fig. 1 pH Change as a Function of Reaction Time

The release of silica, as determined by ICP is shown in Fig.2. This figure represents the cumulative release of Si, made by adding the individual measured time points for the silica released by the samples, remembering that the solutions were changes during the experiments. The total amount of silica release was greater for both of the sol-gel compositions than for the 45S5, and was greater for the 77S sol-gel composition than the 58S glass.

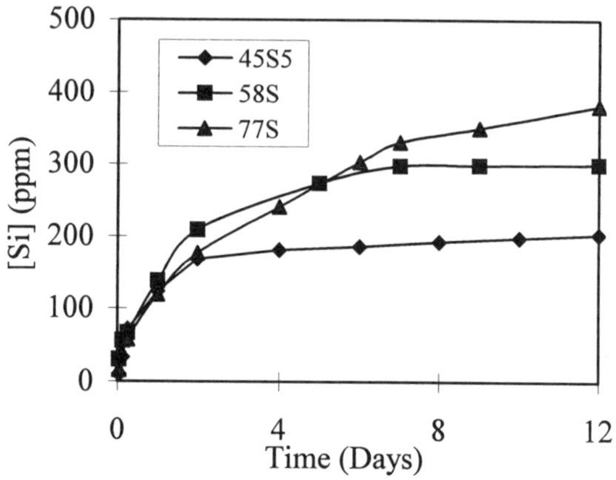

Fig. 2 Cumulative Si Release From 45S5, 58S and 77S in SBF

Fig.3 shows the weight loss as a function of time for the three compositions. The weight loss for each element was calculated from the ionic concentrations in solution at each time point. The sol-gel compositions showed much greater loss of silica than did the 45S5 composition. Note that there was actually a decrease in P concentration in solution as measured by the ICP, which translates into a weight gain for each of the samples. In addition, the concentration of Ca in solution decreased for the 77S sol-gel composition as a function of time, showing as a weight gain.

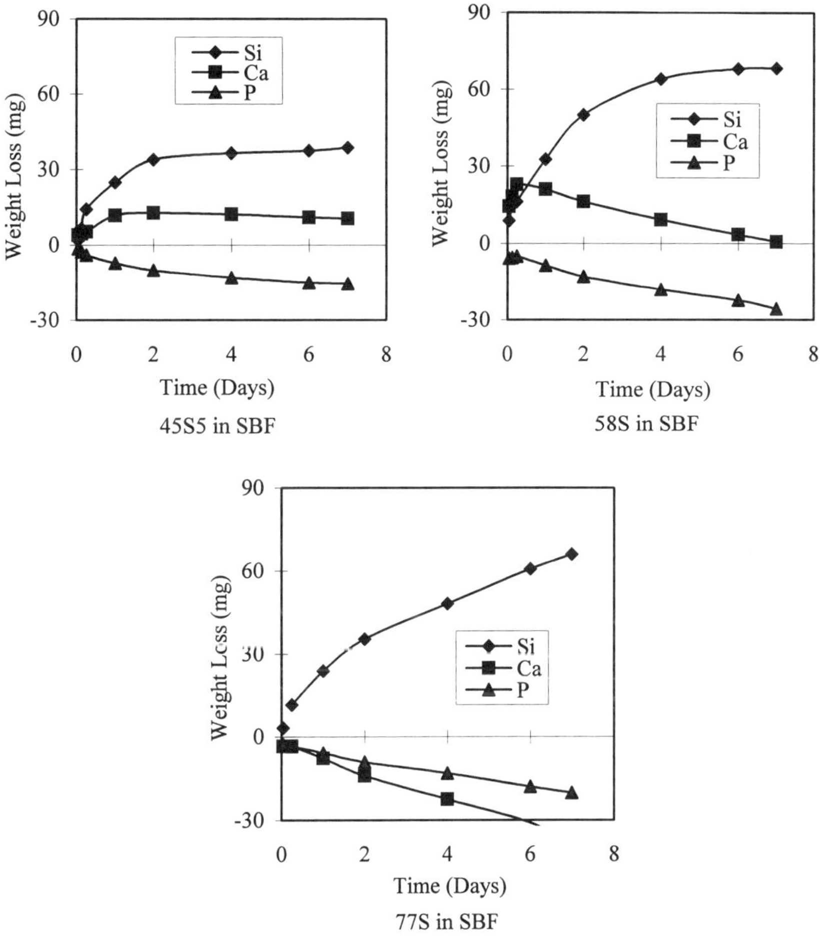

Fig. 3 Weight loss of Si, Ca and P of 45S5, 58S and 77S in SBF

The FTIR data are shown in Fig.4, as a function of time in solution. Of particular interest are the peaks at 550 and 605cm^{-1} and the P-O stretching vibration at 1080cm^{-1}, both of which are indicative of the formation of the HCA layer. By 24 hours of reaction time, all three compositions have formed the bioactive HCA layer. The 58S composition shows the characteristic peaks at 550 and 605cm^{-1} by one hour, which is much earlier than for the other compositions. The Si-OH rocking vibration at about 450cm^{-1}, and the Si-O-Si stretch at 1105cm^{-1}, which are characteristic of the unreacted glass, or the silica rich layer are clearly present at 6 hours of reaction time, an indication that the HCA layer forming on the glass surfaces is still relatively thin.

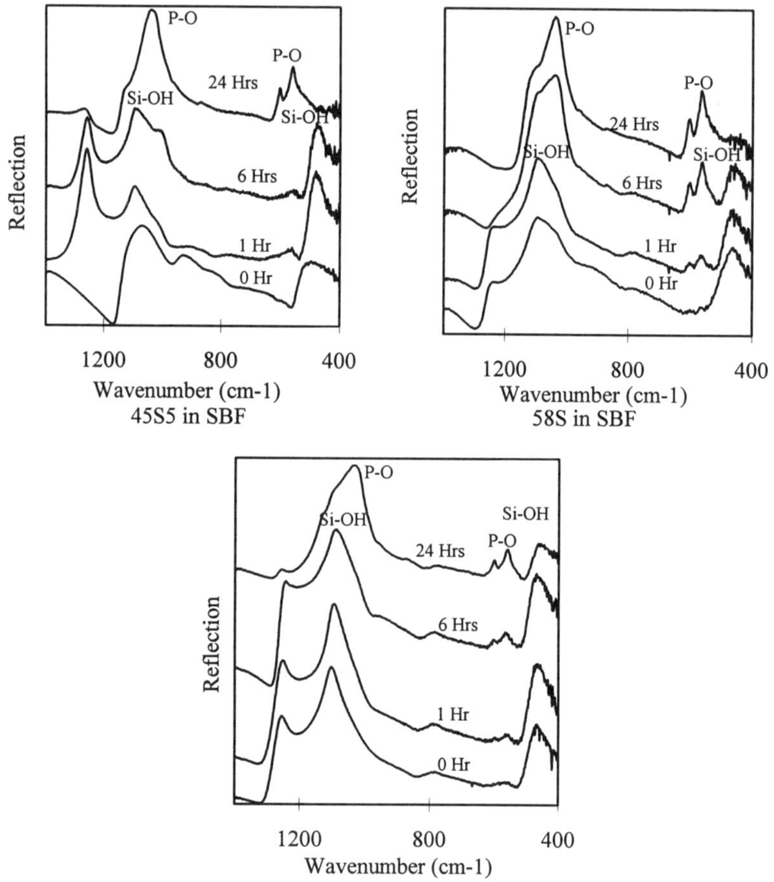

Fig. 4 FTIR Spectra of 45S5, 58S and 77S Reacted in SBF

The histologic results are shown in Fig.5. Fig.5a is a low power section of the 58S taken 4 weeks post-operatively. The darker stained areas are newly formed bone in the defect. Fig.5b shows the 58S glass at 12 weeks. Note that by comparison with Fig.5a, there are many fewer particles. The remaining trabecular bone appears normal. Fig.5c is a higher power micrograph of the 12 week 58S sample. The arrow points to a particle of material which is resorbing. In Fig.5d, the arrow is pointing to the characteristic HCA layer on a particle of 45S5 glass after 12 weeks post-operative.

DISCUSSION

The bone bonding nature of bioactive glasses has been ascribed to the formation of the HCA layer, which in turn occurs, by a series of surface reactions as mentioned earlier in this paper. Recently, the possibility that bioactive glasses can be resorbed has been proposed for the 45S5 composition, when fabricated in a certain particle size range(19-21). Hench clearly demonstrated the well known sequence of surface reactions some 25 years ago, and research has also shown that the surface reactions on the 45S5 glass are limited to about 150μm. Thus if the glass is formed as a particulate, and the size of the particles are sufficiently small, the entire particle can react, and what is left is the HCA layer formed by those surface reactions.

The cumulative loss of Si as a function of time shown in Fig.2 is greater for the sol-gel compositions, which suggests that these glasses are more reactive than the 45S5 composition. In fact, the continued loss of Si for the 58S glass indicates that this composition is truly resorbing, or dissolving. Previous in-vitro studies conducted on the glasses showed that after 2 weeks in solution the remaining material was composed only of calcium phosphate(11). There was no Si present in the recovered particles.

The weight loss data presented in Fig.3 clearly shows that all of the glass compositions are losing Si ions to the solution. However, the weight gain in Ca and P, demonstrates the formation of the calcium phosphate layer on the surface of the glasses. The more dramatic uptake of Ca and P for the sol-gel glasses as shown in Fig. 3 confirms the greater surface reactivity of these materials, compared to the 45S5 glass. The pH increase, shown in Fig. 2, is a result of the sodium-hydrogen ion exchange in the 45S5 glass, and the calcium-hydrogen ion exchange for the sol-gel derived glasses. Thus, it is not surprising that the pH increase is greater for the 45S5 composition than for either sol-gel composition. This is due to the fact that the melt glass contains Na^+ that rapidly diffuse into solution and cause a rapid change of pH. The FTIR data of Fig.4 clearly shows the HCA layer formation for all of the compositions by about 6 hours, and much

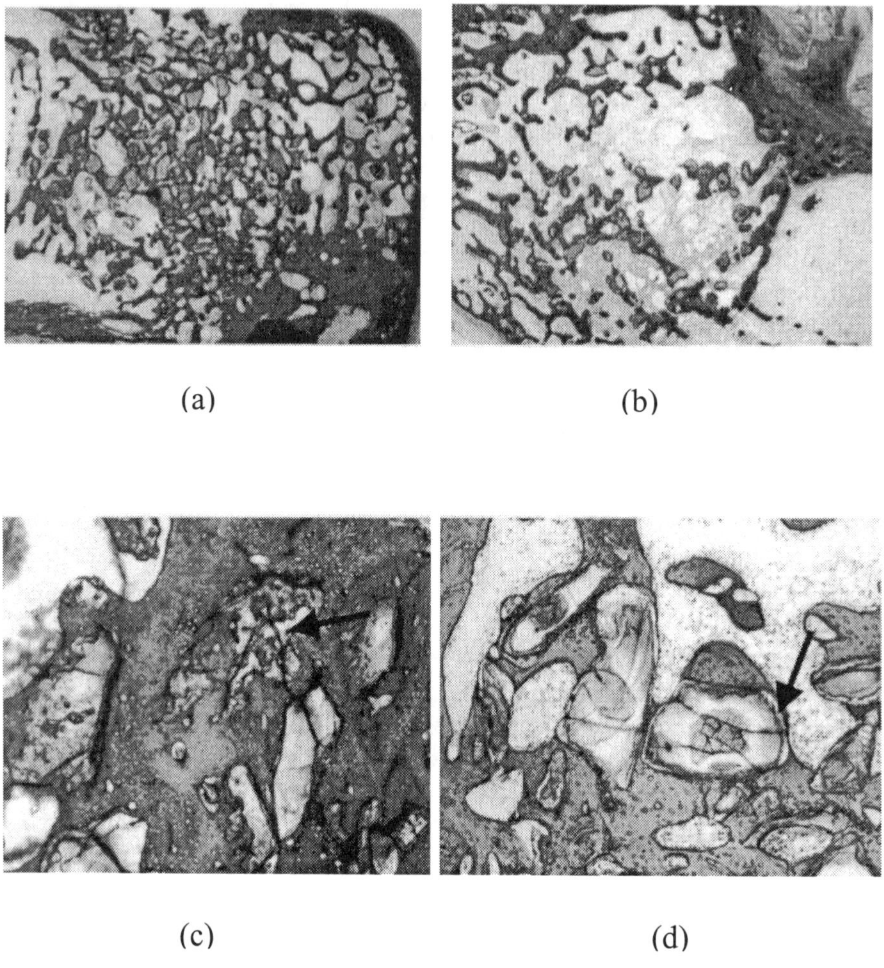

Figure 5. Composite micrographs showing particle resorbability. (a) 58S at 4 weeks, 12.5x. (b) 58S at 12 weeks, 12.5x. (c) 58S at 12 weeks, showing resorption of particles (arrow) and infiltration of new bone, 100x. (d) 45S5 at 12 weeks, showing typical gel layer formation and bone bonding (arrow), 100x.

earlier for the 58S sol-gel glass. This early HCA formation on the sol-gel derived glasses correlates with the weight loss data for these materials, and is in agreement with previous studies of these bioactive glasses(15,17,18).

The histology shown in Fig. 5a and b clearly shows the resorption of the 58S sol-gel glass as a function of time. By 12 weeks, more than 50% of the particles have been totally resorbed. The new trabecular bone that has formed appears normal for this site. In Fig.5c, the arrow points to a particle of the 58S sol-gel material which is being resorbed and filled in by new bone. The arrow in Fig. 5d shows the characteristic HCA layer on the 45S5 Bioglass® that is in contact with the newly formed bone. However, these layers are absent from the 58S glass seen in Fig. 5c. This behavior is different from that previously seen with any other bioactive glass. Although there is evidence of the HCA layer formed in-vitro, it cannot be found in looking at the histology, either at 4, 8, or 12 weeks.

Previous studies on the 58S sol-gel glass in-vivo have demonstrated that by 12 weeks in a rabbit model, the remaining particles in the bone matrix contained Ca and P, but no appreciable Si, as determined by EDXA(15). The data presented here suggests that the high rate of release of Si allows a greater solubility of the entire particle. Rather than forming a surface HCA layer, the ions released from the glass could be taken up by the surrounding tissue, thereby stimulating the formation of new bone directly. It is also likely that osteoclast activity is occurring on the glass surface, increasing the resorption of the material.

It has been stated that the ideal synthetic graft material will be biocompatible, be able to restore form and function, and be completely removed from the body in a reasonably short period of time with no adverse reactions. To date, bioactive ceramics and glasses have demonstrated good compatibility, an ability to allow bone to heal and restore function, and show little or no adverse reactions. However, the issue of resorbability still remains. Melt derived bioactive glasses, even those of relatively small particle size ranges, leave behind the characteristic HCA layer after the surface reactions have been completed. This is clearly visible in previously published papers(1,6,15). It has been surmised that over time all of the glass will be removed from the body by the natural remodeling process(21), but this has still not been proven.

This study has demonstrated that sol-gel derived bioactive glass compositions can be resorbed in-vivo and replaced by bone in a manner which is different from that shown by melt derived glasses. This resorption process does not limit the particle size used to repair bony defects, which is a major advantage, since many defects can be quite large in size. Longer term in vivo studies are necessary in various bony defects to validate the utility of these new bioactive ceramic materials.

REFERENCES

1. L.L. Hench, R.J. Splinter, T.K. Greenlee, and W.C. Allen, "Bonding Mechanisms at the Interface of Ceramic Prosthetic Materials", J. Biomed. Mater. Res., No. 2, pp. 117-141 (1971)

2. L.L. Hench, "Biomedical Applications and Glass Corrosion", Proceedings of the Xth International Congress on Glass, Ceramic Society of Japan, Kyoto, Japan, pp. 30-41 (1974)

3. C.G. Pantano, Jr., A.E. Clark, Jr. and L.L. Hench, "Multi-Layer Corrosion Films on Bioglass®", J. American Ceramic Society 57 (9) pp. 412-413 (1974)

4. A.E. Clark, Jr., C.G. Pantano, Jr. and L.L. Hench, "Compositional Analysis of the Formation of the Bone-Implant Bond", U.S. Army Medical Research and Development Command, Report #6, pp. 39-47 (1975)

5. L.L. Hench, "Bioceramics" Science of Ceramics, Vol. 9, pp. 193-211 (1977)

6. V. Gross, R. Kinne, H. J. Schmitz and V. Strunz, "The Response of Bone to Surface Active Glass/Glass Ceramics", in CRC Crit. Rev. Biocompat. 4, pp. 2-15 (1988)

7. M. Ogino, F. Ohuchi, and L.L. Hench, "Compositional Dependence of the Formation of Calcium Phosphate Film on Bioglass®". J. Biomed. Mater. Res. 14, pp. 55-64 (1980)

8. T. Kokubo, "A/W Glass Ceramic: Processing and Properties" in an Introduction to Bioceramics, pp. 75-88, Editors: L.L. Hench and J. Wilson, World Scientific Pub. Co. Pte. Ltd., Singapore (1993)

9. L.L. Hench. "Bioactive Ceramics for Skeletal Repair", Ceramic Technology International, pp. 29-32 (1997)

10. L.L. Hench and J.K. West, "Biological Application of Bioactive Glasses", Life Chemistry Reports 13, pp. 187-241 (1996)

11. R. Li, A.E. Clark, and L.L. Hench, "An Investigation of Bioactive Glass Powders by Sol-Gel Processing", J. Appl. Biomaterials 2, pp. 231-239 (1991)

12. R. Li, A.E. Clark, and L.L. Hench, "Effect of Structure and Surface Area on Bioactive Powders Made by Sol-Gel Process", pp. 627-633 in Chemical Processing of Advanced Materials, Editors: L.L. Hench and J.K. West, Wiley and Sons, New York (1992)

13. M.M. Pereira, A.E. Clark and L.L. Hench, "Homogeniety of Bioactive Sol-Gel Derived Glasses in the System CaO-P_2O_5-SiO_2", J. of Materials Synthesis and Processing 2 (3), pp. 189-196 (1994)

14. K.D. Lobel and L.L. Hench, "In-Vitro Protein Interactions with a Bioactive Gel-Glass", J. of Sol-Gel Science and Tech. 7, pp. 69-76 (1996)

15. D.C. Greenspan, J.P. Zhong, and D.L. Wheeler, "Bioactivity and Biodegradability: Melt vs. Sol-Gel Derived Bioglass® In-Vitro and in-Vivo", in Bioceramics 11, Editors: R. LeGeros and J. LeGeros, World Scientific Publishing Co., Pte., Ltd. New York, pp. 345-348 (1998)

16. T. Kokubo, H. Kushitani, C. Ohtsuki, S. Saklea, and T. Yamamuro, "Chemical Reaction of Bioactive Glass and Glass-Ceramics with a Simulated Body Fluid", Materials in Medicine 3, pp. 79-83 (1992)

17. J. Zhong and D.C. Greenspan, " Porous Sol-Gel Bioglass® from New Equilibrium Drying", in Bioceramics 2, Vol. 10, Editors: L. Sedel and C. Rey, Elsevier Science, Ltd., pp. 265-268 (1997)

18. D. C. Greenspan, J.P. Zhong, X.F. Chen, and G.P. LaTorre, " The Evaluation of Degradability of Melt and Sol-gel Derived Bioglass® In-Vitro", in Bioceramics, Volume 10, Editors: L. Sedel and C. Rey, Elsevier Science, Ltd., pp. 391-394 (1997)

19. E.J.G. Schepers, P. Ducheyne, P. Barbier, L. Scheoess, "Bioactive Glass Particles of a Narrow Size Range: A New Material for the Repair of Bone Defect", Implant Dentistry 2, pp. 151-156 (1993)

20. A.M. Gatti, P. Ducheyne, P. Piattelli, E.J.G. Schepers, D. Trisi, L. Chianni, "Glass Corrosion Layers on Bioactive Glass Granules of Uniform Size Affect Cellular Function", in bioceramics Vol. 6, Editors: P. Ducheyne and D. Christiansen, Butterworth-Heinemann, Ltd., pp. 295-400 (1993)

21. E. Schepers, P. Ducheyne, "Bioactive Glass Particles of Narrow Size Range for the Treatment of Oral Bone Defects: A 1 to 24 Month Experiment with Several Materials and Particle Sizes and Size Ranges", J. Oral Rehabil. 24, pp. 171-181 (1997)

THE EFFECT OF PORE SIZE ON METAL CLUSTER FORMATION IN SILICA SOL GEL MONOLITHS

Kevin W. Powers
Department of Materials Science and Engineering
University of Florida, Gainesville, FL 32611-6400

Larry L. Hench
Department of Materials
Imperial College of Science, Technology and Medicine,
Prince Consort Road, London SW7 2BP

ABSTRACT

The discovery of nonlinear optical properties in colloidal metal glasses has led to an increased interest in controlling the size, shape, number and environment of metal colloids in optical substrates. The traditional methods of precipitating colloidal metals in glasses have been augmented by ion implantation and the sol gel route to produce such materials. In this study the ability to produce silica monoliths with pore diameters from 2.0 to 40 nanometers is exploited to assess the deposition of gold nanoclusters in the porous glass. The gold doped silica gels are heat treated at various temperatures up to full density to determine the effect of pore size and heat treatment on the optical absorption and scattering of the composites.

INTRODUCTION

Scientists have been interested in the phenomena of colloidal color since the mid 19th century when Faraday investigated the stability of gold sols.[1] Early in this century, Mie used colloidal metal suspensions in the development of his theory of light scattering.[2] Throughout this century colloidal sols and specifically colloidal metals in glasses have been studied primarily to elucidate their optical properties and to further develop Mie's theory of light scattering. Doremus studied the optical characteristics of gold and silver colloidal glasses to determine the effects of particle size and method of preparation.[3,4] His calculations used more reliable optical constants than were available to Mie. He reported that below a particle diameter of 85Å the optical adsorption peak near 530nm becomes proportional to the particle size indicating a deviation from the optical properties of the bulk metal (primarily ε_2). He attributes this to the size of the particles becoming much smaller than the mean free path of the electrons.

It has been found that such glasses exhibit high optical non-linearity and thus are attractive candidates for use in optical devices.[5,6,7] Gold colloid particles, for example, can have a large third order nonlinear optical susceptibility ($\chi^{(3)} = 1.2\times10^{-7}$esu) and a near resonance nonlinear response that is on the order of 50 picoseconds.

THEORY

The colors produced in ruby glasses are attributed to the scattering of light by the small metal particles as described by Mie. However, the phenomenon encompasses more than light scattering. Metal colloids display a complex refractive index $n^{**} = n - jk$ where n and k are the real and imaginary (refractive and absorption) components respectively. The relatively high value of k for metals gives high optical absorption in the visible and hence their high reflectivity. At longer wavelengths both k and n tend to increase linearly with wavelength according to the free electron model of Drude.[8] Since the electrons in the metal are assumed to react together in an electromagnetic field, this resonance is called a plasma (or plasmon) resonance. For most metals there is little spectral variation (dispersion) in the optical constants throughout the visible region and colloidal coloration tends toward neutral shades of brown or gray. Gold, silver and copper, however depart from the norm and display plasmon resonances in the visible region. There is a dramatic shift in the optical constants for the metals that affects both the adsorption and scattering of incident light. This is the source of the red colloidal coloration of gold and copper, and the yellow coloration of silver. It is this same phenomenon that is responsible for the gold and copper coloration of the bulk metals. Figure 1 shows the optical constants for gold as a function of wavelength. In the vicinity of 530nm there is a rapid decrease in n and a concurrent increase in k. This is the gold plasmon band.

Although Mie's classic paper regarding colloidal gold particles dealt primarily with scattering phenomena, it is a combination of scattering and absorption of the colloidal particles that is responsible for the color of these glasses.

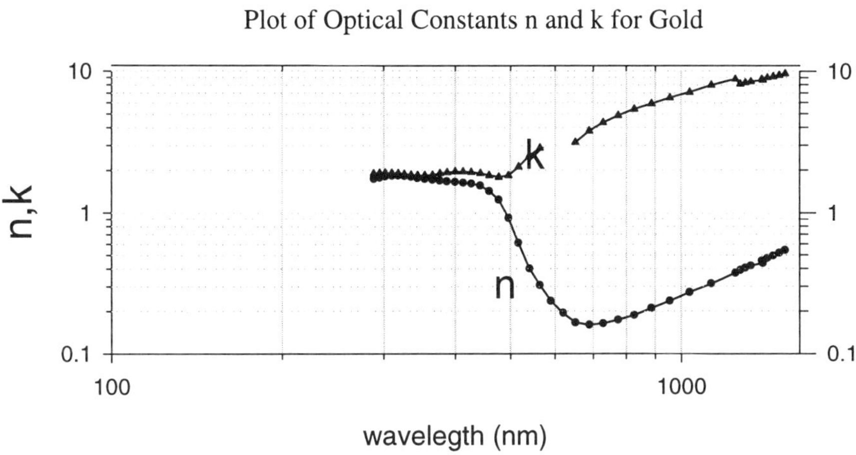

Figure 1. Optical Constants for Gold (Pal80).

If one assumes spherical colloids, and a low concentration of scattering centers (independent scattering events), the scattering and absorption efficiency contributions to the total turbidity can be calculated by equations [1] and [2].[1]

$$Q_{scat} - \mathrm{Re}\left[\frac{8}{3}\cdot(x)^4\cdot\left[\left|\frac{m^2-1}{m^2+2}\right|\right]^2\right] \qquad [1]$$

$$Q_{abs} = -\,\mathrm{Im}\left[4\cdot x\cdot\left[\frac{m^2-1}{m^2+2}\right]+4\cdot\frac{x^3}{15}\cdot\left[\left|\frac{m^2-1}{m^2+2}\right|\right]^2\cdot\frac{m^4+27m^2+38}{2m^2+3}\right] \qquad [2]$$

where

$$x = \frac{2 \cdot \pi \cdot a \cdot n_s}{\lambda} \qquad \text{and} \qquad m = \frac{N - iK}{n_s}$$

a = particle size in nm
λ= wavelength of incident light
N = real part of the refractive index
K = imaginary part of refractive index
n_s = refractive index of the matrix (in this case silica)

At particle sizes less than 10nm, the scattering contribution is negligible and the ruby color is generated by a combination of the plasmon band at 530nm and an slightly increased absorption in the blue. Figure 2(a) shows the theoretical absorption and scattering efficiencies for a 10nm particle using the best available optical constants for gold.[9] In Figure 2(b) the same functions are plotted for a 40nm colloid size. Note that the scattering contribution, which is minimal at 10 nm, increases dramatically as the particle size increases. The total diminuation of transmitted light, represented by the total efficiency factor Q_{total}, is the sum of the scattering and absorption contributions. Figure 3 shows the total attenuation of transmitted light as a function of gold colloid size. As particle size increases, intuition might tempt one to expect the scattering component to eliminate more of the blue due to Rayleigh-Gans type scattering. However, the dispersion of gold is such that its index is relatively close to that of the glass matrix in the blue region and this minimizes the scattering in this area. By the same token, the variation (rapid decrease) in the real part of the refractive index in the vicinity of 530 nm causes much greater scattering near the plasmon band. Finally, the tails of both the scattering peak and the absorption peak are large and extend into the red region of the visible spectrum. Consequently, as particle size grows, both the scattering and absorption components combine and grow larger in the red. The combination of these effects produces the characteristic absorption/scattering spectrum we see. Thus as the colloidal gold particles increase in size either by aggregation or by growth (e.g. "overstriking" of a ruby glass), the result is a progressive change in color from red to violet to blue. If the particles grow large enough to behave as bulk metal (i.e. the particle size becomes much larger than the penetration depth ~15nm for gold), reflection begins to play a role and the glass takes on a muddy color.

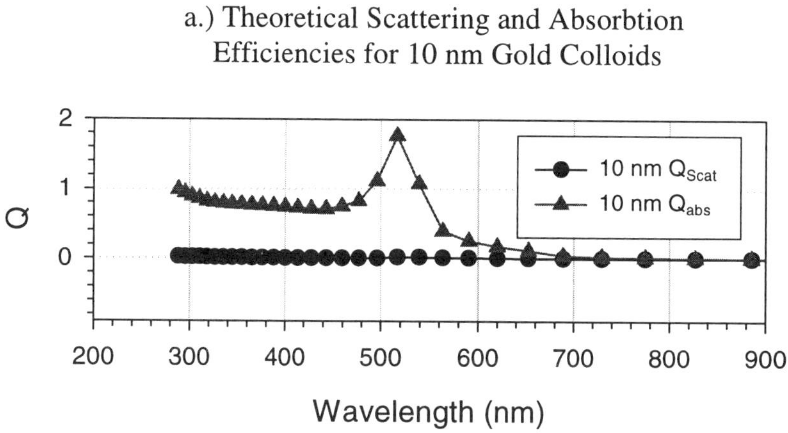

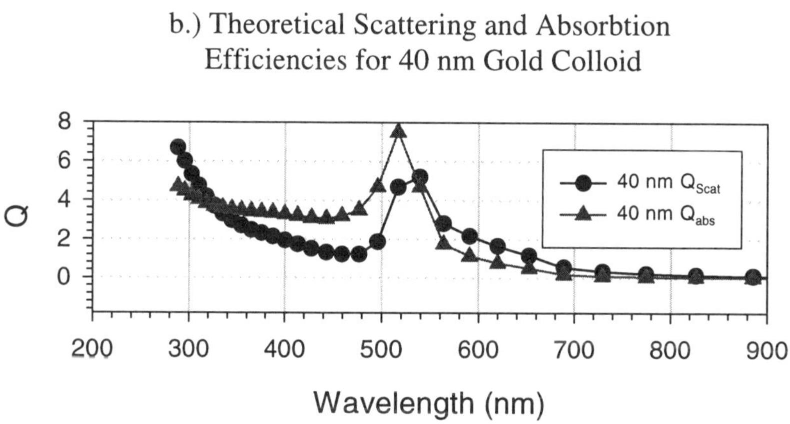

Figure 2. a.) Theoretical absorption and scattering efficiencies as a function of wavelength for 10 nm gold colloids in silica. b.) Same as (a) but calculated for 40 nm particles.

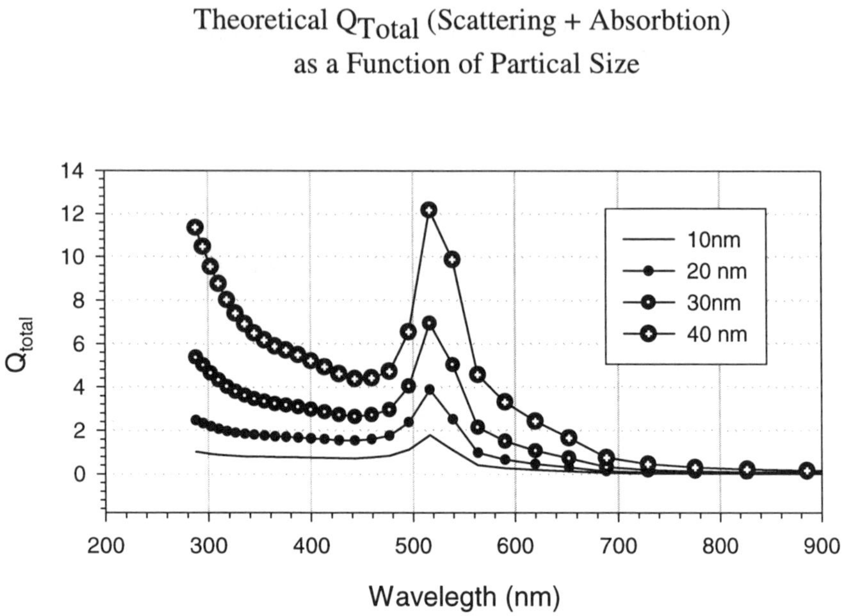

Figure 3. Total attenuation factor for several different particle sizes.

In summary, as a gold sol is formed from very small spherical particles it develops a ruby red color due primarily to absorption in the blue and green. As the particles grow the color develops to a wine red as scattering and absorption tails grow into the red part of the spectrum. Larger particles or aggregates cause a rapid increase in scattering in the red allowing only blue light through. These gels appear blue in transmission and reddish or violet when viewed from the side. As particles become very large, the solution takes on a reddish brown or "muddy" look as reflection begins to contribute.

Scattering in Sol Gel Derived Porous Glasses

For most dense glasses there is very little scattering internal to the glass itself due to the fine scale of homogeneity in the glass. For porous sol gel glasses, however, there is inhomogeneity due to the empty pores (n=1.0 for air in pores and n=1.46 for silica glass). The scattering of pure silica gels of various pore for most mesoporous gels is dominated by Rayleigh scattering which is proportional to $1/\lambda^4$. The scattering is greatest in the ultraviolet and blue and there is significant loss in the visible spectrum as the pore size increases. Figure 4 shows the attenuation of a 200Å radius gold doped silica gel monolith plotted against $1/\lambda^4$. The linear region is indicative of Rayleigh scattering caused by the porosity of the gel substrate. In an undoped gel, the scattering remains linear with respect to $1/\lambda^4$ throughout the visible spectrum. It is clear that there is an additional source of absorption and/or scattering that is responsible for the ruby red color of the colloidal gold glass.

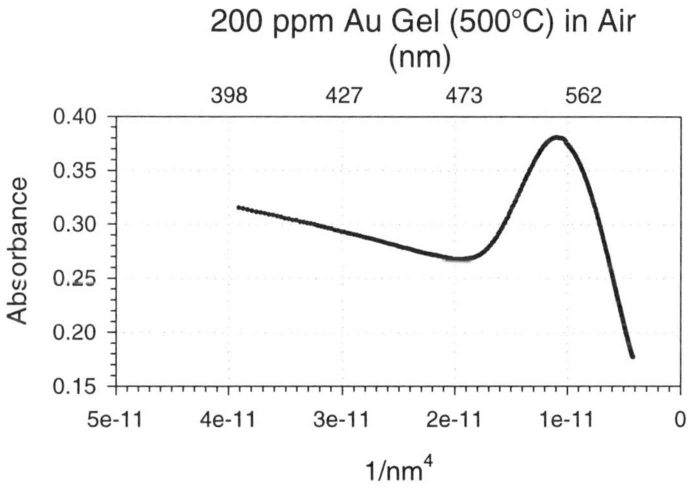

Figure 4. Plotting the absorbance vs. $1/\lambda^4$ clearly shows the contribution of the plasmon band versus the matrix (Rayleigh) scattering contribution.

Thus when analyzing the optical properties of gold colloids in porous glasses, both the Rayleigh scattering of the glass and the scattering from the colloidal particles must be considered. In dense ruby glasses, there are no pores and the scattering contribution of the glass matrix is negligible.

EXPERIMENTAL

The preparation of small metal clusters in sol gel derived silica has received scant attention, perhaps due to the difficulty in drying (and/or densifying) samples large enough to analyze or in producing stable thin films. Clusters in porous samples are a necessary precursor to dense samples but may have applications of their own, particularly in the area of metal catalysts.

In this work, small gold colloids are produced in monolithic porous glass of three different pore sizes. The difficulties of drying and densifying monoliths can be overcome by the proper formulation and careful control over the drying and densification conditions.[10] Gold can be introduced into the silica gel by adding the desired amount to the gel formulation in the form of a metal salt (usually chloride). The salt remains in the pore liquor and is deposited on the interior surfaces of the pores as the liquid is evaporated during drying. Colliodal metal particles are produced by the decomposition of the salt during subsequent heat treatments. Small sol gel monoliths can be produced with pore radii ranging from less than 10Å to 200Å. It is postulated that the pore size will constrain the growth of the colloids producing different sizes of metal clusters suitable for analysis.

10Å Radius Gels

Dried sol gel monoliths with 10Å pore radii are produced by using a low R-ratio (R= mole ratio of water to TMOS) in a heated bath under highly acidic conditions. The high temperature and acidic conditions insure rapid hydrolysis of the TMOS and promote the evaporation of methanol formed by the reaction. The elimination of the methanol is an important feature of this scheme for several reasons. Highly acidic sols promote re-esterification of silicic acid back to the methoxy precursor. Reducing the concentration of methanol should shift the reaction towards more complete hydrolysis by LeChatelier's principle and increase the relative percentage of water. This enhances the charge stabilization of the condensed species. Finally, the removal of methanol reduces the overall volume of the gel producing a higher initial gel density (increased solids loading). The higher gel density correlates to smaller pore size which increases capillary pressure (and hence syneresis) during drying. At the same time, it imparts greater strength to the wet gel helping it survive the drying process. The ultimate result is a microporous gel with an average pore radius of approximately 1 nanometer.

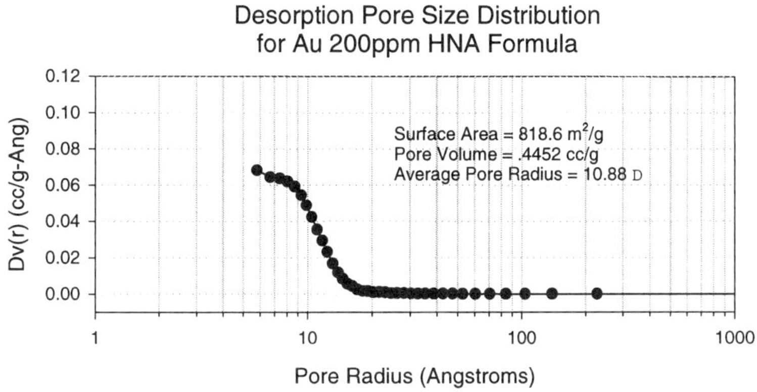

Figure 5. Desorption isotherm and pore size distribution for microporous (10Å) 200 ppm gold gel.

The BET data for a typical, as dried, gold doped microporous gel is depicted in Figure 5. The average pore radius is 10.88Å but the Kelvin pore size distribution indicates substantial porosity below 10Å. (Although micropore radius determined by the Kelvin equation is usually understated by a significant margin it is clear that a significant proportion of micropores are present).

30Å Radius Gels.

Gels with an average pore radius of 30Å are prepared at room temperature by using an R ratio of 16:1 under acidic conditions (pH=2). A small quantity of HF (.001 moles HF:1 mole TMOS) is added as a catalyst. Figure 6 shows the desorption pore size distribution data for these gels. The isotherm (not shown) exhibits the typical Type E hysteresis loop of mesoporous silica gels indicating ink-bottle type pores.

200Å Radius Gels

The formula used to produce a 200Å radius gel uses an R ratio of 16:1, acidic conditions (pH=2) and is also catalyzed with HF. For the larger pore gel a higher concentration of HF is required (.006 moles HF: 1 mole TMOS). In order to slow down the reaction, the sol is diluted with an equal volume of methanol and cooled to 5°C before the addition of the TMOS. Figure 7 shows the BET pore size analysis of the large pore gel doped with 250 ppm Au.

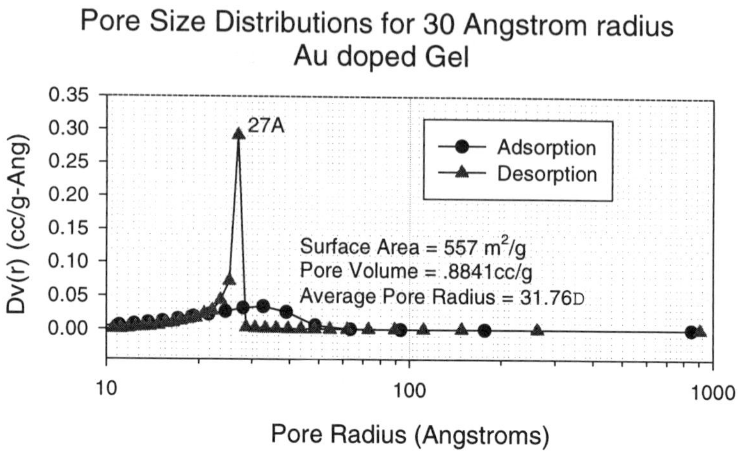

Figure 6. Pore size distribution for 30Å Au doped gel.

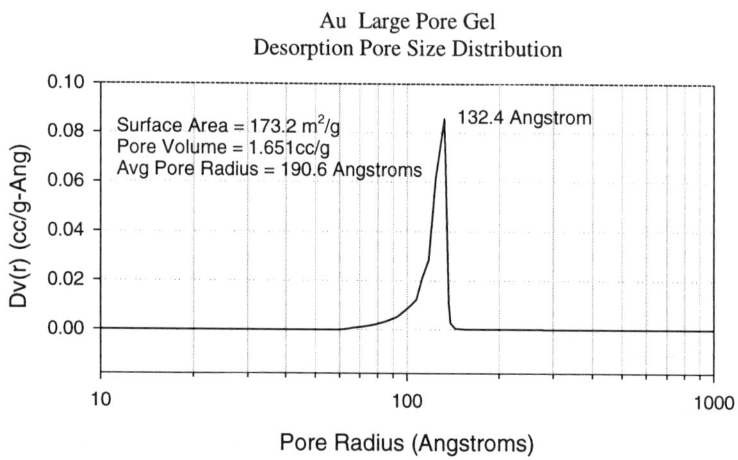

Figure 7. Pore size distribution for large pore (200Å) silica gel monolith doped with 250ppm Au.

Doping and Heat Treatment

All of the gels in this experiment were pre-doped. The gold was added to the sol prior to gelation in the form of an ICP standard composed of AuCl in 2% nitric acid. The concentration of gold in the standard solution was 1000μg/ml. The final concentration of gold in ppm was calculated with reference to the dry weight of the stoichiometric silica in solution. (e.g. 100ppm Au indicates 100 parts Au per million parts silica by weight).

The gels were all aged and dried identically with a final drying temperature of 180°C. Some of the gels showed a slight coloration after drying indicating some formation of gold particles had already begun. Full formation of the colloids was complete after heat treating at 400°C. UV-Vis spectra were taken at this temperature, at 800°C and after achieving full density for the largest pore size (1200°C for the 200Å). The smaller pore sizes could not be brought to full density due to the evolution of water at high temperatures (bloating).

UV-Vis spectra were taken with a Perkin-Elmer Lambda 9 UV-Visible-NIR spectrophotometer. Spectra were taken of the gels after treatment at 400°C, 800°C, and, in the case of the 200Å radius gels at 1200°C after achieving a fully dense glass. A peak at the plasmon frequency of ~530nm is indicative of well dispersed particles and a spherical morphology. The presence of a broad plasmon band in the vicinity of 600 nm is indicative of large irregularly shaped particles or large aggregates.

RESULTS AND DISCUSSION

Figures 8 and 9 show the UV-Vis spectra for the three types of porous glass at several different annealing temperatures. The 10Å gels all appeared blue in transmission with clearly visible scattering of red when viewed from the side. Figure 8(a) confirms this with a very broad band in the red which extends into the near infrared. This is most likely a product of aggregation as the particle size of the gels should be limited by the pore size of the glass. Linear or fractal aggregates can form along the pore network even with small pore sizes and give rise to the broad red shifted band.[11] The 30Å glass (Figure 7(b)) also appeared bluish after treatment at 400°C indicating either large irregularly shaped particles or aggregates. After treatment at 800°C, however, the plasmon band shifted back to 530nm and narrows. This same effect was reported by Doremus[4] upon heat treatment of his "purple" sols and can be attributed to the particles becoming more spherical or the sintering of aggregates. Oddly, the 200Å gels showed the opposite effect upon densification at 1200°C. Whereas they started out with a plasmon peak at 529nm, the peak red shifts and broadens as the gel becomes dense. This may be a function of the large degree of shrinkage in these gels (~400% by volume, ~40% linearly) during densification.

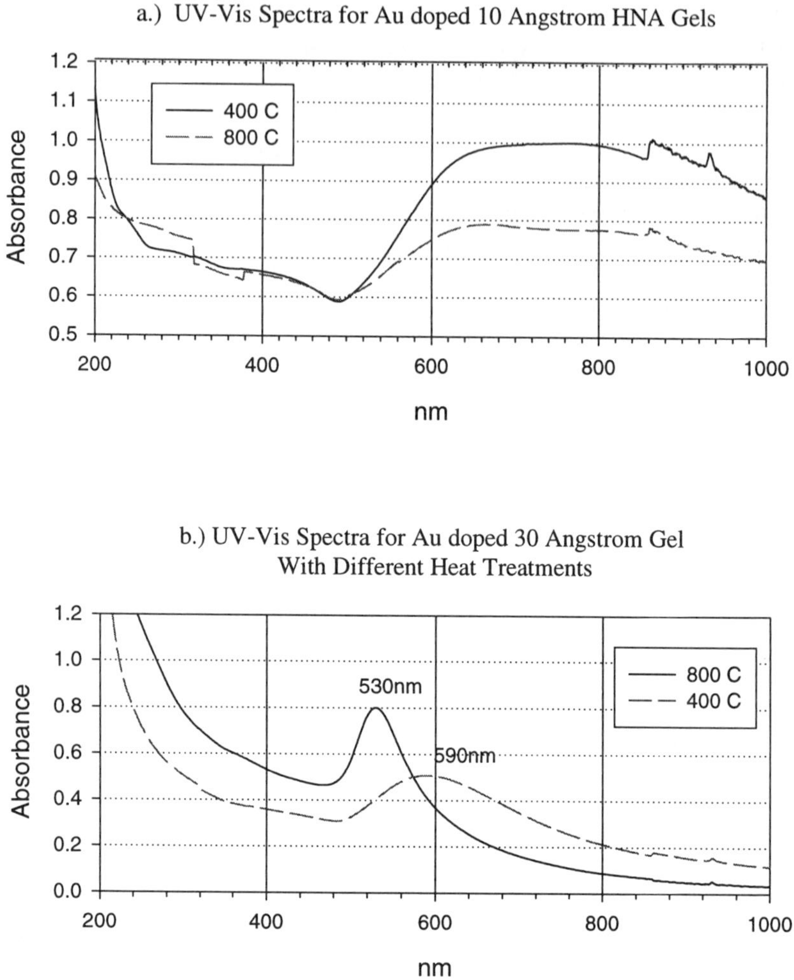

Figure 8. UV- Visible absorption spectra for 10Å and 30Å Au doped gels (200ppm) at different stabilization temperatures. a.) UV- Visible absorption spectra for 10Å gel doped at 200ppm. b.) UV- Vis absorption spectra for 30Å Au doped gel (200ppm).

Thus the gold clusters are brought into closer proximity and may form aggregates. It should also be pointed out that the final densification temperature for the large pore gels is above the melting point of gold (1063°C) giving the gold a high degree of mobility through the pore network as densification occurs.

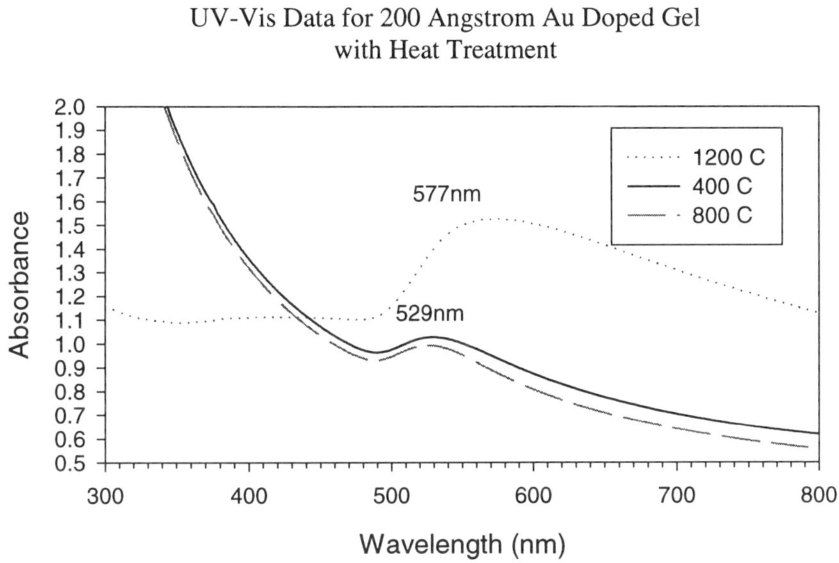

Figure 9. UV-Visible absorption spectrum for 200Å Au doped gels for three different stabilization temperatures (250ppm).

CONCLUSIONS

It is evident that the size and optical properties of the gold clusters is not a simple function of pore size but rely also on the heat treatment and probably the morphology as well. If the goal is the production of dense silica glass with well dispersed and spherical clusters, the 30Å or 200Å gel formulas are promising. However, the 10Å formula, which might be expected to produce very small colloids by constraining growth, produces a blue glass resembling an "overstruck" glass containing large or irregular clusters. This is likely due to an irregular morphology in the confines of the pore network. The size and morphology of the clusters needs to be established by TEM and compared with the porosity and spectroscopy data. The doping parameters can then be further explored in order to control the size and distribution of the metal clusters.

References

1. Bamford, C. R, Colour Generation and Control in Glass, Elsevier Scientific Publishing Company, New York, N.Y., (1977).
2. Mie, G., Ann Physik, 25, (1908), 377-445.
3. Doremus, R. H., J. Chem Phys., 40, (1964), 2389.
4. Doremus, R. H., J. Chem Phys., 42, (1965), 414.
5. Inagaki, T, Arakawa, E. T., and Williams, M. W., Phys Rev. B, 23 (1981), 5246.
6. Kadono, K., Sakaguchi, T., Miya, M., Matsuoka, J., Fukumi, T., and Tanaka, H., Journal of Materials Science:Materials in Electronics, 4, (1993), 59-61.
7. Ahmadi, T. S., Logunov, S. L., and El-Sayed, M. A., J. Phys Chem., 100, (1996), 8053-8056.
8. Hummel, R. E., Electronic Properties of Materials 2nd Ed., Springer Verlag, Berlin, (1993).
9. Palik, E.D., Editor, Handbook of Optical Properties of Solids, Academic Press Inc., (1985).
10. Powers, K.W., The Development and Characterization of Sol Gel Substrates for Chemical and Optical Applications, PhD Dissertation, University of Florida, Gainesville, FL, (1998).
11. Chen, W.P., and Chen, J.M., Journal of the Optical Society of America, 71, (1981), 189.

IN VITRO SURFACE REACTIVITY OF MELT VERSUS SOL-GEL DERIVED BIOGLASS®

J.P. Zhong and D.C. Greenspan
USBiomaterials Corp., One Progress Blvd., Alachua, FL 32615, USA

ABSTRACT

Two different processing methods have been used to make bioactive glasses: melt and sol-gel processing. The melt processing produces dense glass while the glass derived by the sol-gel method is normally stabilized and possesses a porous structure and high surface area. In this study, melt and sol-gel derived glasses were reacted in various buffer solutions, the pH was monitored, and the rate of HCA formation was determined by Fourier Transform Infrared spectroscopy (FTIR). The glasses were compared in terms of their bioactivity.

INTRODUCTION

It has been well documented that Bioglass® develops a bond with bone and stimulates bone tissue growth through a series of surface reactions[1,2]. Recently, a hypothesis has been proposed that the biocompatability and bioactivity of 45S5 Bioglass® is most probably derived from its rapid formation of a hydroxy-carbonate apatite (HCA) layer on surface and soluble silicon released to surrounding area after implantion[3]. As summarized in Table I Hench first proposed, the 11 reaction events that take place upon implantation[1].

Two different processing methods have been developed to synthesize bioactive glasses: 1) melt processing and, 2) sol-gel method[4,5]. The melt processing is a traditional technology that produces dense glass. By contrast, the sol-gel technology requires lower processing temperatures, and the resulting glasses possess a porous structure and high surface area after stabilization.

45S5 melt-derived Bioglass® has over 10 years of clinical success. It has been reported that when the silica content for the melt-derived bioactive glasses reaches more than 55 wt%[1], the ability to bond with bone will be lost, and no early HCA formation can be observed on the glass surface *in vitro*. This loss of bone bonding is attributed to the low surface reactivity in which there will not be enough surface area developed[6], nor enough silanol groups (Si-OH) to initiate the

HCA formation. In addition, there is not sufficient soluble silicon leaching to stimulate bone growth.

Table I Surface Reaction Stages on Bioglass®

1	Ion exchange of Na+ with H+ from surrounding environment;
2	Dissolution of soluble silica to surrounding environment and formation of Si-OH;
3	Polycondensation of SiOH+SiOH →Si-O-Si;
4	Adsorption of amorphous $Ca+PO_4+CO_3+OH$;
5	Crystallization of HCA;
6	Adsorption of biological moities in HCA layer;
7	Action of macrophages;
8	Attachment of stem cells;
9	Differentiation of stem cells;
10	Generation of matrix;
11	Crystallization of matrix.

It has been assumed that sol-gel bioactive glasses with their porous structure and high surface area would possess a much higher degree of surface reactivity. A number of recent studies have reported that a sol-gel glass showed an early induction and formation of HCA and also demonstrated a high level of resorbability[7,8]. Additionally, the porous structure and partial stabilization of the sol-gel network allows extension of the bone bonding composition range well above the 55 wt% SiO_2 limit found in melt derived bioactive glasses. In fact, early HCA formation has been demonstrated in sol-gel bioactive glasses with SiO_2 content as high as 80 wt%[4,9].

In this work, a sol-gel bioactive glass is compared with 45S5 Bioglass® and a non-bonding composition 60S. These were tested in vitro and their surface reactivity, specifically the HCA layer, evaluated. This information provides more evidence for the effects of surface structure on in vitro bioactivity.

MATERIALS AND METHODS

45S5 and 60S melt-derived dense glasses and 58S sol-gel derived glass were used in this work, and their compositions are listed in Table II. In the processing of the melt-derived glasses, the oxide compounds are well mixed, melted and homogenized at 1350°C for 5 hours. The glass frit was obtained by water quenching to room temperature, and the frit was ball milled and then separated through sieves to produce the glass particles with size ranging from 710-300μm.

Table II The Compositions of Bioactive Glasses

Sample ID	Composition, Mol%			
	SiO_2	CaO	P_2O_5	Na_2O
45S5	46.1	24.4	2.6	26.9
60S	60.1	19.6	2.6	17.7
58S	60.0	36.0	4.0	0

For sol-gel glass[5], the gel was prepared by mixing D.I. water, HCl, tetraethoxysilane (TEOS, $Si(OC_2H_5)_4$), triethylphosphate (TEP, $OP(OC_2H_5)_3$) and calcium nitrate, tetrahydrate (CN, $Ca(NO3)_2{\bullet}4H_2O$) in order. It took 2 hours to complete the mixing and the mixture was then cast into a polypropylene mold for aging at 60°C for 55 hours. After aging, the pore liquor was removed and the gels were transferred to a drying vessel. The drying process was conducted from room temperature to 180°C for 3 days under 95% relative humidity, and then the dried gel was heated to 700°C for stabilization. The stabilized gel was then ground to 710-300μm particles for testing.

The surface structure of the samples was analyzed using an NOVA 2000 nitrogen absorption/desorption BET system (Quantachrome, FL). The analysis provided data on specific surface area, average pore diameter, and pore volume.

For the in vitro testing, the particulates were tested dynamically in tris buffer and a simulated body fluid (SBF)[10] at 175 RPM at 37°C. For these tests, 0.2g particles in 200ml solution[7] has been used. The tris buffer used was a 0.2 N solution and the SBF is the tris buffer containing similar ionic concentrations as human blood plasma[10]. The pH of the reaction solutions was monitored during the reaction and the reacted particles were separated with solution by vacuum filtration through 8 μm filter paper after each reaction time period, dried with acetone and surface characterized for HCA formation by FTIR [4,7,9,11].

RESULTS

The surface structure of the samples from the BET analysis are listed in Table III. 45S5 and 60S are dense glasses with very low surface area, while 58S is a porous material with nearly $5x10^3$ greater surface area than dense glasses.

The pH of the reaction solutions versus time for the three materials are shown in Fig. 1. The 45S5 Bioglass demonstrated a rapid pH increase, and 60S increased slightly while 58S showed a moderate pH rise.

The FTIR reflection spectra of the three materials reacted in tris and SBF are presented in Fig. 2. In both solutions, no HCA layer was observed on the 60S melt derived glass even in 20 days. In tris, 45S5 showed faster formation of HCA on its surface than 58S in 24 hrs. In SBF, however, 58S showed faster formation of HCA than 45S5.

Table III The Pore Structure of Bioactive Glasses

Sample ID	Ave. Pore Diameter(Å)	Surface Area (M^2/g)	Pore Volume (cm^3/g)
45S5	Dense	0.027	0
60S	Dense	0.027	0
58S	68	151	0.45

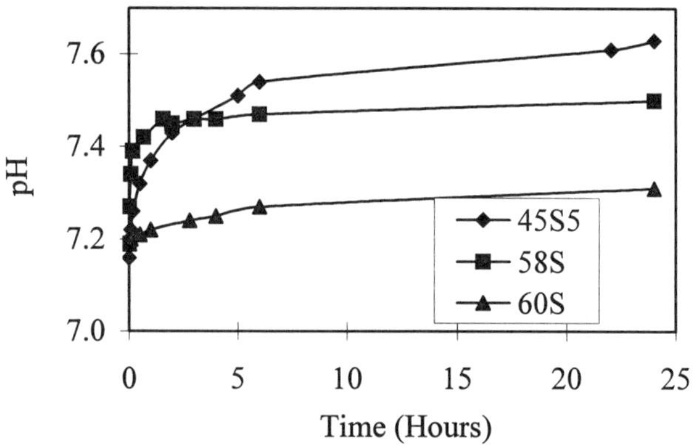

Fig. 1 pH of Reaction Solution of Bioactive Glasses vs. Time

DISCUSSION

It is well known that melt derived bioactive glasses follow the reaction sequence (steps 1-5, Table I) as proposed by Hench. The resulting porous structure, with the high surface area and high silanol concentration can induce the HCA formation[1,6,12,13]. 45S5 with a low silica content and more non bridging bonds has a more open network. Its surface reaction occurs rapidly which results in the fast pH increase by the ion exchange of H^+ in sulotion with Na^+ inside the glass and rapid formation of HCA on surface. The HCA formation in conjunction with the soluble silicon release has been shown to be responsible for its clinical success as a bone graft implant.

60S with a higher silica content has a tight glass network which gives rise to a slow surface reaction with a slow ion exchange and network dissolution. Its surface area and concentration of silanol groups are never sufficient to induce the HCA formation, even at long time periods[6]. In addition, the pH changed only

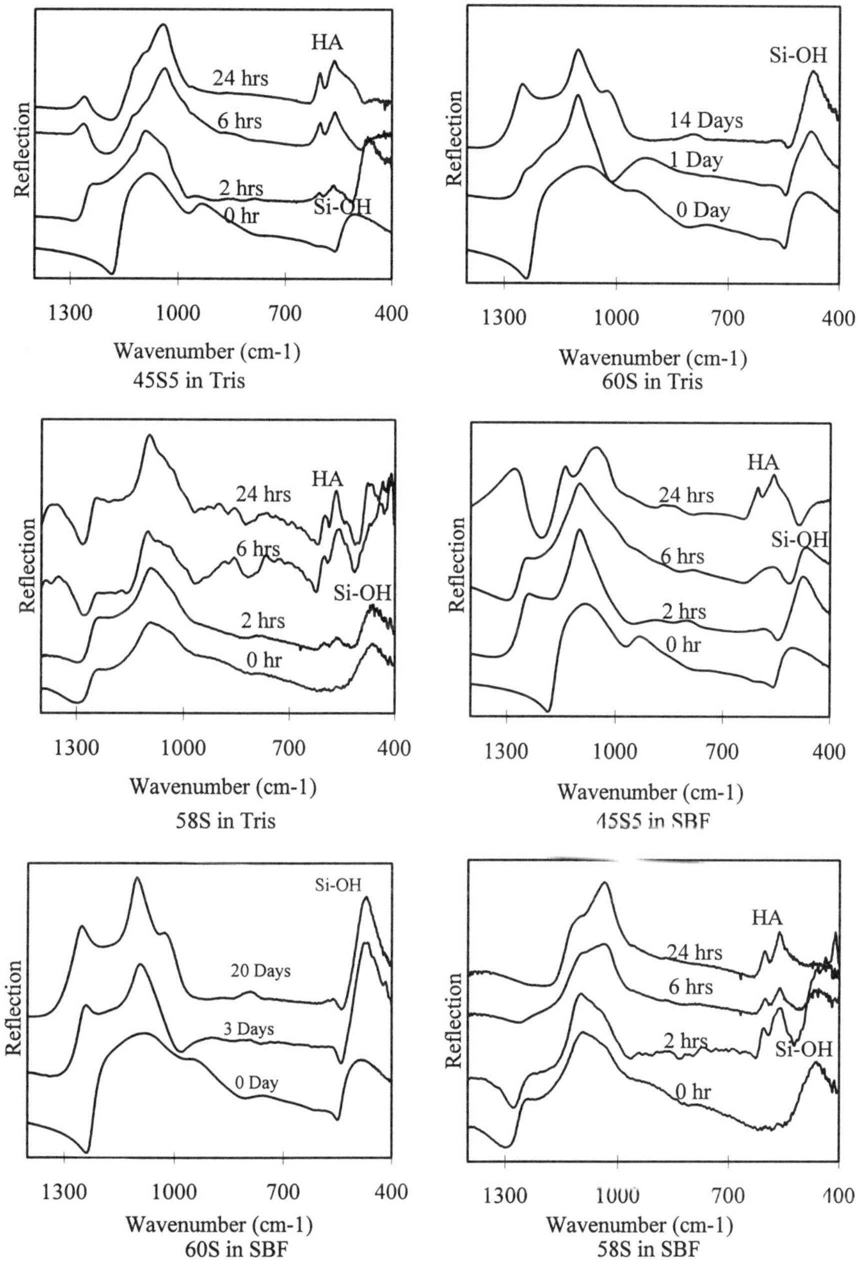

Fig. 2 FTIR Spectra of 45S5, 60S and 58S in Tris and SBF

slightly for 60S glass, indicating little alkali leaching (stages 1-3) from the glass.

Although 58S contains the same mol% of silica as the non-bonding 60S glass, its initial porous structure and high surface area provide a high concentration of silanol groups and thus, the HCA layer can form rapidly, provided the Ca^{2+} and PO^{4-} groups are available from the solutions. As can be seen in Fig. 2, the HCA formation rate was similar on both 58S and 45S5 in tris. However, although the pH change was modest for 58S due to the absence of Na^+ and lack of fast initial ion exchange, the HCA formed within two hour in SBF on 58S, more rapidly than 45S5 (6 hours).

Clearly, a slow surface reaction has been observed on 45S5 in SBF. It can be assumed that the early precipitation of Ca^{2+} and PO^{4-} from SBF solution inhibited the further surface reaction and the development of surface area[14]. Thus, the formation of thick and well developed HCA layer was delayed. Since SBF is more close to the *in vivo* conditions, more rapid surface reaction could be expected on 58S *in vivo*. That, in addition to the modest pH change, and its soluble-Si associated resorbability as reported previously[7,8], should enable the sol-gel glass an ideal bone graft material.

CONCLUSION

The initial fast accumulation of Ca^{2+} and PO^{4-} groups for melt-derived 45S5 in tris solution gave rise to a rapid formation of HCA on surface. The melt-derived 60S with 60 mol% silica content did not show any bioactive response *in vitro*. Although it contains a high concentration of silica, the sol-gel bioactive glass with initial porous structure and high surface area demonstrated the formation of HCA more rapidly than 45S5 in SBF with only modest pH changes. Those characteristics of the sol-gel bioactive glass with its great resobability should make it an ideal bone graft material.

REFERENCE

1. L.L. Hench, "Bioceramics: From Concept to Clinic," J. Am. Ceram. Soc., 74[7], 1487, 1991
2. L.L. Hench, "Bioactive Ceramics: Theory and Clinical Applications," Bioceramics 7, 1-14, Andersson, Happonen, YliUrpo, eds., Butterworth-Heinemann, 1994
3. L.L. Hench, "Bioactive Glass and Tissue Regeneration," Bioceramics 9, 3-6, T. Kokubo, T. Nakamura, and F. Miyaji, eds., Elsevier Science, 1996
4. R. Li, A.E. Clark, and L.L. Hench, "An Investigation of Bioactive Glass Powders by Sol-Gel Processing," J. Appl. Biomater., 2, 231-39 (1991)
5. J.P. Zhong, and D.C. Greenspan, "Porous Sol-Gel Bioglass From Near-

Equilibrium Drying," Bioceramics 10, 265-68, L. Sedel and C. Rey, eds., Elsevier Science, 1997

6. D.C. Greenspan, J.P. Zhong and G.P. LaTorre, "The Evaluation of Surface Structure of Bioactive Glasses In-Vitro," Bioceramics 8, 477-81, Hench, Wilson, Greenspan, eds., Elsevier Science, 1995
7. D. C. Greenspan, J. P. Zhong, X. F. Chen, and G. P. LaTorre, "The Evaluation of Degradability of Melt and Sol-gel Derived Bioglass *in-vitro*," Bioceramics 10, 391-94, L. Sedel and C. Rey, eds., Elsevier Science, 1997
8. D.C. Greenspan, J.P. Zhong, and D.L. Wheeler, "Bioactivity and Biodegradability: Melt vs. Sol-gel Derived Bioglass *In Vitro* and *In Vivo*," Bioceramics 11, 345-52, R. LeGeros and J. LeGeros eds., World Science, 1998
9. M.M. Pereira, et al., "Effect of Texture on the rate of Hydroxyapatite Formation on Gel-Silica Surface," J. Am. Ceram. Soc., 78[9] 2463-68, 1995
10. T. Kokubo et al., 'Solutions able to reproduce *in vivo* surface-structure changes in bioactive glass-ceramic A-W," J. Biomed. Mater. Res., 24, 721 -34, 1990
11. M.R. Filgueiras, G.P. LaTorre, and L.L. Hench, "Solution Effects on the Surface Reactions of a Bioactive Glass," J. Biomed. Mater. Res., 27, 445-53, 1993
12. T. Kokubo, et al., "Dependence of Bone-like Apatite Formation on Structure of Silica Gel," Bioceramics 7, O. Andersson, Happonen, YliUrpo, eds., , 49-54, 1994
13. P.J. Li, C. Ohtsuki, and T. Kokubo etc., "Apatite Formation Induced By Silica Gel in a Simulated Body Fluid," J. Am. Ceram. Soc., 75[8], 2094-97, 1992
14. J.P. Zhong, D.C. Greenspan, "Bioglass Surface Reactivity: From Invitro to In vivo", Bioceramics 11, 415-18, R. LeGeros and J. LeGeros eds., World Science, 1998

HIGHLY POROUS SILICON OXYCARBIDE GLASSES PRODUCED BY POLYMER PYROLYSIS

Harald Schmidt, Dietmar Koch, Georg Grathwohl
Ceramic Materials and Components, University of Bremen,
IW3 / Am Biologischen Garten 2, 28359 Bremen, Germany

ABSTRACT

Microporous Silicon Oxycarbide (SiOC) materials are fabricated by pyrolysis treatment of polysiloxane precursors. The thermal treatment of the precursors was carried out in a temperature range from 500 °C to 1000 °C in flowing nitrogen atmosphere. A high degree of microporosity is reached depending on the temperature applied; maximum BET specific surface areas of nearly 450 m^2/g are achieved at a pyrolysis temperature of 600 °C. These large surface areas can be realized in high volume components if the formation of orientated pore channels is additionally provided by incorporation and first step decomposition of polymer fibers. Heat treatment to higher temperature leads, however, to micropore collapsing and loss of inner surface area. This degradation can efficiently be hindered by the modification of the polymer. For this purpose various alkylsilane additives and a further portion of prepyrolized filler to be incorporated in the green body are used. The temperature of pore collapsing can be shifted to higher values and large inner surface areas are still being kept at a level of higher than 160 m^2/g after heat treatment at 1000 °C. Mercury (Hg)-porosimetry and gas adsorption measurements are applied to characterize the porosity and the high BET surface areas and DFT derived pore distributions.

INTRODUCTION

Porous ceramics and glasses are used in a large number of applications as filters, catalysts and membranes but also in thermal and acoustic barrier systems or heat exchangers for severe operating conditions. Depending on their structures, compositions and porosity characteristics these inorganic porous components are fabricated by partial sintering of suitable powders, sol-gel technology, sintering of composites with combustible phases, annealing and etching of systems with a solubility gap or by pyrolytic decomposition of polymeric precursors. These inorganic non-metallic components allow use of their functional porosity in separation, purification, adsorption and reaction processes where elevated temperatures, aggressive media or biological environments are to be considered. Some further advantages compared e.g. to conventional polymer membranes can be seen in the improved mechanical properties, higher stiffness, wear resistance and the higher pressures to be applied during operation.

The use of silicon-organic polymers as precursors for ceramic materials offers an interesting alternative route to produce ceramic components; this is at present a favourite process to fabricate ceramic matrix composites. A large variety of preceramic precursors based on polycarbosilanes, polysilanes, polysiloxanes or polysilazanes can be used for the production of ceramic components via shaping processes as being typical for organic polymers and pyrolysis treatment at elevated temperatures and controlled atmosphere [1]. Structure, degree of crystallinity and microstructural parameters of the inorganic component being formed are evolving under control of the pyrolysis conditions and in dependence of the chemical system being used.

Silicon oxycarbide (SiOC) matrices can easily be prepared by pyrolysis of polysiloxane precursors at temperatures above 600 °C in flowing argon or nitrogen atmospheres. Polysiloxanes have the advantage over low-oxygen polymers in that they are much easier to handle and process under ambient environmental conditions. During the pyrolysis treatment various constituents of the polymer structure are going to be released depending on temperature and gas pressure applied. The result of this thermal degradation is the polymer-ceramic transformation accompanied by the formation of open porosity being influenced by the gas evolution. Modification of polysiloxane precursors by alkylsilanes of various filler systems incorporated in the polymer can be used to produce highly porous SiOC-materials [2,3]. The thermal and chemical stability of the resulting pyrolytic decomposition product in conjunction with the high specific inner surface area and the pore network is very important for use as filters in aggressive environments [4].

In this work porous polymer derived SiOC materials are produced via the polymer pyrolysis route. The aim of this development was to define the conditions of the generation of the resulting micro- and mesoporous structures. In addition, orientated macropore channels were intentionally formed in the early states of thermal degradation in order to influence the gas releasing process. The pore size distribution was measured by Hg-porosimetry as well as by N_2 and CO_2 gas adsorption analysis. The thermal stability of the filler containing SiOC was investigated at temperatures up to 1000°C in flowing nitrogen.

POLYMER PYROLYSIS PROCESS AND EVOLUTION OF MICROSTRUCTURE

The pyrolysis of Si-containing polymers to a ceramic material leads to a significant increase of density (e.g. *polymer phase:* $\rho_P \approx 1{,}0\ \mathrm{g\ cm^{-3}}$; *ceramic phase:* $\rho_C \approx 2{,}2 - 2{,}6\ \mathrm{g\ cm^{-3}}$). Due to the resulting volume reduction of up to 60 % and the simultaneous mass loss through gaseous release of volatile components (oligomers, reaction gases) pores and cracks which may result in catastrophic failure of the components are created. During pyrolysis of preceramic polymers between 100 °C and 450 °C cross-linking and structural reorganisation start combined with the release of volatile low molecular Si-containing oligomers. In the range of 600 °C to 800 °C metalorganic bonds are decomposed which lead to the formation of various hydrocarbons (mainly CH_4) [5]. Between 650 °C and 1200 °C additional hydrogen is released. Finally above 1200 °C, depending on the particular polymer system, amorphous silicon oxycarbide starts to crystallize leading to SiO_2, SiC and graphitic carbon. At even higher temperatures above 1400 °C CO is released through the carbothermal reduction [6].

During the pyrolysis of Polymethylsiloxane *PMS*, e.g. the color of the specimens changes from transparent (250 °C) to orange (600 °C), brown (800 °C) and black (1000 °C) which can be correlated with the reorganization and the thermal degradation of the polymer. Furthermore the change in color is a result of the formation of free carbon which starts at 600 °C and increases with further rising pyrolysis temperature. Additional network forming carbon emerges. Above 1000 °C turbostratic graphite is observed.

EXPERIMENTAL SETUP AND PROCEDURE

To investigate the polymer-ceramic transformation two polymers are used: an addition polymerizing Methylphenylvinylhydrogenpolysiloxane *MPVHS* (trade name H62C, Wacker, Germany) and a condensation polymerizing Polymethylsiloxane *PMS* (trade name NH2100, Hüls, Germany).

The liquid MPVHS was filled in teflon cylinders at normal pressure to form the green bodies. Additionally electrostatic loaded polyester short fibers with a diameter of 20 µm and a length of about 10 mm were shot in parallel orientation into a thin *MPVHS* –layer of 500 µm by an electrical high voltage field of 50 kV (Figure 1). Then the orientated fibers were infiltrated with liquid *MPVHS*. After cross-linkage and outgassing at 200 °C in vacuum a green body with orientated fiber arrangement was formed. Through repeated infiltration and fiber embedding various fiber distributions are possible [7]. By the pyrolysis process the fibers are decomposed and leave open pore channels with defined diameter.

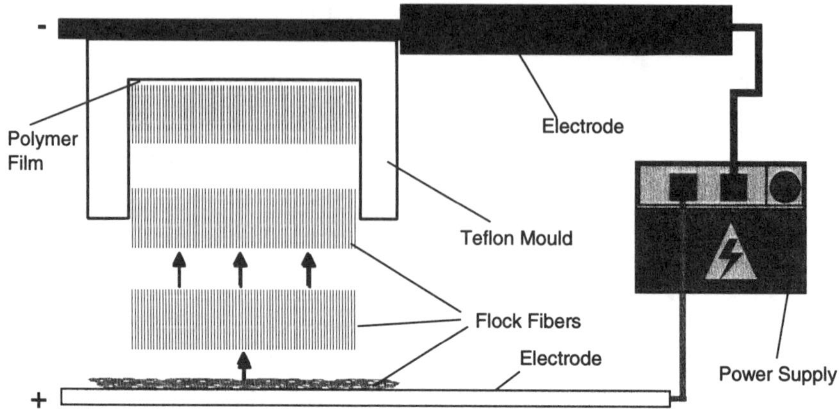

Figure 1: Schematic view of the flocking equipment for orientated embedding of fibers in the *MPVHS* matrix.

A further process route is applied to prepare samples of powdered *PMS*: first *PMS* is interlaced at 200 °C in vacuum, second it is pyrolized in N_2-atmosphere up to 600, 800 and 900 °C, respectively. The prepyrolized material is milled in an attritor up to a bimodal particle size distribution with peak frequencies of 2 and 4 µm. This so-called filler material is then mixed, homogenized and hydrolized in the mass ratio 1:1 with aliphatic alkylsilanes with different chain lengths (Methyltrimethoxysilane *MTMS* and Propyltrimethoxy-silane *PTMO*, delivered by ABCR, Germany). The precursor was hydrolized in a water-ethanol-solution at pH 7. After drying the agglomerates are milled in a ball mill. The green bodies were produced by uniaxial pressing at 20 MPa. The process for manufacturing the porous ceramic bodies is shown in Figure 2.

MPVHS and filler containing *PMS* precursors were pyrolized in a furnace under controlled N_2 gas flow at temperatures between 500 °C and 1000 °C. Adsorption-desorption-isothermal plots for determination of the inner surface area and the pore size distribution were measured using N_2 at -196 °C as the analysis gas (ASAP 2010, Micromeritics, USA). In the case of highly microporous bodies, CO_2 at -78 °C was also used. Inner surface was measured at relative pressures between $p/p_0 = 0{,}05$ - $0{,}3$ using BET analysis. Pore distribution was calculated from the measured adsorption isotherm using DFT (Density Functional Theory) [8] for pores between 0.8 nm and 400 nm. The macropores which occurred due to thermal decomposition of the incorporated fibers were measured with Hg-infiltration method up to a pressure of 2000 bar (Pascal 240, Porotec, Italy). The thermal stability of the pyrolyzed filler containing SiOC was investigated up to 1000 °C in flowing nitrogen.

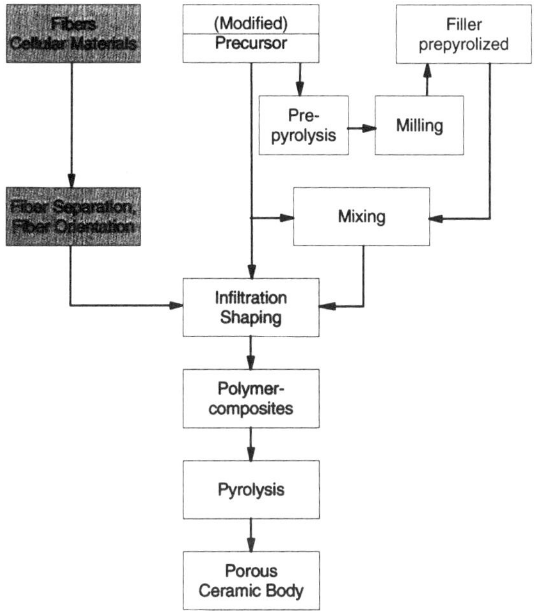

Figure 2: Polymer pyrolysis as production process for porous ceramics.

RESULTS AND DISCUSSIONS

As the first step of the pyrolysis the polymer fibers embedded in the *MPVHS* precursor are decomposed providing open cylindrical pore channels within the matrix. The pore channels allow a controlled release of the further gaseous reaction products arising from the pyrolysis process of the matrix constituents. The radius distribution of the cylindrical macropores in a pyrolized *MPVHS*-siloxane after heat treatment up to 1000 °C was measured using Hg-intrusion porosimetry (Figure 3a). The pore radii varied depending on the fibers used between 5 and 20 µm. The accessory SEM-picture shows the typical distribution of these macropores (Figure 3b). Decomposition gases as hydrocarbons and hydrogen which are mainly produced in the temperature range between 600 °C and 800 °C by the pyrolysis process of the metalorganic compounds of the matrix escape easily through the orientated pore channels. Thus, a crack free porous body with large dimensions is formed with ceramic structure after heat treatment above 1000 °C [9].

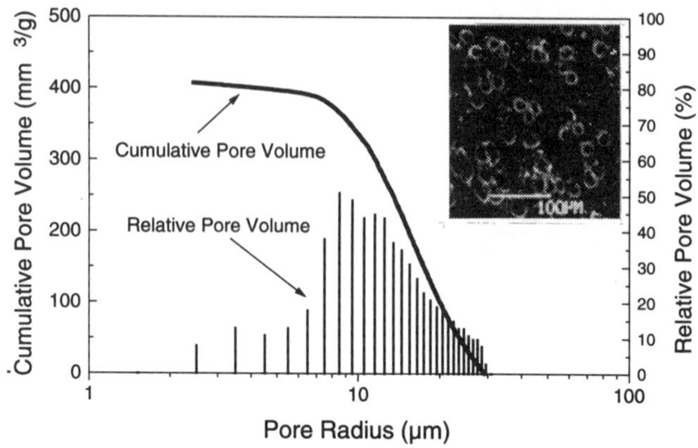

Figure 3: Channel pore radius distribution in a *MPVHS*-polymer ceramic composite after decomposition of the polyester fibers and following pyrolysis at 1000 °C in N_2 gas flow (10^{-3} m³/min) during a soaking time of 4 h; Mercury (Hg)-porosimetry and corresponding SEM-picture.

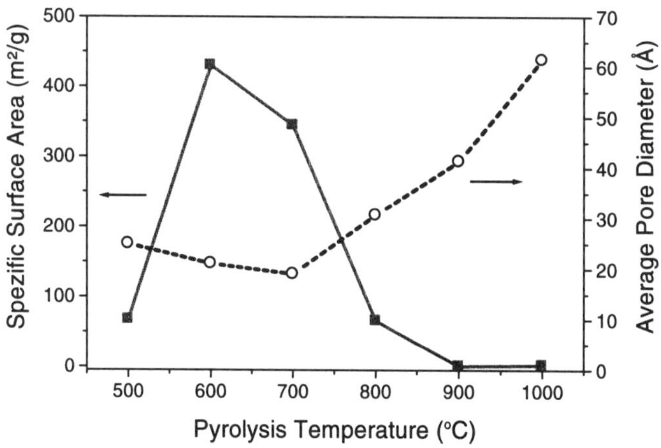

Figure 4: Specific surface area and average pore diameter as calculated from BET results of pure PMS-Precursor as function of pyrolysis temperature.

The formation of the micro pore network as a function of pyrolysis temperature is investigated using gas adsorption analysis. In Figure 4 the inner surface area and the calculated average pore diameter of pure polysiloxane dependent on increasing temperature are shown. Under flowing gas conditions the decomposition and gas release increase significantly at 600 °C; large inner surface areas of nearly 450 m^2/g (measured with N_2 as analysis gas) and 360 m^2/g (measured with CO_2 as analysis gas) are observed. Above 900 °C mesopores and micropores start to collapse due to viscous flow ($\eta \approx 10^{14}$ Pas) and due to the significant reduction of release of decomposition gases.

In Figure 5 and Figure 6 the BET specific surface areas and the pore size distribution of modified *PMS* precursors after pyrolysis up to 1000 °C are shown in comparison to pure *PMS*. Unmodified PMS exhibits no remarkable micro- and mesopores as shown in Figure 6. Modifying the precursor by use of *PMS* filler which was prepyrolized up to 600 °C and of short-chain methyltrimethoxysilane (*MTMS*) the amount of macropores is reduced but there is no significant increase in smaller pore sizes. However, using long-chain propyltrimethoxysilane (*PTMO*) instead of *MTMS* marked micro- and mesoporosity is measured. The specific surface area (Figure 5) which is too small to be measured via gas adsorption in pure *PMS* increased to 12 m²/g with addition of *MTMS* and even to 65 m²/g with addition of *PTMO*. Additionally, the temperature of prepyrolysis of the filler was changed from 600 to 800 and 900 °C, respectively. The mixtures of PTMO and these fillers lead to even higher specific surface areas with 150 and 163 m²/g combined with an increase of the volume fraction of micropores. Thus, by modification of *PMS* with prepyrolized fillers and alkylsilanes the thermal decomposition can be shifted and prolonged to higher temperatures above 1000 °C. With this step the pore skeleton is supported and pore collapsing is avoided to a great extent.

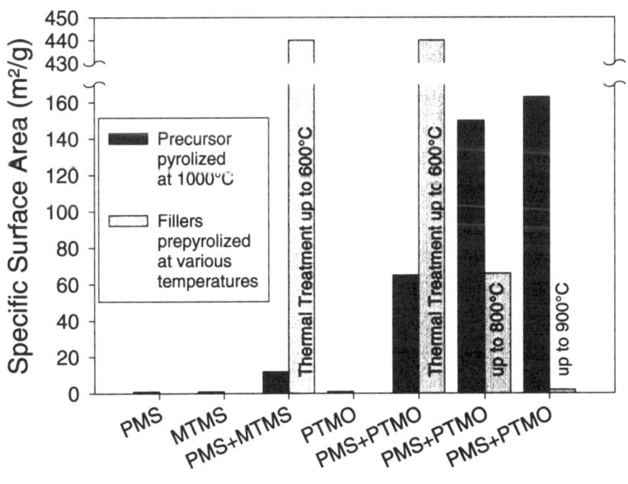

Figure 5: Specific surface area of precursors modified with different alkylsilanes (methyltrimethoxysilane *MTMS* and propyltrimethoxysilane *PTMO*) and filler particles after pyrolysis at 1000 °C in N_2 gas flow. The fillers were prepyrolized up to different temperatures (600, 800, 900 and 1000 °C).

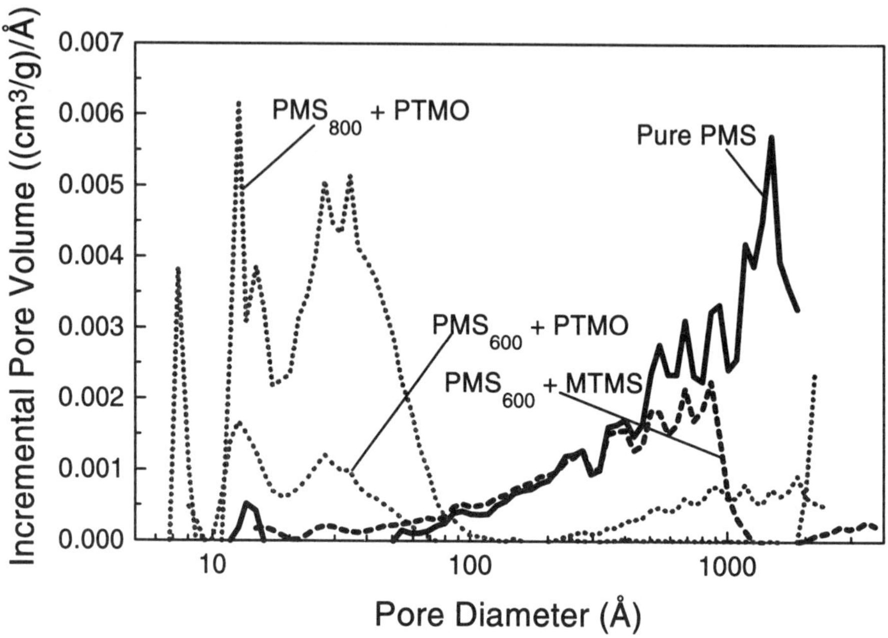

Figure 6: Comparison of pore distribution of *PMS* modified with filler particles prepyrolyzed at 600 °C and 800 °C and with different alkylsilanes (methyltrimethoxysilane *MTMS* and propyltrimethoxysilane *PTMO*) after pyrolysis at 1000 °C in N_2 gas flow as evaluated by gas adsorption according to DFT.

To explain the phenomenon of the influence of the different chain lengths on the formation of pore size distribution, the decomposition processes have to be considered. Decomposition products, e.g. radicals, result from bond splitting and lead to formation of defects. Within the polymer structure the Si-C- (69 kcal/mol) and the C-H-bonds (99 kcal/mol) are the easiest ones to be cracked [10]. The cracking of hydrocarbon bonds leads to the release of reaction gases such as CH_4, C_6H_6, H_2 etc. and to the formation of free carbon which is stabilized in carbon clusters. The occurring radical reactions are summarized in Figure 7. The use of alkylsilanes with longer chains (e.g. *PTMO*), however, results in an increased gas separation of hydrocarbon groups (i.e. propyl groups). The pore network is additionally stabilized by the prepyrolized filler particles as they reduce the viscosity processes [11].

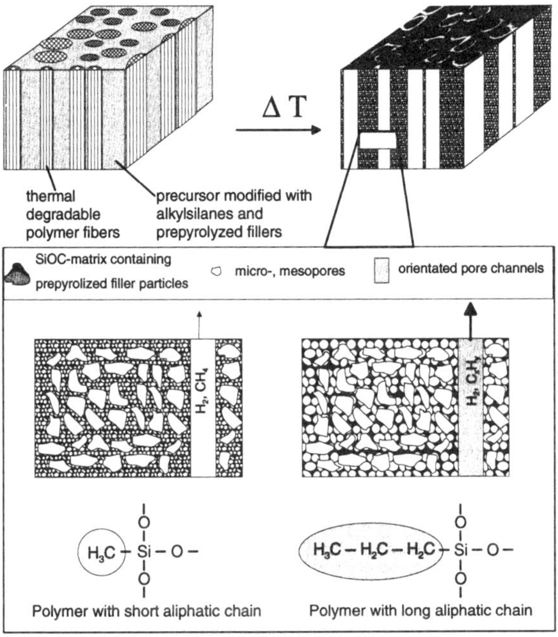

Decomposition reactions:

$$\equiv Si - CH_3 \rightarrow \equiv Si \bullet + \bullet CH_3$$

$$\equiv C - H \rightarrow \equiv C \bullet + \bullet H$$

$$\equiv C - H + \bullet CH_3 \rightarrow \equiv C \bullet + CH_4 \uparrow$$

$$\equiv C - H + \bullet H \rightarrow C \bullet + H_2 \uparrow$$

Figure 7: Scheme describing the optimized formation of orientated channel pores in combination with alkylsilanes to build up thermostable microporous matrix (left) and possible pyrolysis decomposition reactions (right).

CONCLUSION

The use of thermal degradable polymer fibers allows the formation of ideally orientated channel pores within a pyrolized matrix made from Si-organic polymers. Emerging decomposition gases are easily released through these channels and therefore they do not induce fatal cracks. Thus, the construction of high volume parts with functional microporosity is possible which otherwise is restricted to small scale volumes as e.g. layers, membranes. Induced by pyrolysis of the matrix, a micropore network is established in the ceramic body which can be adjusted by using different alkylsilane additives with different chain lengths in combination with semi-reactive prepyrolized filler particles. The modification of the polymer leads to the stabilization of the pore network that emerges at temperatures from about 600 °C to temperatures of 1000 °C and higher. Using propyltrimethoxysilane as a modifier, large inner surface areas (163 m²/g) can be reached after pyrolysis up to 1000 °C compared to 1 m²/g with pyrolized polymer without any modification. Obviously the pore structure is also stabilized through the filler particles as they reduce viscous flow processes. Furthermore, the increased gas separation by decomposition of long chain alkylsilanes supports the pore structure. The combination of both stabilization of the pore pattern and increased gas separation lead to a bimodal pore distribution in compact micro-porous ceramic bodies with orientated macropores.

REFERENCES

[1] R. W. Rice, "Ceramic from Polymer Pyrolysis, Opportunities and Needs-A Material Perspective", *Ceramic Bulletin*, **62** [8] 889 - 92 (1983).

[2] C. Liu, H. Zhang S. Komarneni and C.G. Pantano, "Porous Silicon Oxycarbide Glasses from Organically Modified Silica Gels of High Surface Area", *J. Sol-Gel Sci. Techn.,* **1** 141 - 51 (1994).

[3] Z. Li, K. Kusakabe, S. Morooka, "Preparation of thermostable amorphous Si-C-O membrane and its application to gas separation at elevated temperature", *J. Membrane Sci.*, **118** 159 -68 (1996).

[4] P.W. Lednor, "Synthesis, stability and catalytic properties of high surface area silicon oxynitride and silicon carbide", *Catalysis Today*, **15** 243 - 67 (1992).

[5] H. Schmidt, P. Buhler and P. Greil, "Pyrolytic Conversion of Poly (Methyl-siloxane) to Silicon (Oxy) Carbide Ceramics", *4th Euro Ceramics, Basic Science - Development in processing of advanced ceramics, Part **I**, [1],* 299 - 306 (1995).

[6] G.T. Burns, R.B. Taylor, Y. Xu, A. Zangvil and G.A. Zank, "High-Temperature Chemistry of the Conversion of Siloxanes to Silicon Carbide", *Chem. Mater.*, **4** [6] 1313 - 22 (1992).

[7] T. Schmedders, S. Poehnitzsch, H. Schmidt and G. Grathwohl ,"Texturierte Porosität in oxidkeramischen Systemen" in Werkstoffwoche 1998, Symp. 9a, III. Sintern und Eigenschaften, Willey-VCH, 89-94, (1999).

[8] J.P. Olivier, W.B. Conklin and M. v. Szombathely, "Determination of Pore Size Distribution from Density Functional Theory: A Comparison of Nitrogen and Argon Results, Characterization of Porous Solids III", *Studies in Surface Science and Catalysts*, **87** 81 - 89 (1994).

[9] G. Grathwohl, M.A. Mohammed, H. Schmidt and D. Koch, "Verfahren zur Herstellung von Keramiken mit Mikroporenkanälen und nanoporöser Matrix durch Pyrolyse präkeramischer Precursoren mit eingelagerten Polymerfasern", Deutsches Patentamt (OS-DE 198 38 263 A1), Aug. 22, 1998.

[10] L. Bois, J. Maquet, F. Babonneau, H. Mutin and D. Bahloul, "Structural Characterization of Sol-Gel Derived Oxycarbide Glasses. 1. Study of the Pyrolysis Process", *Chem. Mater.*, **6** [6] 796 - 802 (1994).

[11] G. M. Renlund, S. Prochazka, R. H. Doremus, Silicon Oxicarbide Glasses: Part I. Preparation and Chemistry; Part II. Structure and Properties, *J. Mater. Res.* **6** [12] 2716 -34 (1991).

TiO_2 CERAMIC SURFACES WITH CATALYTIC REACTIVITY IN ULTRAPURE WATER TECHNOLOGY

Dietmar Scherer*, Rainer Gadow
University of Stuttgart
Institute for Manufacturing Technologies of Ceramic Components and Composites
Allmandring 5b
D-70569 Stuttgart
GERMANY

Farhang Shadman
University of Arizona
Department of Chemical & Environmental Engineering
P.O. Box 210011
Tucson, AZ 85721-0011

ABSTRACT

The reactivity of titanium dioxide (TiO_2) surfaces, including TiO_2 / polymer composites, for oxidative removal of trace organic compounds from ultrapure water by a catalytic pervaporation process is demonstrated in this study. The membrane matrix material used is polydimethylsiloxane (PDMS), a ceramic precursor material, which forms the membrane matrix. PDMS is selective for the removal of organic compounds from water. In the PDMS matrix catalytically active TiO_2 particles with anatase type crystal structure are imbedded. On the surface of the TiO_2 particles hydroxyl and perhydroxyl radicals are formed in aqueous environment upon exposure to UV light which leads to oxidation of organic compounds. Thus by running pervaporation experiments with these catalytic membranes, oxidation of organic compounds at the feed/membrane interface as well as inside of the membrane is achieved. Isopropyl alcohol (IPA) and acetone are used as organic model compounds. As a result, partial and complete oxidation of IPA and acetone as well as simultaneous permeation of the oxidation products and the feed compounds could be observed during the catalytic pervaporation experiments.

INTRODUCTION

Large amounts of ultrapure water are used in the semiconductor industry. The required standard for the quality of ultrapure water shifts continuously towards lower levels of tolerable impurities in the water. One issue that becomes more and more important due to ecological and economical considerations is recycling and reutilization of the used ultrapure rinse water back into the polishing loop during manufacturing of semiconductor devices. In order to achieve recycling of the rinse water back into the production line, different trace organic contaminants that were used in the semiconductor production line have to be removed from the water.

In this research a novel catalytic pervaporation process was evaluated with possible application to the removal of these trace organic contaminants from ultrapure water. Catalytic pervaporation combines membrane separation by pervaporation with photocatalytic oxidation of organic compounds. Isopropyl alcohol (IPA) and acetone were used as model trace contaminants since both compounds are typically found in used rinse water and both compounds are highly recalcitrant, i.e. hard to remove.

CONCEPT OF CATALYTIC PERVAPORATION

Pervaporation is a membrane process used to separate mixtures of dissolved solvents. In pervaporation processes a membrane is placed as a permselective layer between a liquid and a gaseous phase. A liquid mixture is maintained at atmospheric pressure on the feed side of the membrane and one of the components permeates preferably to the permeate side of the membrane by solution and diffusion processes [1,2]. At the permeate side the component is evaporated and removed by applying a vacuum or by using an inert sweeping gas. In this study polydimethylsiloxane (PDMS), a ceramic precursor material, was used as membrane matrix material since PDMS shows high permeabilities for organic/water mixtures and is widely used for the removal of organics from water.

The goal of this research is not only the removal of organic compounds from ultrapure water, but also the immediate oxidation of these trace organic contaminants by photocatalytic oxidation. It has been demonstrated that many organic compounds are decomposed in aqueous solutions in the presence of TiO_2 powder as photocatalyst illuminated with UV-light and this process can be used for water purifying [3]. In this photocatalytic process, the absorption of a photon by TiO_2 promotes an electron from the valence band to the empty conduction band and thus produces an electron-hole pair [4,5]. In the presence of water and oxygen, highly reactive hydroxyl radicals and perhydroxyl radicals are generated on the catalyst surface by interaction with the catalyst. The radicals react readily with organic contaminants and oxidize them. This process leads to the degradation (partial or complete mineralization) of the organic contaminant and finally results in the formation of carbon dioxide and water if the reaction proceeds to completion. Anatase is the more active phase in comparison to rutile, which is probably due to differences in the extent and nature of the surface hydroxyl groups present in the low temperature anatase structure. In addition, the photoactivity enhancement can be related to the Fermi level of anatase which is about 0.1 eV higher than that of rutile.

To achieve a combination of selective removal of organic compounds by pervaporation and photocatalytic oxidation of these organic compounds, TiO_2 photocatalyst particles with anatase type crystal structure were imbedded within a PDMS matrix during the membrane fabrication process. Since for organic/water separation by pervaporation both the organic compounds as well as water permeate through the PDMS membrane, hydroxyl and perhydroxyl radicals are formed on the surface of the catalyst particles inside of the membrane and at the feed/membrane interface upon exposure to UV light. These radicals can react with organic compounds and partially or completely oxidize them. The oxidation products are successively removed on the permeate side.

In Fig.1 a schematic pathway of the catalytic pervaporation process for the removal of organic contaminants is shown with alcohol being the representative for a possible organic contaminant.

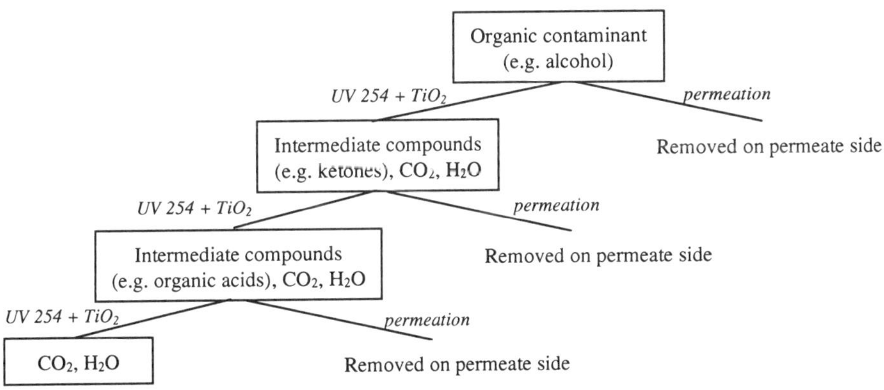

Fig. 1 Schematic pathway of catalytic pervaporation

Whereas one part of the organic contaminant, i.e. in this case alcohol, is partially or completely oxidized, another part permeates through the membrane without being oxidized and is removed on the permeate side. Partial oxidation of the organic compound leads to the formation of a large number of intermediate compounds (e.g. ketones, organic acids). These intermediate compounds are either oxidized further to yield other intermediate compounds as well as CO_2 and H_2O or permeate through the membrane to the downstream side without being oxidized or diffuse back to the feed liquid. The permeation rates of the intermediate compounds formed are usually higher than the permeation rate of the "original" organic compound itself. An exact description of the complete path of oxidation is nearly impossible.

EXPERIMENTAL SETUP

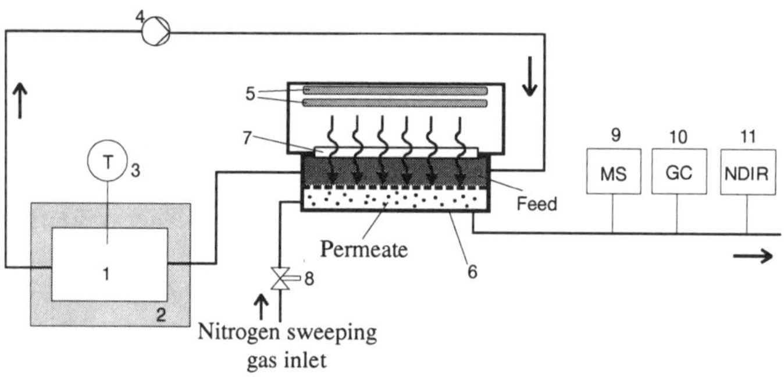

1 Feed reservoir
2 Water bath
3 Temperature control
4 Circulation pump
5 UV 254 source
6 Membrane holder
7 Quartz window
8 Flowmeter
9 Mass spectrometer
10 Gas chromatograph
11 NDIR CO_2 analyzer

Fig. 2 Experimental setup

An experimental setup as shown in Fig. 2 was provided to run and evaluate the catalytic pervaporation experiments. A commercially available disc filter holder (Millipore 316 Stainless Filter Holder, 90 mm) was modified in order to match the requirements for running catalytic pervaporation experiments. A quartz window was placed on top of the upstream compartment to allow exposure of the catalytic membranes to UV light during the catalytic pervaporation experiments. The effective surface area of the membrane holder was 68.9 cm^2.

For all experiments, nitrogen was used as sweeping gas on the permeate side to create the driving force for the pervaporation process. The use of a sweeping gas allows on-line analysis of the concentrations of the permeated compounds by means of a mass spectrometer (Dycor M 200 Quadrupole Gas Analyzer), a gas chromatograph (Varian Model 3700 Gas Chromatograph) and a NDIR (non dispersive infrared) CO_2 Gas Analyzer (Horiba VIA-510 Gas Analyzer).

EXPERIMENTAL PROCEDURE

Experiments for the catalytic pervaporation process were run using PDMS membranes containing 1 wt. % TiO_2 particles of the anatase form as photocatalyst

[6]. The membrane thickness of the fabricated membranes was 5 mil (125 µm). IPA and acetone were used as organic contaminants with feed concentrations of 360 ppm and 90 ppm, respectively. Low nitrogen sweeping gas flow rates (0.75 cc/s) were applied on the permeate side of the membrane holder in order to yield concentrations of the permeated compounds well above the detection limits of the analytical equipment used. The sweeping gas flow rate was kept constant for all experiments performed. Since CO_2 is one of the main products formed by the oxidation of organic compounds, the CO_2 concentration on the permeate side was continuously monitored and used as an indicator for photocatalytic oxidation of the organic contaminant within the membrane.

In order to yield a photocatalytic effect, the catalytic membrane was exposed to UV 254 nm. This wavelength was used since it does not have the potential to directly oxidize organic molecules, as e.g. UV 185 nm does. Thus all of the CO_2 produced during the experiments can be related to photocatalytic oxidation of organic compounds within the catalytic membrane.

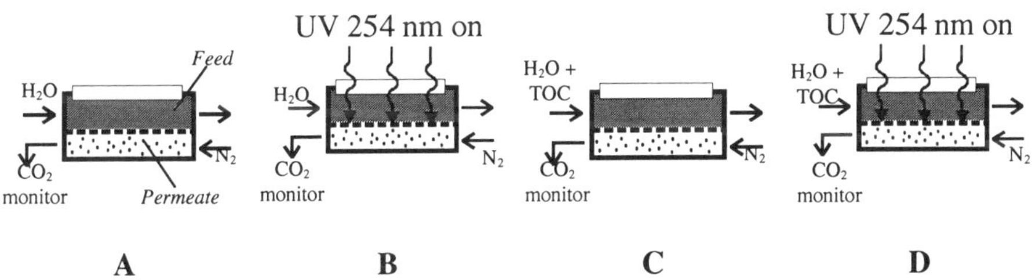

Fig. 3 Experimental procedure to evaluate the catalytic pervaporation process

All experiments were run in a procedure of four steps as shown in Fig. 3. During the first two steps, A and B, the background of the system was determined by feeding pure ultrapure water, not containing any organic model contaminant yet, into the membrane separator. In step A the UV source was off and in step B the UV source was turned on. In steps C and D an organic contaminant (TOC) was added to the ultrapure water and this mixture (H_2O + TOC) was fed into the system. Whereas in step C the UV source was off, in step D the UV source was turned on. During all process steps the concentration of CO_2 in the sweeping gas on the permeate side was monitored. Since in steps B and D the catalytic membrane was exposed to UV light, photocatalytic oxidation of organic compounds could occur within the membrane and at the feed/membrane interface. If thus an increase of the CO_2 concentration on the permeate side is detected during step B or D it can be related to photocatalytic oxidation of organic compounds.

RESULTS AND DISCUSSION

In Fig. 4 a plot of the CO_2 concentration on the permeate side as monitored by the NDIR CO_2 Analyzer is shown over time for the different steps of the catalytic pervaporation experiment. The organic model contaminant in water is IPA with a concentration of 360 ppm.

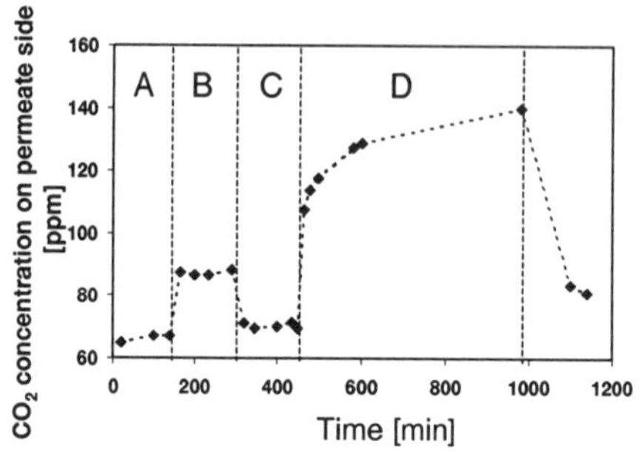

Fig. 4 Catalytic pervaporation for IPA/ water mixture (c_{IPA}=360 ppm); CO_2 concentration on permeate side determined with NDIR CO_2 analyzer

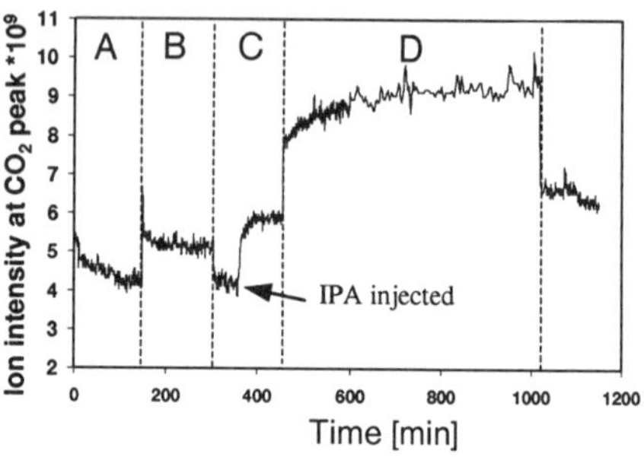

Fig. 5 Catalytic pervaporation for IPA/ water mixture (c_{IPA}=360 ppm); ion intensity at CO_2-peak on permeate side determined with mass spectrometer

During step A ultrapure water without IPA was used as feed liquid and the UV source was turned off. All of the ultrapure water which was used for the experiments was removed from an ultrapure water system. Since it was exposed to air prior to feeding it into the membrane holder, some CO_2 got dissolved in the water prior to the start of the experiment. Since part of this dissolved CO_2 is continuously removed from the feed during step A due to the high permeability of the PDMS membrane for CO_2, a certain background concentration of CO_2 (approx. 65 ppm) is monitored on the permeate side during step A.

In step B the UV source is turned on. This causes an increase of the CO_2 concentration on the permeate side. This increase is due to different reasons. One reason is that trace organic compounds were already in the experimental system as background contaminants. These "background organics" are oxidized in step B due to the photocatalytic effect of the membrane with CO_2 being a reaction product. Another reason for the increase of the CO_2 concentration on the permeate side is that the local temperature at the feed side of the membrane holder increases due to exposure to UV light. Due to this temperature increase the equilibrium concentration of CO_2 in water becomes lower. Thus CO_2 is driven out of the water on the feed side and CO_2 permeates through the membrane and causes a higher CO_2 concentration on the permeate side.

After the UV source is turned off in step C, the concentration of CO_2 decreases back to the level of step A. IPA is subsequently injected into the feed liquid leading to a final concentration of c_{IPA}=360 ppm. No increase in the CO_2 concentration is observable due to the injection of IPA as long as the UV source is turned off. But by turning the UV source on in step D a sharp increase of the CO_2 concentration on the permeate side is visible well above the CO_2 concentration level observed during step B which represented the background CO_2 concentration in the ultrapure water. Thus the difference in the CO_2 concentration on the permeate side between step B and step D can be related to actual photocatalytic oxidation of IPA inside of the membrane and at the membrane/feed interface which leads to the formation of CO_2.

During step D there is an increase of the CO_2 concentration visible over time. This is due to back-diffusion of CO_2 from the membrane to the feed. The rate of oxidation is high at the feed/membrane interface, since the concentration of IPA as well as the intensity of the UV light are high in this region. Therefore a large amount of CO_2 is released at the feed/membrane interface as reaction product and part of the CO_2 diffuses back into the feed liquid instead of permeating through the membrane. Thus the concentration of CO_2 in the feed increases continuously which leads to higher CO_2 transmembrane fluxes and therefore to a higher CO_2 concentration in the permeate.

As soon as the UV source is turned off after step D, the CO_2 concentration decreases sharply. The CO_2 concentration does not decrease back to the same level as in step C again. This is another indicator of back-diffusing CO_2 during step D. The elevated concentration of CO_2 in the feed is still present after the UV source is turned off.

Besides monitoring the CO_2 concentration on the permeate side by means of a NDIR CO_2 analyzer, the CO_2 concentration was also monitored independently by a mass spectrometer. The relative concentration of CO_2 on the permeate side, as represented by the measured ion intensity at mass-to-charge ratio 44, for the different process steps is shown in Fig. 5. The same major changes of the CO_2 concentration during the different steps can be observed as in Fig. 4. In addition, there is obviously an increase of the CO_2 concentration visible after injection of IPA. But this increase is due to the fact that there is an overlap of the main response peaks of CO_2 at mass-to-charge ratio 44 and IPA at mass to charge ratio 45. Since after injection of IPA some IPA permeates through the PDMS membrane, it is detected in the sweeping gas on the permeate side by the mass spectrometer.

In addition to monitoring the CO_2 concentration in the sweeping gas as an indicator for photocatalytic oxidation, also the IPA concentration on the permeate side was measured with a gas chromatograph. This gas chromatograph was also used to detect possible intermediate compounds formed by the oxidation of IPA as well. As a result, a decrease of the IPA peak and thus a decrease of the IPA concentration in the permeate can be detected on the permeate side during step D in comparison to step C since part of the permeating IPA was oxidized inside of the membrane. An additional peak is visible in the plot of the gas chromatograph during step D (see Fig. 6).

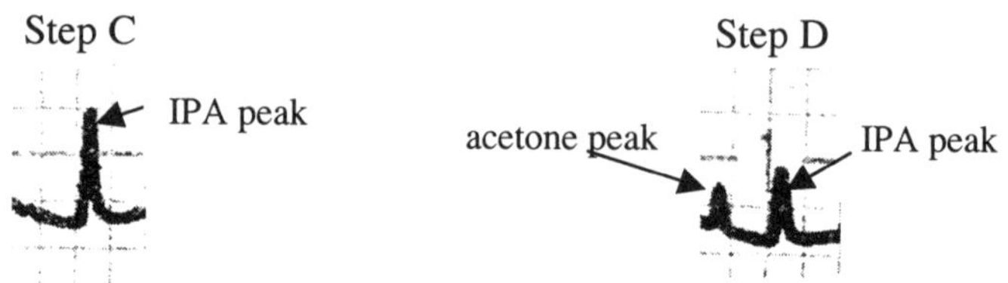

Fig. 6 Gas chromatograph recording of detected organic compounds in sweeping gas during step C and step D

This peak corresponds to acetone which is an intermediate compound formed by the oxidation of IPA. Thus during the catalytic degasification process both complete oxidation, as represented by the increase in CO_2 concentration on the permeate side, and partial oxidation of IPA, as represented by the formation of acetone, occurred. This verifies the assumption of the schematic pathway of the catalytic degasification process as shown earlier in Fig. 1.

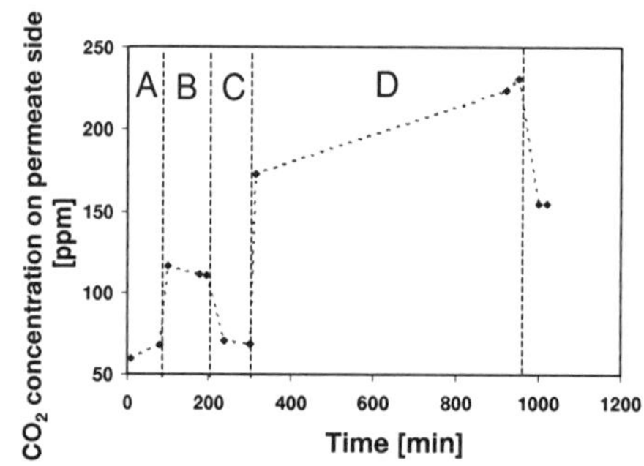

Fig. 7 Catalytic pervaporation for acetone/water mixture ($c_{Acetone}$=90 ppm); CO_2 concentration on permeate side determined with NDIR CO_2 analyzer

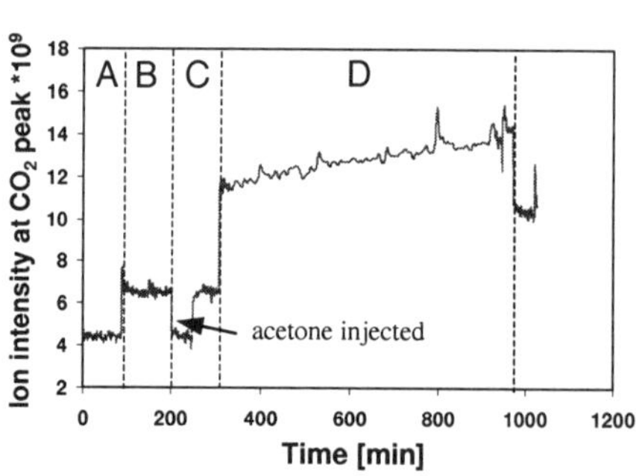

Fig. 8 Catalytic pervaporation for acetone/water mixture ($c_{Acetone}$=90 ppm); ion intensity at CO_2-peak on permeate side determined with mass spectrometer

The same experimental procedure was run using acetone as model organic contaminant ($c_{Feed\ Acetone}$ =90 ppm). The results are shown in Figs. 7 and 8. It is obvious that also acetone is oxidized by means of the photocatalytic membrane leading to a higher CO_2 concentration on the permeate side in step D in comparison to the background reading of the system during step B.

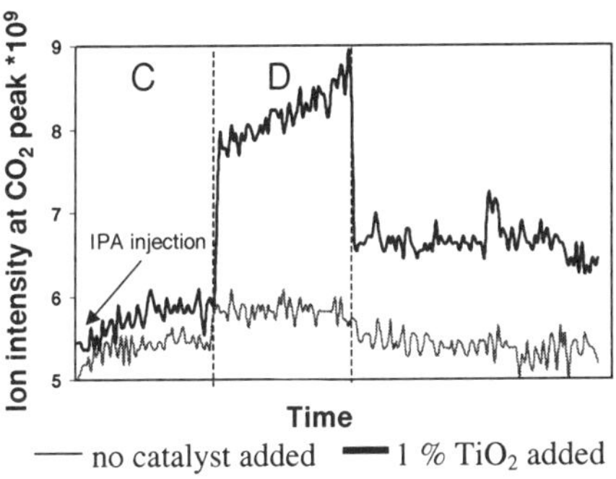

Fig. 9 Comparison using non-catalytic and catalytic PDMS membranes for IPA/water mixture (c_{IPA}=360 ppm)

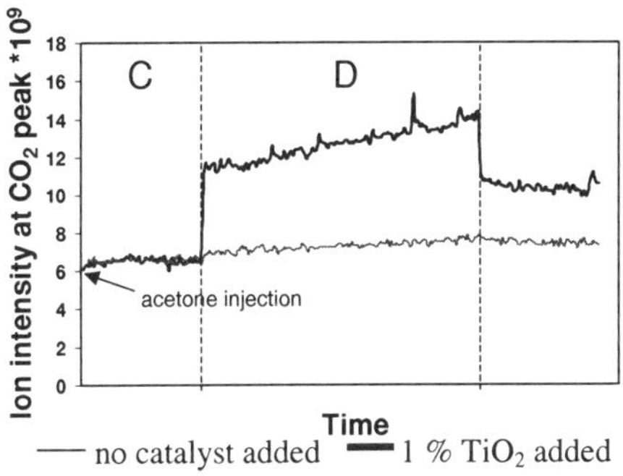

Fig. 10 using non-catalytic and catalytic PDMS membranes for acetone/water mixture ($c_{acetone}$=90 ppm)

The last set of experiments was run using a PDMS membrane of the same thickness but without the TiO_2 catalyst particles. This was done to rule out the possibility that the increase in CO_2 concentration on the permeate side was due to other effects than oxidation of the organic model compounds. A comparison of the performance of the non catalytic ("pure" PDMS) membrane and the performance of the catalytic (PDMS with TiO_2) membrane is given in Figs. 9 and 10. The relative concentration of CO_2 in the permeate as detected by the mass spectrometer is shown over time for the two different membranes. Only the last two steps, i.e. step C and D, of the experimental procedure are shown.

In Fig. 9 IPA is used as organic model compound. At the beginning of the experiment (step C) the IPA/water mixture is fed into the membrane holder with the UV source being turned off. It can be seen that the initial concentration of CO_2 in the permeate is essentially the same using the catalytic or the non-catalytic membrane. In step D the UV source is turned on. Exposure of the non-catalytic membrane to UV 254 nm does not change the CO_2 concentration in the permeate whereas exposure of the catalytic membrane causes a significant increase of the CO_2 concentration in the sweeping gas. Thus the increase of the CO_2 concentration which can be observed by using the catalytic membrane can be related to photocatalytic oxidation of IPA. The slight increase of the CO_2

concentration observable by using the non-catalytic membrane is most likely due to a local increase in temperature inside of the membrane separator due to exposure to UV light and thus the equilibrium concentration of CO_2 in the water is changed. The same comparison is shown in Fig. 10 for an acetone/water mixture.

CONCLUSIONS

The experiments performed demonstrate the feasibility of the concept of catalytic pervaporation, which combines two separate processes, selective removal of trace organic compounds and photocatalytic oxidation, in a synergistic way. It was shown that partial oxidation and removal of IPA and acetone, two highly recalcitrant organic compounds, is possible. Further research has to be done to optimize this process, e.g. with regard to membrane configuration and membrane material.

REFERENCES

[1] R.C. Binning, R.J. Lee, J.F. Jennings, E.C. Martin, "Separation of liquid mixtures by permeation," *Industrial and Engineering Chemistry*, Vol. 53 , No.1 (1961) 45-50.

[2] P. Aptel, J. Neel, Pervaporation. In *Synthetic Membranes: Science, Engineering, and Application*, ed. P.M. Bungay, H.K. Lonsdale, M.N. dePinho; D. Reidel Publishing Company, Dordrecht, Holland, 1986.

[3] R.W. Matthews, "Purification of water with near-U.V. illuminated suspensions of titanium dioxide," *Water Resources*, Vol. 24, No. 5 (1990) 653-660.

[4] A. Wold, "Photocatalytic properties of TiO_2 ," *Chemistry of Materials*, 5 (1993) 280- 283.

[5] R.W. Matthews, "Photo-oxidation of organic material in aqueous suspensions of titanium dioxide," *Water Resources*, Vol. 20, No. 5 (1986) 569-578.

[6] D. Scherer, "Removal of organic contaminants from ultrapure water by membrane processes, Master's Thesis, University of Arizona, 1996.

Coatings and Composites

STRUCTURAL AND BIOACTIVITY ANALYSIS OF POROUS TITANIA FILMS PRODUCED BY AN ELECTROCHEMICAL METHOD

Marivalda M. Pereira and Wander L. Vasconcelos
Dep. of Metallurgical and Materials Eng., UFMG
Rua Espirito Santo n° 35, Centro - Belo Horizonte, MG 30160-030 - Brazil
mpereira@demet.ufmg.br

ABSTRACT

Thick titania films can be produced on titanium substrates by an electrochemical method, through anodic oxidation of the substrate in a methanolic solution. This film can increase the stability of the substrate and may also make it bioactive, depending on the structural characteristics and surface chemistry of the oxide formed. In this work titania films were produced under a fixed current density and processing time, in methanolic solutions with different water content. The bioactivity of the films was evaluated upon immersion in simulated body fluid. The structural characteristics of the films and its bioactivity was evaluated by X-Ray Diffraction, Infrared Spectroscopy and Nitrogen Adsorption. The surface reactions in simulated body fluid was discussed in the light of the structure of the films and their surface chemistry.

INTRODUCTION

Titanium and its alloys are largely used as implant materials in orthopedics and dentistry. Among titanium alloys the Ti6Al4V has specially interested researchers due to its high fatigue resistance, an elastic modulus close to that of bone and its biocompatibility. It is believed that its biocompatibility is related to a titanium oxide layer formed on the surface of the metal, which limits metallic ion release. However, the natural oxide layer is very thin and may be destroyed with time. Recent advances have been in the direction of modifying the metal surface by the formation of a thicker oxide layer (1,2).

Besides their biocompatibility some materials present a specific behavior called bioactivity, generally considered as the ability of the material to chemically bond to bone through the formation of a hydroxyapatite (HA) layer on their surface in the physiological environment (3). The advantage of this behavior is the promotion of the implant stabilization in the short term and the permanent fixation of prosthesis.

The titanium oxide is a potential bioactive material. Damen et al (4) showed that titania particles induce the precipitation of HA from solutions containing Ca^{2+} and HPO_4^{2-} ions. Li et al (5) also showed the precipitation of HA on titania films produced by sol-gel processing. However, it is not known which characteristics of the titania make it bioactive. Possible features that control the bioactivity of oxides are the presence of a large concentration of MOH (M= Metal) on the surface and the presence of a microporosity, as demonstrated for silica (6,7). These features are typical of oxides formed by the sol-gel process.

In the case of titanium oxide, thick layers (up to 50 μm) can be produced by an electrochemical method (8,9). The author suggests that the formation of the oxide involves the direct synthesis of titanium metoxide by anodic oxidation of titanium in a methanolic electrolyte followed by its rapid hydrolysis and conversion to oxide in the presence of water. The advantages of the process are its simplicity, uniformity of the oxide layer obtained and the control over the thickness of the layer. We think that it may also be possible to control the texture of the oxide formed, if it really involves a sequence of hydrolysis and condensation reactions, similar to the sol-gel process.

In this work titania films were produced under a fixed current density and processing time, in methanolic solutions with different water content. The bioactivity of the films was evaluated upon immersion in simulated body fluid. The structural characteristics of the films and its bioactivity was evaluated by x-ray diffraction, infrared spectroscopy and nitrogen adsorption. The surface reactions in simulated body fluid were discussed in the light of the structure of the films and their surface chemistry.

EXPERIMENTAL PROCEDURE

Substrate Preparation

The substrate used was a titanium alloy Ti6Al4V. Samples in the shape of discs (14,5 mm in diameter x 1 mm thickness) were ground, polished in diamond paste, cleaned in acetone in an ultrasonic cleaner and dried.

Film Preparation and Characterization

The electrolyte solutions were prepared using absolute methanol and sodium nitrate in the concentration 10 g/l. The electrochemical process was conducted with constant current density, in a potentiostat/galvanostat (Radiometer, PGP201) in the galvanostat mode. The electrolytic cell was composed of a platinum auxiliary electrode, a calomel reference electrode and the titanium alloy as the working electrode. The titania films were obtained with current density of 10 mA/cm^2. Films were prepared from methanolic solutions with and without water addition (0.5, 1 and 2% volume). All reagents used were analytic grade.

The structure of the films obtained was analyzed by scanning electron microscopy - energy dispersive spectroscopy (SEM-EDS), x-ray diffraction (XRD) and Fourier transform infrared spectroscopy (FTIR) using the diffuse reflectance attachment. The surface area and porosity of the samples were evaluated by nitrogen adsorption.

Bioactivity Analysis

Previous studies indicate that there is a direct correlation between the *in vitro* formation of an HA layer on the surface of a material and its ability to bond to bone (3). Therefore, the HA formation was evaluated after soaking samples with and without titania films in simulated body fluid (SBF). The SBF solution was prepared by dissolving reagent grade sodium chloride, sodium bicarbonate, potassium chloride, dibasic potassium phosphate, magnesium chloride, calcium chloride and sodium sulfate in deionized water. The solution was buffered at pH 7.25 with hydrochloric acid and tris hydroxymethyl aminomethane (THAM). The ion concentration of SBF is presented in Table I. Samples were suspended in the solutions at 37°C for various periods of time, the specimens were removed from solution and dried in air. The reacted surfaces were analyzed by FTIR over the range of 400 to 1400 cm^{-1}.

Table I. Ion concentration (mM) of SBF and human blood plasma

	Na^+	K^+	Mg^{2+}	Ca^{2+}	Cl^-	HCO^{3-}	HPO_4^{2-}	SO_4^{2-}
SBF	142	5.0	1.5	2.5	147.8	4.2	1.0	0.5
Blood Plasma	142	5.0	1.5	2.5	103.0	27.0	1.0	0.5

RESULTS AND DISCUSSION

Figure 1 presents an SEM micrograph of a titania film produced with current density 10 mA/cm^2 and reaction time of 24 minutes and the SEM-EDS spectra of the substrate and film. A comparison of the EDS spectra showed a higher oxygen content in the film, indicating the formation of an oxide. The variation of the film thickness with time is presented in Table II. It should be noted that the film thickness is not homogeneous and the values in the table represent a range in thickness values.

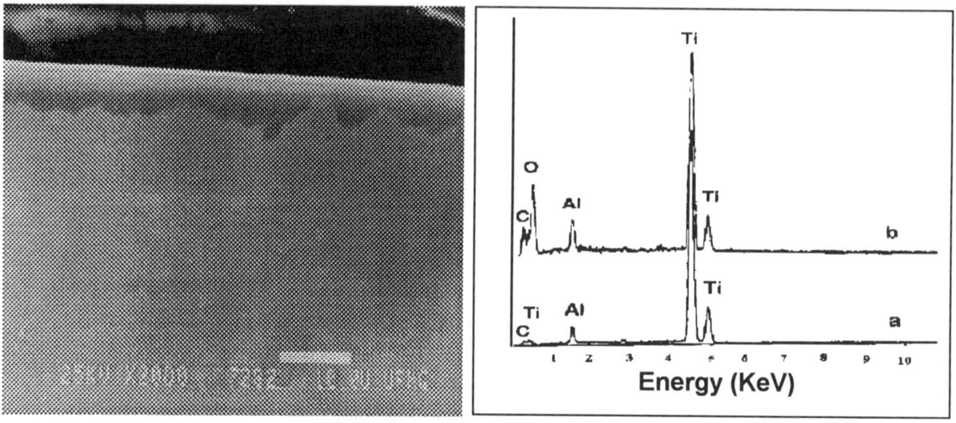

Figure 1 - SEM micrograph of the transverse section and EDS of (a) substrate and (b) film obtained with methanolic solution, current density 10 mA/cm^2; time = 24 minutes

Table II. Thickness of films as a function of time (current density of 10 mA/cm^2).

Time (minutes)	Thickness (μm)
10	0,2
24	3 to 8
32	4 to 9
35	6 to 10
40	5 to 12
60	10 to 18

The films were identified by XRD analysis. The XRD spectra of as prepared and heat treated samples are illustrated in Figure 2. The diffraction band observed in the samples submitted to the electrochemical process, that do not appear in the substrate spectra (not shown), indicates that the film produced is amorphous. When the sample is heat treated at 600°C the XRD spectrum also shows this band with lower intensity, indicating that part of the film is still amorphous. Strong peaks corresponding toTiO_2 anatase structure were observed, indicating the partial crystallization of the amorphous titania film. Peaks related to the substrate are also observed. These results are similar to the ones obtained by Shirkhanzadeh (9). In his work he showed also that, after heat treating the samples at 800°C peaks corresponding to both the rutile and anatase structures were observed. We also analyzed a sample of the substrate only heat treated at 600°C, and not submitted to the electrochemical process (not shown here). Peaks corresponding to rutile were

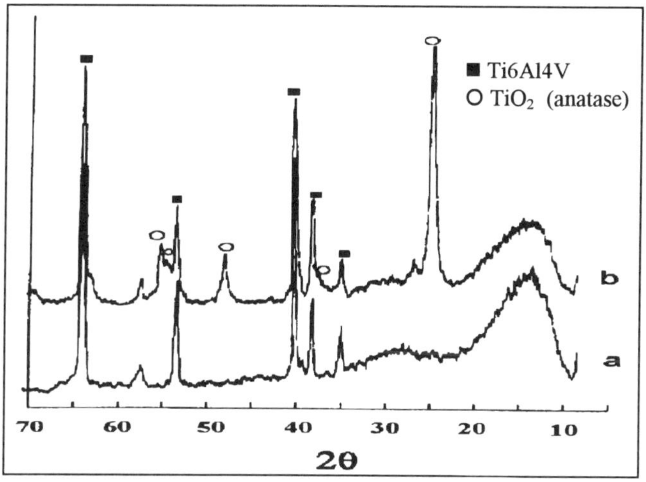

Figure 2 - X-Ray diffraction spectra of films a) as prepared (methanol, 10 mA/cm^2, 24 minutes) and b) after heat treatment at 600^{0}C for 1 hour.

observed but they were of much smaller intensity than the peaks observed in the samples submitted to the electrochemical process.

When the methanolic solution was prepared with water addition a significant difference was observed in the film structure. The film obtained without water addition was homogeneous and adherent. For 1% water addition the film presented a similar aspect. However, for 2% water addition, only scattered, non-adherent agglomerates were observed. The XRD spectrum of samples produced with 2% water addition does not show the amorphous diffraction band, being very similar to the spectrum of the pure substrate.

Table III shows the results obtained with the nitrogen adsorption analysis for films obtained with and without water addition. The calculation of surface area and pore volume was based on an estimate of the film weight, considering its thickness and the density of titania. Values for samples not submitted to the electrochemical process could not be obtained because the adsorption volume is very small, indicating a very low surface area. The results show that the films obtained present a high surface area. This high surface area is associated to a microporosity of the film, as illustrated in the pore size distribution curve shown in Figure 3. It is also observed that a small water addition increases the microporosity and the surface area of the film. The values for a silica gel with pore radius of 12 Å is presented for comparison.

The bioactivity test results are presented in Figure 4. Two absorption bands are observed at around 980 and 520 cm^{-1} for the titania film as obtained. After immersion in SBF the same bands are still apparent and the bands characteristic of hydroxyapatite formation are not observed. Samples immersed for longer times, up to 60 days, show similar behavior. Based on these results it can be said that the titania films obtained in the conditions used in this work are not bioactive, since they do not induce the formation of HA upon immersion in SBF.

It has been shown that silica and titania substrates produced by sol-gel processing induce the formation of HA (5-7). As discussed by Li et al (5,6) and Pereira et al (7), two characteristics of the substrates that make it bioactive are the presence of a large microporosity and a high degree of hydrolysis, giving origin to

Table III. Nitrogen adsorption results for samples with titania films produced in the indicated solutions (current density = 10 mA/cm^2, time = 2 hours) and of a silica-gel

Electrolyte Solution	Surface Area (m^2/g)	Pore Size (Å)	Pore Volume (cm^3/g)
Methanol	21	23	0.024
Methanol + 0.5% Water	32	10	0.016
Silica-Gel	748	11.5	0.431

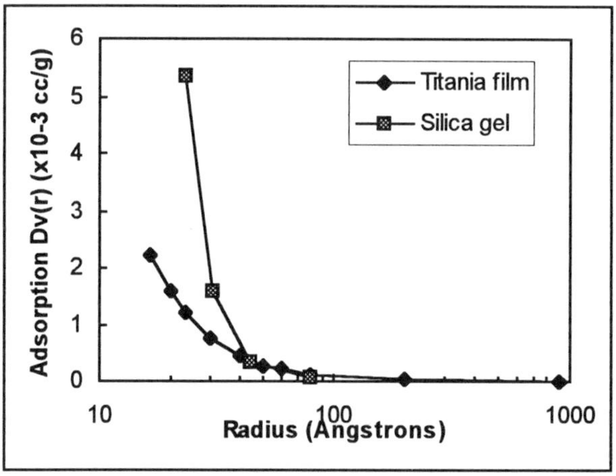

Figure 3 - Pore size distribution, obtained by nitrogen adsorption, of a sample with titania film obtained in methanol (10 mA/cm^2, 2 hours), and of a silica gel.

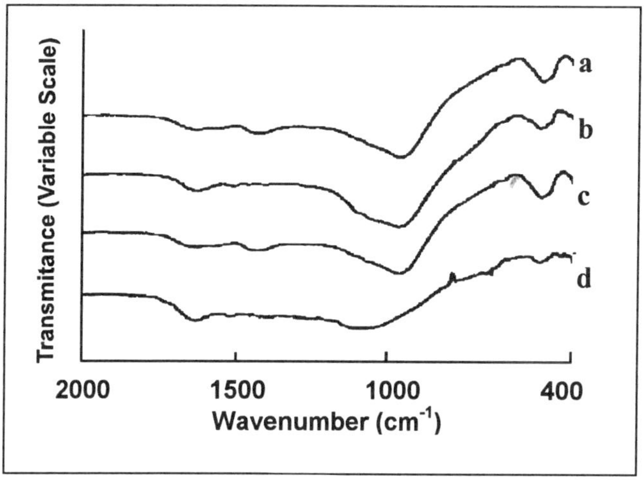

Figure 4 - FTIR spectra of samples with titania films a) as preparared (methanol, 10 mA/cm^2, 2 hours); b) sample as obtained in a and immersed in SBF for 15 days c) as preparared (methanol with 0,5% water, 10 mA/cm^2, 2 hours) d) sample as obtained in c and immersed in SBF for 15 days.

a negative surface charge. Both characteristics are present in titania and silica gels which have a large surface area, associated with the microporosity, and a large concentration of hydroxyl groups on the surface, responsible for a negative charge at the pH 7.25 (above the isoelectric point).

As pointed out before, it is suggested (8) that the reaction sequence that occurs in the electrochemical process used to obtain the titania films in this work are similar to the sol-gel reactions. Therefore, it was expected that the films obtained would be porous and have a large degree of hydrolysis on the surface, having the characteristics that would make them bioactive. In fact the films obtained are porous, as shown by the BET results. However, the surface area of the films, for the preparation conditions used in this work, are smaller than the surface area associated with silica-gel samples. It should be pointed out though, that the surface area was only estimated, since the actual film weight is not known. At this point we can not say anything about the degree of hydrolysis of the surface.

It should also be considered that for silica-gels it was discussed (7) that the pore size may also play an important role in the HA nucleation process. As the pore size decreases restricted diffusion of ions into the pores becomes a limiting factor and nucleation rate decreases. As seen on the pore size distribution curves (Figure 3) a large fraction of the pores present in the titania films are smaller than

20 Å and this may be a contributing factor to the inability of the films to induce the nucleation of HA.

Samples with an electrochemical titania film, produced at a current density of 10 mA/cm^2 and 40 minutes, were submitted to a treatment with 5M NaOH solution, at 60°C for 1 hour. After the treatment the samples were cleaned in water, dried, and immersed in SBF, at 37°C for 20 days. The samples were then analyzed by FTIR and the results are shown in Figure 5. The spectrum of the sample treated with NaOH is very similar to the untreated sample. After immersion in SBF bands are observed at 602 and 559 cm^{-1}. These two bands, assigned as bending vibrations in the [PO_4] tetrahedra (11), are characteristic of an amorphous hydroxyapatite phase formed. The formation of this layer is shown in the SEM micrograph in Figure 6. These results show that it is possible to change the characteristics of the film formed with a simple treatment in NaOH. However, it remains to be investigated what modifications occur in the film, that make it capable of inducing the formation of a hydroxyapatite layer. This modifications may be either in the pore structure and in the surface chemistry.

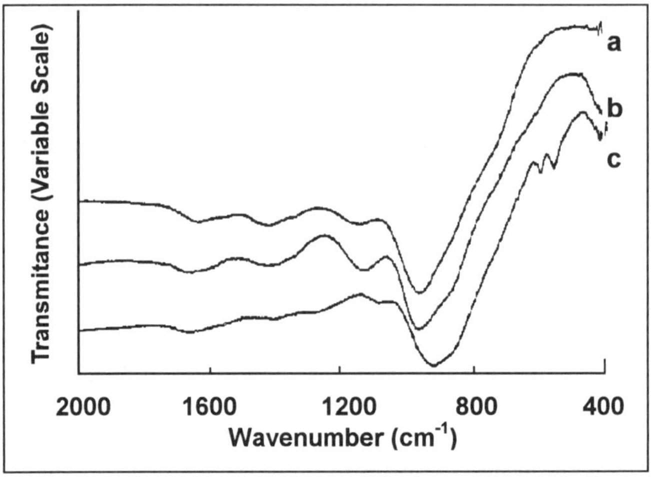

Figure 5 - FTIR spectra of eletrochemical film (methanol, 10 mA/cm^2, 40 minutes) as obtained (a), treated with NaOH 5M (b), and immersed in SBF for 20 days (c).

CONCLUSIONS

Titania films with thickness varying up to 20 μm were obtained on Ti6Al4V alloy by the electrochemical process. The films obtained were initially amorphous,

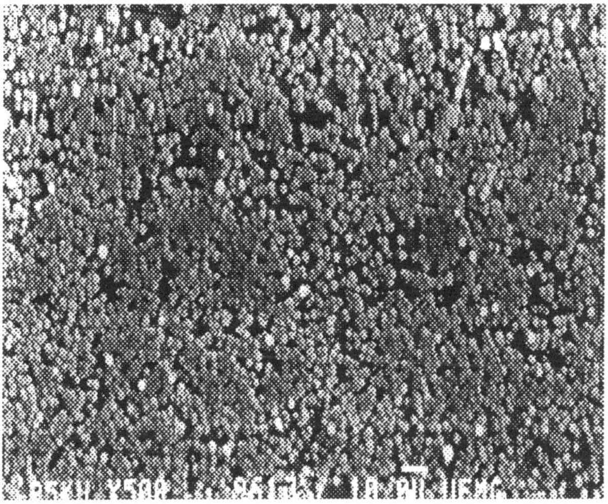

Figure 6. SEM micrograph of a sample with eletrochemical film (methanol, 10 mA/cm^2, 40 min) after treatment with NaOH (60°C, 1 hour) and immersion in SBF.

partially crystallizing to anatase upon heat treatment at 600°C. Regular and adherent films were obtained in methanolic electrolyte with water addition up to 1%. Higher water contents lead to the formation of scattered agglomerates. The titania films present a high surface area associated with a large microporosity. The films produced do not induce the formation of hydroxyapatite in SBF solution. After treatment of the films with NaOH it was observed the formation of hydroxyapatite upon immersion in SBF solution.

REFERENCES

[1]M. Browne, P.J. Gregson and R.H. West, "Characterization of Titanium Alloy Implant Surfaces with Improved Dissolution Resistance," *Journal of Materials Science: Materials in Medicine*, 7, 323-329 (1996).

[2]D.B. Haddow, P.F. James and R. Van Noort, "Characterization of Sol-gel Surfaces for Biomedical Applications," *Journal of Materials Science: Materials in Medicine*, 7, 255-260 (1996).

[3]L.L. Hench, "Bioactive Ceramics," *Annals of New York Academy of Science*, **523**, 54-71 (1988).

[4]J.J.M. Damen, J.M. Ten Cate, J.E. Ellingsen, "Induction of Calcium Phosphate Precipitation by Titanium Dioxide," *Journal of Dental Research,* **70** [10] 1346-1349 (1991).

[5]P. Li, K. de Groot and T. Kokubo, "Bonelike Hydroxyapatite Formation Initiated by Sol-gel Derived Titania Coating on a Titanium Substrate," *Journal of the American Ceramic Soc*iety, **75**, 2094-2097 (1992).

[6]P. Li, "In Vitro and In Vivo Calcium Phosphate Induction on Gel Oxides," Ph.D. Thesis, Leiden University, Leiden, Holanda, 1993.

[7]M.M. Pereira, A.E. Clark and L.L. Hench, "Effect of Texture on the Rate of Hydroxyapatite Formation on Gel-Silica Surface," *Journal of the American Ceramic Soc*iety, **78** [9] 2463-2468 (1995).

[8]M. Shirkhanzadeh, "Eletrochemical Preparation of Protective Oxide Coatings on Titanium Surgical Alloys," *Journal of Materials Science: Materials in Medicine* **3**, 322-325 (1992).

[9]M. Shirkhanzadeh, "Fabrication and Characterization of Alkoxy-derived Nanophase TiO_2 Coatings," *NanoStructured Materials*, **5**, 33-40 (1993).

[10]V.C. Farmer, "Infrared Spectra of Minerals," *Mineralogical Society*, London, 1974.

[11]B.O. Fowler, "Infrared Studies of Apatites, I. Vibrational Assignments for Calcium, Strontium, and Barium Hydroxyapatites Utilizing Isotopic Substitution.," *Inorganic Chemitry*, **13**, 194-207 (1974).

DEVELOPMENT OF UNIFORM HYDROXYAPATITE COATING ON POROUS Ti METAL BY ELECTRODEPOSITION TECHNIQUE

S. Bhaduri, S.B. Bhaduri, and J.B. Zanotti
Department of Metallurgical Engineering
University of Idaho
Moscow, ID 83844-3024.

ABSTRACT

Hydroxyapatite (HAp) is one of the most appropriate ceramic materials for prosthetic applications due to its excellent biocompatibility and bioactivity. HAp can be used as a coating material on metal implants. The idea is to combine mechanical advantages of metallic materials with the excellent biocompatibility of HAp. Thick films of HAp were deposited on porous Ti metal discs using an electrodeposition technique. An electrolyte containing $Ca(NO_3)_2$, $(NH_4)_2HPO_4$ was used for deposition with the disc acting as the anode. After annealing at 200°C, the as-deposited thick films were identified to contain HAp (and CaO) by x-ray diffraction. The coating was further characterized by SEM in order to observe the microstructure. Results showed that dense and uniform coatings of Hap (with CaO) formed, which had final thickness of up to 500 μm.

INTRODUCTION

It is well known that hydroxyapatite (HAp) is an important constituent of the bone. In the literature, many studies showed the potential benefits of HAp as a prosthetic material. However, HAp is a ceramic and hence a brittle material [1].

The fracture toughness of artificial HAp is less than 1 MPa.√m (even though human bone has fracture toughness values between 2-12 MPa.√m [2]). Strategies for utilizing HAp as a successful prosthetic material can be several. These include addition of other ceramic reinforcements to HAp, coating a biocompatible metal with HAp or making HAp/polymer composites. All of these strategies have their own niche applications.

This paper is about coating HAp coating on porous Ti substrates. The innovation here is our serendipitous discovery for a simple process for making porous Ti, which itself is a prosthetic material [3]. In order to work as a successful biomaterial, HAp coating should be approximately 40-200 μm thick, should be porous and should not contain impurities. The first requirement is to resist resorbability of HAp, the second requirement is to facilitate bone growth and the third one is related to degradation. In the literature there are many references for coating metallic substrates. These include plasma spraying [4-6], sol-gel [7], flame spraying [8], MOCVD [9], magnetron sputtering [10], laser ablation [11], bio mimicking [12,13] and electrodeposition [14-20]. Plasma spraying is the most common process for producing HAp coatings. Plasma spraying has many advantages as well as disadvantages. Since plasma spraying produces high temperature, HAp particles melt and form strong bonds to the substrate. It is also a fast process for building up layers. For spraying HAp no protective atmosphere is needed. However, prolonged high temperature may alter the microstructure of the substrate. Plasma spraying being a line-of-sight process, coating does not penetrate the pores. Yet another drawback with plasma spraying is that there may be amorphous phases present due to the non-equilibrium nature of the technique. In particular, the amorphous phases tend to resorb somewhat early. Therefore, other techniques need to be examined. Flame spraying generates higher particle velocities and lower temperatures using a high velocity oxy-fuel (HVOF). This allows coating of complex contours. Sol-gel, magnetron sputtering and laser ablation are slow processes for building up the requisite layers of suitable thickness. Both bio mimicking and electrodeposition have been reported in the literature and are relatively easier to use for thick coatings. The electrodeposition technique is simpler than the bio-mimicking process in that a glass of specific composition is needed in the latter. Therefore, an electrodeposition technique was used in the present study. Broadly speaking, the electrodeposition technique can be further classified into electrophoretic deposition and electrocrystallization. In the first case, the crystalline HAp powders are dispersed in a proper liquid medium. The substrate to be deposited is connected to an electric potential to create an electric field. This causes electrophoresis to occur and HAp is deposited on the substrate [18]. After deposition, the as-coated films need sintering. As opposed to electrophoresis, the electrocrystallization technique deposits the amorphous or weakly crystalline coatings from a solution [15-17, 19,20]. These coatings require a low temperature calcination process to crystallize the HAp. This not only cuts down a processing step but also has enhanced capability to coat porous substrates without clogging the pores. Since the emphasis here is to coat a porous substrate, the electrocrystallization technique was chosen.

EXPERIMENTAL PROCEDURE

The porous substrates were prepared by pressing 6mm discs of titanium powder (-325 mesh, Alfa Aesar, Ward Hill, MA). The substrates were about 80% of the theoretical density with average pore size of 50 μm. This allows for excellent handling strength and the penetration of the coating into the pores and cavities. The electrolyte used for the fabricating of coatings was made by mixing 0.042M $Ca(NO_3)_2$ $4H_2O$ (99.6% pure, Fisher chemical, Fair lawn, NJ) and 0.025M $NH_4H_2PO_4$ (98% pure, Alfa-Aesar, Ward Hill, MA) solutions. The solution was prepared by mixing 50 % ethanol with 50 % deionized water. The coating process was carried out in a 1L beaker at 69°F, using the porous titanium discs as the anode, and a stainless steel cathode (approximately 20 cm in length and 1mm in diameter). A magnetic stirrer was used throughout the coating process so that no precipitate can deposit at the bottom of the beaker. An Elanco precision (model XP-650) dual variable power supply was used to control the voltage and current. Voltages between 15 and 25 V, and currents of 0.5-1.5 amps were applied. The deposition of HA precursor electrolytes was carried out for 20 minutes, and resulted in thick coatings. The coated discs were then allowed to air dry over night and calcined at 100°C to 400°C. The calcination stage is the most crucial in the entire procedure. After several trial runs temperature was optimized 200°C where no oxidation of Ti occurred, with the predominant phase identified as HAp.

RESULTS AND DISCUSSIONS

Figure 1 represents the x-ray diffraction pattern of the as heat-treated coating. Nearly the entire pattern can be attributed to the presence of HAp [21]. The only other peak is due to the presence of CaO. Because of the very low heat-treatment temperature, the peaks were broad. This indicates the presence of fine crystallites, probably of the order of nanometers. This is similar to the naturally occurring HAp in bones and teeth. The presence of CaO indicated the Ca/P ratio is greater than 1.67. It is also not clear if any other phases are present due to the broad peaks of the pattern. A higher temperature heat-treatment would have produced sharper peaks but would have oxidized the Ti substrate.

The presence of other phases in electrodeposited coating is quite common. Refs 19 and 20 report the formation of Brushite as the other phase present. Ref. 18 reports the presence of α and β-tricalcium phosphate (TCP). The presence of TCPs enhances the resorbability, which enhances delamination. This also means that an optimum assembly of phases by the electrodeposition technique has not been found yet and our future goals will be directed towards achieving that.

Figure 2 shows elemental maps of Ca and P taken with a scanning electron microscope to identify the extent of coating deposition. The coating appears brighter than the substrate. Figure 3 shows a side view of the coated substrate, which formed in about an hour. As seen in the figure, the coating thickness is about 435 μm. In comparison, Chen et al. [20] reported a coating thickness of 800 μm during a deposition time of 3 hours. Thus the current deposition rate was faster than theirs. The faster deposition rate is believed to be due to the synergism between addition of alcohol and the presence of interconnected pores. Chen at al. [20] argued that the coating formation takes place via precipitation from the solution followed by absorption. The latter of the two is the rate controlling step. They argued alcohol addition improved the absorption. The presence of interconnected pores improved further penetration of the coating. A planar view of the dense coatings along with porous feature of the substrate is clearly seen in Figure 4. Pores are visible because the coating can penetrate them without clogging them up. This is the factual evidence for a faster coating rate.

CONCLUSIONS

The paper describes an electrodeposition technique for coating porous Ti substrates for possible prosthetic applications. The coatings were heat-treated at 200°C for 2 hours to crystallize HAp and CaO. This indicated that the coatings were off-stoichiometry (Ca/P ratio greater than 1.67). Previous reports indicated assemblages of different phases and work needs to be done to optimize their presence. Due to the low heat-treatment temperature, the coatings consisted of fine crystallites. The presence of macro-pores in the substrate enabled the coatings to penetrate porous Ti without clogging the pores and relatively thick coatings of up to 435 μm could be deposited.

ACKNOWLEDGEMENT

This work is supported by a seed grant from the University of Idaho.

REFERENCES

1. A. Posner, A. Perloff, and A.F. Dionio, Acta Cryst., 11 (1958) 308.
2. T. Nonami and F. Wakai, J. Ceram Soc. Japan., 103 (1995) 639.
3. S.B. Bhaduri and S. Bhaduri, Invention disclosure.
4. G.L. de Lange and K. Donath, Biomater., 10 (1989) 121.
5. W. Tong, J. Chen, and X-D. Zhang, Biomater., 16 (1995) 829.
6. J.C. Knowles, K. Gross, C.C. Berndt, and W. Bonfield, Biomater., 17 (1996) 639.
7. S.W. Russel, K.A. Luptak, C.T. A. Suchicital, T.A. Alford, and V.B. Pizziconi, J. Am. Ceram. Soc., 79 (1996) 837.
8. J.D. Haman, L.C. Luca, and D. Crawmer, Biomater., 16 (1995) 229.
9. G. Spoto, E. Ciliberto, and G.C. Allen, J. Mater. Chem, 4 (1994) 1849.
10. J.G.C. Wolke, K.van Dijk, H.G. Schaekenk, K. de Groot, and J.A. Jassen, J. Biomed. Mater. Sci., 28 (1994) 1477.
11. S. Sardin, M. Varela, and J.L. Morenza, pp. 225 in Hydroxyapatite and Related Compounds, edited by P.W. Brown and B. Constanz, CRC Press, Cleveland, OH and Boca Raton, FL (1994).
12. M.M. Pereira, A.E. Clark, and L.L. Hench, J. Am. Ceram Soc., 78 (1995) 2463.

13. K. Hata, T. Kokubo, T. Nakamura, and T. Yamamuro, J. Am. Ceram. Soc., 78 (1995) 1049.
14. J. Redepenning and J.P. McIsaac, Chem. Mater., 2 (1190) 625.
15. M. Shirkhanzadeh, J. Mater. Sci. Lett., 10 (1991) 1415.
16. M. Shirkhanzadeh, J. Mater. Sci. Lett., 12 (1993) 16.
17. M. Shirkhanzadeh, M. Azadegan, V. Stack, and S. Scherer, Mater. Lett., 18 (1994) 211.
18. X. Ding, K. Yamashita, and T. Umegaki, J. Ceram. Soc. Japan, 103 (1995) 854.
19. J. Redepenning, T. Schlessenger, S. Burnham, L. Lippiello, and J. Miyano, J. Biomed. Mater. Res., 30 (1996).
20. J-S. Chen, H-Y. Juang, and M-H. Hon, J. Mater. Sci. Mater. Med., 9 (1998) 297.
21. JCPDS file no. 9-0432.

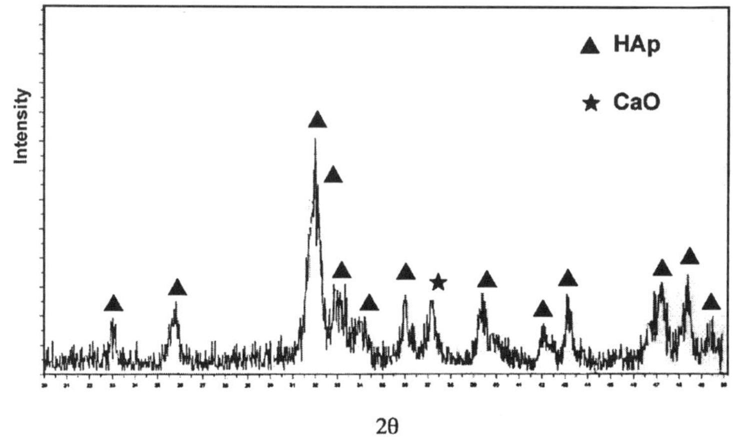

Figure 1. X ray diffraction of as heat-treated HAp coating.

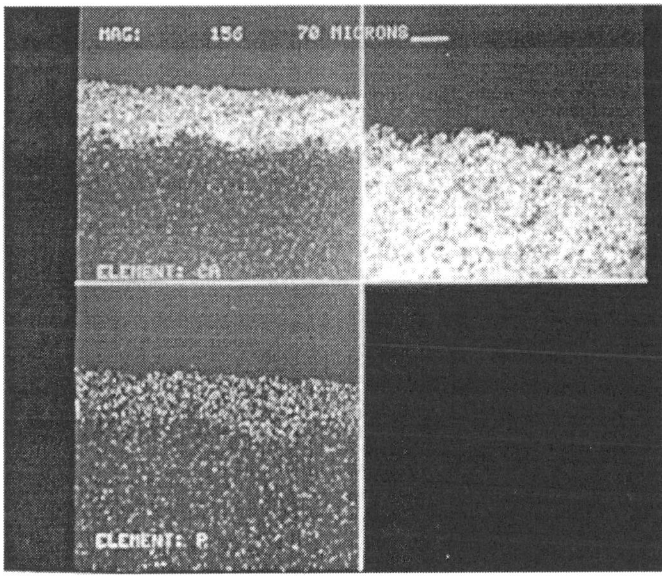

Figure 2. Elemental maps of Ca and P within the coating.

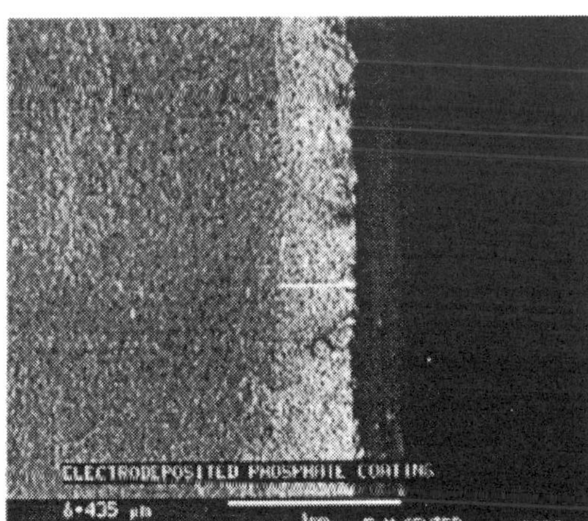

Figure 3. A low magnification side view of the coating.

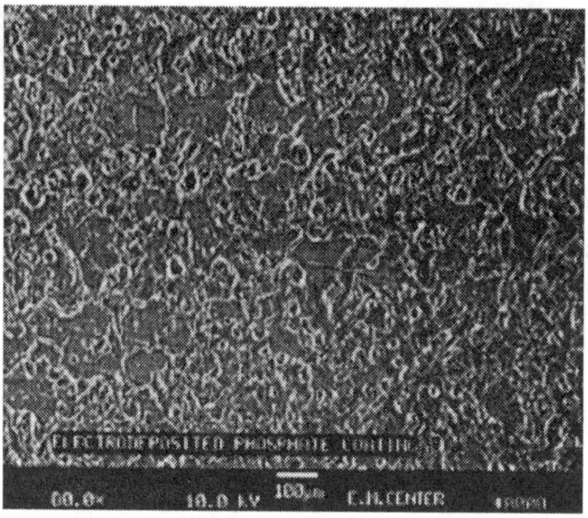

Figure 4. A low magnification planar map of the coating shows the porous substrate underneath.

MECHANICAL PROPERTY CHARACTERIZATION OF 45s5 BIOGLASS®/PLG COMPOSITES

G. DiPasquale* and J.J. Mecholsky, Jr., University of Florida, Department of Materials Science & Engineering, P.O. Box 116400, Gainesville, FL 32611; N. Leatherbury, OsteoBiologics, Inc., University Business Park, 12500 Network, Suite 112, San Antonio, TX 78249; D.C. Greenspan, USBiomaterials Corp., One Progress Blvd., Box 23, Alachua, FL 32615.

ABSTRACT

Bioglass® particulate has been documented to form a hydroxy-carbonate apatite (HCA) layer within 24 hours of exposure to a buffer solution.[1] This layer facilitates the bond formation between the implant and surrounding tissue. Polylactide-co-glycolide (75/25 wt.%) is a biodegradable polymer that is being developed for orthopedic fracture fixation. For this study, three-point flexure was used to determine the strength and fracture toughness for each composite group. The 45s5 Bioglass®/PLG composites containing 20, 40, and 60 vol.% Bioglass® particulate in a PLG matrix were tested to failure in 3-pt bending. The composites containing 20 vol.% and 40 vol.% Bioglass® particulate prereacted for 20 hours in tris buffer produced the highest average failure strength of 71±23 MPa and 66±9 MPa, respectively. Direct critical flaw size measurement of the fracture surfaces showed that the composites containing 20 vol.% and 40 vol.% Bioglass® particulate prereacted for 20 hours in tris buffer produced the highest average K_{IC} of 2±0.6 MPa$\sqrt{m}$ and 2.1±0.4 MPa$\sqrt{m}$, respectively. Weibull statistical analysis for the strength distributions showed that the composites containing 40 vol.% Bioglass™ that was prereacted 0 hours and 20 hours produced the best Weibull modulus values, which were 9 and 7, respectively. Further advances in this field should result in the development of tougher degradable fixation devices.

INTRODUCTION

The 45s5 Bioglass®-75/25 PLG composite system is under consideration as a viable alternative to traditional metal implants used for structural applications. The ceramic component is 53-38 μm 45s5 Bioglass® particulate and

and the matrix material is PLG, which is a thermoplastic copolymer comprised of 75% polylactic acid (PLA) and 25% polyglycolic acid (PGA). The components devising this ceramic-polymer composite offer several advantages compared to a traditional metallic implant.

Currently, applications for bioactive glasses and ceramics include radiation therapy, hearing restoration, and dental/maxillofacial implants, all of which are non-load bearing applications. The use of these materials can be greatly extended if they can be made to be load bearing. Bioactive ceramics are preferred over their metal counterparts because they are much more corrosion resistant. Unlike metals, the 45s5 Bioglass® particulate in this study does not release potentially harmful heavy metal ions into surrounding tissue.[2] The PLG matrix is advantageous because as it degrades in the body, its by-products are lactic acid, carbon dioxide, and water. They are removed through normal respiratory functions of the body and usually do not adversely affect surrounding tissue. This is of particular interest because when combined with a bioactive ceramic, such as Bioglass®, the PLG allows sufficient time for the ceramic to bond with the surrounding tissue before it completely degrades.

Orthopedic clinical applications of this copolymer include fracture fixation for malleolar fractures of the ankle, elbow intra-articular fractures, plates, and screws.[3] Suture material, such as Vicryl®, is another application of the PLG copolymer (90% PGA/10% PLA).[4] The primary reason that the commercialization of PLG devices for fracture fixation has been limited is due to inferior mechanical properties compared to its metal and long bone counterparts.[5] In previous studies, composites involving the PLG copolymer have been reinforced with carbon, PLA, or PGA fibers.[6]

The development of specifically designed materials for use in the body, which will be bioactive or resorbable, will be critical in solving future problems. Most, if not all, of the bioactive materials developed to date are bioactively superior to inert materials but structurally inferior. Thus, the objective of this research is to measure the mechanical properties of a toughened, load-bearing, bioactive/biodegradable composite of ductile matrix material (PLG) with a bioactive ceramic- reinforcing phase (45s5 Bioglass®).

EXPERIMENTAL METHODS

The composites used for this study contained 53-38 μm 45s5 Bioglass® particulate (provided by USBiomaterials Corporation, Alachua, Florida) in a 75/25 PLG matrix. The composite material was fabricated in a compression mold and then cut into 1mm x 2 mm x 25 mm test bars (OsteoBiologics, Inc., San Antonio, TX). The particulate used for this study was classified into three different bioactivity categories before fabrication into the composite: 0 hours reacted in tris-hydroxy methyl amino methane (tris) buffer, 1 hour reacted in tris buffer, and 20 hours reacted in tris buffer, respectively.[7] Each of these particulate categories was then incorporated into the PLG matrix in 20 vol.%, 40 vol.%, and 60 vol.% amounts, respectively. Table I shows the complete matrix of test specimens used for this study.

Table I: Test Specimen Matrix

TIME PREREACTED IN TRIS BUFFER	20 VOL % BIOGLASS® PARTICULATE	40 VOL % BIOGLASS® PARTICULATE	60 VOL % BIOGLASS® PARTICULATE
0 HRS	n = 26	n = 28	n = 21
1 HR	n = 29	n = 30	N/A
20 HRS	n = 30	n = 30	n = 27

Note: The 'n' denotes test bars that were fabricated and tested using these solids loading conditions.

The flexure strength was determined using an Instron strength testing machine fitted with a three-point bend fixture, which was chosen due to its simplicity. The average thickness of each specimen was 1 mm and the average width of each specimen was 2 mm. The lower support span was 18 mm and the crosshead speed was at a rate of 5 mm/minute.

The fracture toughness was calculated from the measured fracture stress and flaw size. The failed specimens from the flexure testing were placed under a light microscope for measurement of the critical flaw size, c. Equation 1 was used to calculate the corresponding fracture toughness value, K_{IC}, for each specimen (units are in MPa√m).[8] The geometric factor, Y, was assumed to be 1.4 for a corner "quarter" crack and 1.24 for a semi-elliptical shaped flaw at the surface.

$$K_{IC} = Y\sigma_f\sqrt{c} \qquad (1)$$

Ceramic strength can vary over wide ranges and Weibull analysis is often used in studying the statistics of strength. The Weibull modulus, *m*, is a measure of the variation of data, which is determined by the distribution of strengths after a set of identical specimens has been broken. It is based on each sample in a test group containing many flaws of varying severity. The severity of the "worst" flaw varies among the samples and each of these samples fails from its "worst" flaw in the highest stressed (tensile) region. The distribution of failure stresses for different samples indicates how the severities of the "worst" flaw differ.

RESULTS AND DISCUSSION

Table II shows the average strength and crosshead displacement values for each test group. The sample groups that had the best mean flexural strength were the composites that contained 20 and 40 vol.% Bioglass® particulate which was prereacted for 20 hours. We hypothesize that the formation of the hydroxycarbonate apatite (HCA) layer on the surface of the particulate during pre-reaction acts to bond the glass and PLG. This increased bonding makes the overall composite stronger.

The 40 vol.% composites, with particulate prereacted 1hour, produced a lower than expected strength value (26.3 ± 21.6), most likely due to many large flaws present in the samples. Figure 1 (a) through (c) show the difference among the surfaces of each sample group. Compared to the 20 vol.% (a) and 60 vol.% (c) surfaces, the surfaces of the 40 vol.% composites with particulate prereacted 1 hour (b), are not smooth. Areas of agglomerated glass (white) show that the glass was not homogeneously dispersed in the PLG during processing. Composite 1 in Figure 1 (b) shows a region missing toward the upper edge of the sample. These preexisting defects would be expected to produce lower strength samples (40 vol.%-prereacted 1 hour). The three composites shown in this micrograph are typical representations of the entire group of samples for this testing condition.

Table II: Average 3-point Flexural Strength and Crosshead Displacement Data

Sample Group	n	Cross-head Displacement (mm)	Average Load (Kg)	Average Failure Strength (MPa)
20%-0 hours	26	1.1 ± 0.4	0.4 ± 0.1	52.3 ± 15.1
40%-0 hours	28	0.6 ± 0.2	0.4 ± 0.1	59.8 ± 7.2
60%-0 hours	21	0.3 ± 0.1	0.1 ± 0.1	14.2 ± 5.1
20%-1 hour	29	1.9 ± 0.4	0.5 ± 0.1	66.0 ± 15.1
40%-1 hour	30	0.8 ± 0.2	0.4 ± 0.1	26.3 ± 21.6
20%-20 hours	28	1.6 ± 0.5	0.4 ± 0.2	70.7 ± 23.4
40%-20 hours	30	0.8 ± 0.3	0.5 ± 0.0	65.8 ± 8.8
60%-20 hours	24	0.4 ± 0.2	0.3 ± 0.1	33.4 ± 11.1

Using the SEM, micrographs were taken of the fracture surfaces for samples, which exhibited high and low fracture strengths, for each sample group. Figure 2 (a) shows the fracture surface of a 40 vol.% composite with particulate prereacted 1 hour that had a high fracture strength (84.6 MPa) relative to the strengths of other composites in the same group. Compared to its counterpart with a low fracture strength (15.6 MPa), shown in Figure 2 (b), it can be seen that

(a) does not have large voids throughout the volume of the sample. The corresponding average fracture toughness values for this sample group support this reasoning because it follows the general fracture toughness trend (Table III), which was increasing as prereaction time increased.

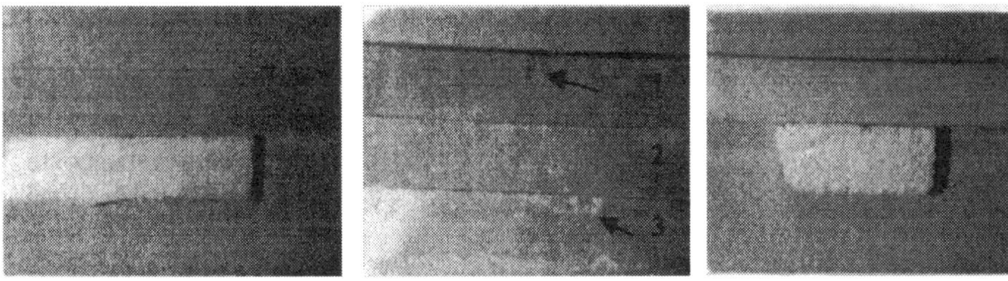

(a) (b) (c)

Figure 1: A comparison between the surfaces of the (a) 20 vol.%, (b) 40 vol.%, and (c) 60 vol.% composites after fabrication. The samples in (b) have more visible and large defects than either (a) or (c).

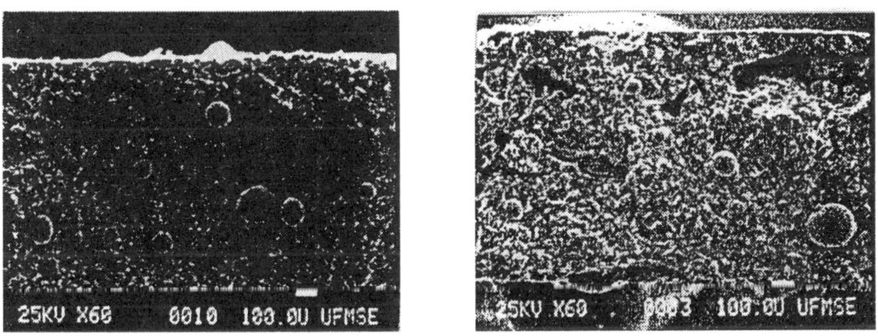

(a) (b)

Figure 2: SEM micrographs of 40 vol.%-prereacted 1 hour composites that exhibited (a) high fracture strength (84.6 MPa) and (b) low fracture strength (15.6 MPa). (Large "spherical" bubbles are due to replication anomalies and are not part of the microstructure or fracture process.)

As expected, the samples that contained 60 vol.% Bioglass® particulate prereacted 0 and 20 hours produced the lowest flexural strength values (14.2 ± 5.1 and 30.8 ± 17.2, respectively). This was most likely due to the high volume of glass particulate, which inhibited the formation of an interconnected polymer network that could provide sufficient bonding among the glass particles. Overall, the strength values could be further increased if better handling and processing procedures are followed. For example, flaws are created when the sample is pulled out of its mold because some material is pulled out of the sample due to sticking. Flaws are also introduced into the composites when they are cut into their final shape and they can be reduced if the cutting blade is properly dressed between cuttings.

The fracture surfaces of the failed specimens were observed under a light microscope and the crack sizes were measured. Figure 3 shows the typical critical flaws, (a) semi-elliptical at the surface and (b) corner, which caused failure in the composites. The average crack size, c, for each composite group is shown in Table III. As the prereaction time for the particulate increased, the measured flaw size increased. The reason for the increased flaw size is unknown. In general, the smallest crack sizes were measured in the sample groups containing 20 vol.% particulate. This can be attributed to the polymer matrix acting to shield the advancing crack tip; thus, minimizing crack size. The largest critical flaw size (841 μm $\pm$ 39 μm) was measured in the 60 vol.% composites containing particulate prereacted for 20 hours.

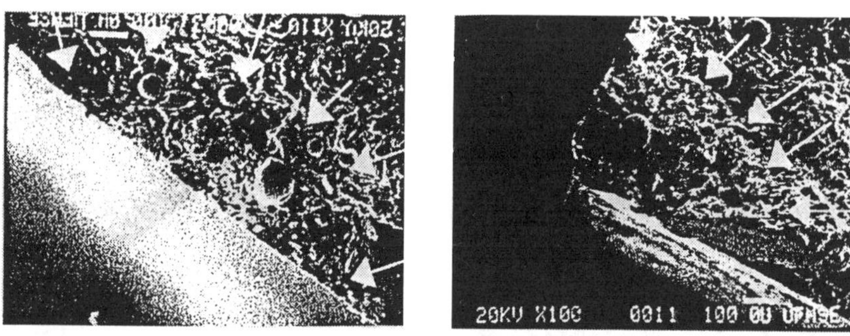

(a) (b)

Figure 3: SEM micrographs showing typical critical flaws found in failed specimens. (a) Semi-elliptical geometry at surface and (b) corner flaw. (Large "spherical" bubbles are due to replication anomalies and are not part of the microstructure or fracture process.)

As the prereaction time of the particulate increased, the K_{IC} values increased for all of the sample groups. Again, this result can be supported by the formation of the HCA layer on the surface of the glass particulate. The bond that was formed with the PLG matrix acted to toughen the overall composite since it required more energy to break than the weak interface between unreacted particulate in the PLG. The 20 vol.% and 40 vol.% sample groups that contained

Bioglass® prereacted 20 hours produced the best average K_{IC} values of 2.0 ± 0.6 MPa√m and 2.1 ± 0.4 MPa√m, respectively. The K_{IC} value of a material is directly related to its microstructure, which is seen in Figure 2. Due to the very porous structure in (b), the K_{IC} decreased drastically from 2.2 MPa√m in the particular specimen showed in Figure 2(a) to 0.33 MPa√m for the particular specimen shown in Figure 2(b). As expected, K_{IC} deteriorated for the 60 vol.% Bioglass® composites because the majority of the composition was glass; therefore, the samples were very brittle.

Table III: Average Crack Size and K_{IC} Determined by Direct Measurement

Sample Group	n	Average Crack Size (μm)	Average K_{IC} (MPa√m)
20%-0 hours	21	209 ± 77	1.0 ± 0.2
40%-0 hours	28	210 ± 76	1.1 ± 0.2
60%-0 hours	21	160 ± 45	0.2 ± 0.1
20%-1 hour	18	278 ± 79	1.4 ± 0.3
40%-1 hour	23	367 ± 120	1.3 ± 0.6
20%-20 hours	17	567 ± 147	2.0 ± 0.6
40%-20 hours	12	688 ± 173	2.1 ± 0.4
60%-20 hours	11	841 ± 39	1.2 ± 0.4

Weibull statistical analysis for strength and crack size distributions was done on the composites that were flexural tested. Figure 4 shows a comparison of the Weibull modulus for distribution of strengths, m, among all of the composite groups in this study. The composites containing 40 vol.% Bioglass® that were prereacted 0 hours and 20 hours produced the best modulus values, 9 and 7, respectively. Although the 20-vol.% Bioglass® composites that contained particulate prereacted 20 hours had a high average flexural strength, the Weibull modulus was low (2). These low values indicate that there was a large variation of strength values within each test group. As expected, the 40 vol.% Bioglass® composites that contained particulate prereacted 1 hour had the lowest Weibull modulus, 2. This was most likely caused by the inhomogeneous microstructure and presence of flaws on the surface of the composites in that group. Overall, the Weibull modulus values for strength distributions were low and could be enhanced if better processing techniques were implemented.

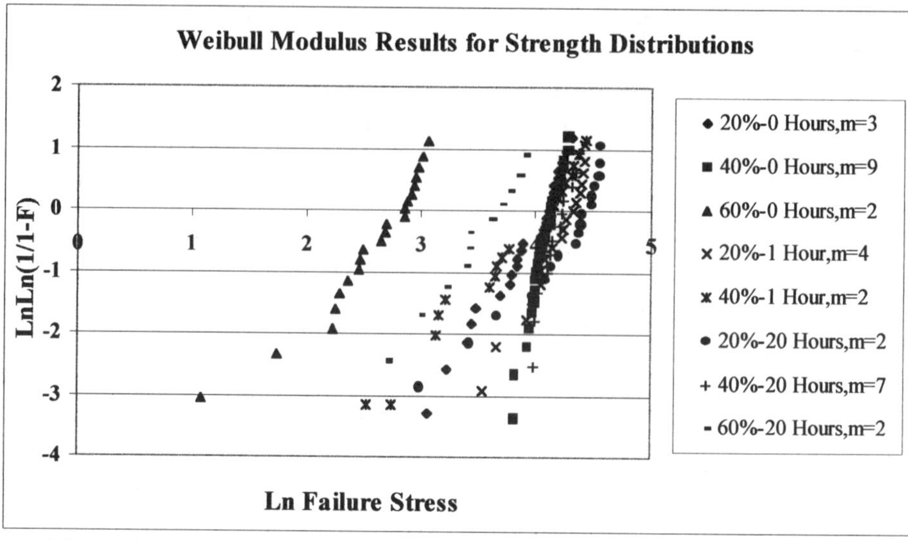

Figure 4: Weibull plot for strength distributions of all composite groups.

CONCLUSIONS

The mechanical properties of the 45s5 Bioglass®-75/25 PLG composite system were investigated. Fracture toughness was calculated from flaw size and most critical flaws emanated from the surface and corners. Generally, the smallest critical flaw sizes were measured in the composites containing 20 vol.% Bioglass® particulate. The composites that contained 20 vol.% and 40 vol.% Bioglass® particulate that was prereacted 20 hours produced the highest average K_{IC} values of 2.0 ± 0.6 MPa√m and 2.1 ± 0.4 MPa√m, respectively. There was a direct correlation between increased particulate prereaction time and increased K_{IC} values. Generally, the K_{IC} of the 20 and 40 vol.% Bioglass® composites containing particulate prereacted 20 hours was 3 times higher than plain Bioglass®. The composites that contained 20 vol% and 40 vol.% Bioglass® particulate that was prereacted 20 hours produced the highest average flexural strength values of 70.7 ± 23.4 MPa and 65.8 ± 8.8 MPa, respectively. The group that produced the poorest average flexural strength (26.3 ± 21.6 MPa) contained 40 vol.% Bioglass® particulate that was prereacted 1 hour. This was attributed to inhomogeneous microstructure as well as fabrication-induced flaws on the surfaces of the samples as opposed to any intrinsic inferiority of composition or processing.

Weibull statistical analysis for strength distributions showed that the composites containing 40-vol.% Bioglass® that was prereacted 0 hours and 20 hours produced the highest Weibull modulus values, 9 and 7, respectively. As expected after analysis of the flexural strength results, the 40 vol.% composites that contained Bioglass® particulate prereacted 1 hour, produced the lowest Weibull modulus of 2. Although the composites that contained particulate prereacted 20 hours produced the highest flexural strength and toughness, they had a low modulus (2), which indicated that there is a large variation in strength values. Overall, compared to plain Bioglass®, PLG added to 45s5 Bioglass® particulate enhanced the flexural strength and K_{IC} values. The reliability of this material would be greatly enhanced if better processing techniques were employed.

REFERENCES

[1] Hench, L.L., Ö.H. Andersson, and G.P. LaTorre, "The Kinetics of Bioactive Ceramics Part III: Surface Reactions for Bioactive Glasses Compared with an Inactive Glass," Bioceramics, Volume 4, Butterworth-Heinemann, Ltd., London, 155-162, 1991.

[2] Ibid.

[3] Böstman, O., "Current Concepts Review: Absorbable Implants for Fixation of Fractures," J. Bone and Joint Sur. 73-A (1), 148-153, 1991.

[4] Nakamura, T., S. Hitomi, S. Waranabe, Y. Shimizu, K. Jamshidi, S.H. Hyon, and Y. Ikada, "Bioabsorption of Polylactides With Different Molecular Properties," J. Biomed. Mater. Res. 23, 1115-1130, 1989.

[5] Athanasiou, K.A., G.G. Niederauer, C.M. Agrawal, and A.S. Landsman, "Applications of Biodegradable Lactides and Glycolides in Podiatry," Clin. Podiatric Medic. Surg. 12 (3), 475-495, 1995.

[6] Christel, P., F. Chabot, J.L. Leray, C. Morin, and M. Vert, "Biodegradable Composites for Internal Fixation," Biomaterials 1980, John Wiley & Sons, New York, 271-280, 1982.

[7] Gomori, G., "Preparation of Buffers for Use in Enzyme Studies," Methods in Enzymology, Volume I, Academic Press, New York, 138-146, 1955.

[8] Tsai, Y.L. and J.J. Mecholsky Jr., "Fracture mechanics description of fracture mirror formation in single crystals,"Int. J. Fract. 57, 169, 1992.

MULTILAYER BIOACTIVE COMPOSITE FOR STRUCTURAL APPLICATIONS

D.C. Clupper and J.J. Mecholsky, Jr.
Department of Materials Science and Engineering
University of Florida
Gainesville, FL 32611

ABSTRACT

A tape-cast, multilayer, laminate, bioactive composite has been developed to act as a model system for structurally tough, biologically compatible materials. A Bioglass®-metal composition has been developed which has strength and toughness values that can be tailored to structural applications in the body such as bone and other hard tissue replacement. Strengths over 100 MPa and work of fracture values one order of magnitude greater than that of monolithic Bioglass® have been achieved.

INTRODUCTION

There are a limited number of biomaterials which are able to form strong bonds with bone and soft tissue. Materials with such qualities are useful in orthopaedic and dental surgery, as adequate implant-tissue fixation is necessary for long-term implant success. Bioglass® has been shown to bond chemically to both bone and soft tissue. Specifically, the 45S5 composition (45% SiO_2, 24.5% Na_2O, 24.5% CaO and 6% P_2O_5) bonds to bone more rapidly than any other known bioactive material; e.g. apatite-wollastonite glass-ceramic and hydroxyaptite.[1]

Upon exposure of Bioglass® to blood plasma or a salt solution approximating the same, Na^+ ions leach from the glass surface and are replaced by water molecules. A hydrated silica network forms and acts as a site for calcium phosphate nucleation.[2,3] This layer becomes more crystalline with time, forming hydroxyapatite [$Ca_{10}(PO_4)_6(OH)_2$], the mineral phase of bone.

Bioglass® is used clinically in dental surgery to fill the gap created by tooth extraction. Packing the void with particulate Bioglass® helps ensure that the jawbone structure is maintained.[1] Bioglass® is also used in middle ear surgery to restore the ossicular chain. However, extended clinical use has been suppressed because of low strength and toughness. The present research was designed to produce a bioactive composite with higher strength and toughness values than monolithic Bioglass® (approximately 70 MPa and 0.7 MPa $m^{1/2}$, respectively). Such a composite could potentially be used clinically in load bearing areas; e.g. in long bones or vertebrae.

Chen and Mecholsky[5] used tape-casting methods to produce alumina-metal laminates which possessed higher strength and toughness values than the monolithic alumina. Similar methods were used in this research to improve the strength and toughness of Bioglass®. Tape-casting offers the potential for producing complex shapes which would be required for clinical use.

MATERIALS AND METHODS

Bioglass® 45S5 particulate (>120 microns) was obtained from USBiomaterials, Alachua, FL. The powder was further decreased by milling in ethanol to a size range of 1-30 microns. Tape-casting slurries were then formulated by mixing an organic binder, plasticizer, solvent and milled Bioglass® particulate for 12 hrs. Slurries were cast 3 ft/min using a doctor blade height of 200 microns. The dried tape thickness ranged from 70-110 microns. A variety of metals were also tape-cast. Bioglass® tapes were then punched from the dried tapes and laminated at 150°C and 20 MPa pressure to produce green bodies approximately 2.5 mm thick. Metallic tapes were also laminated with the Bioglass® tapes to produce green composites. Organics were then removed below 600°C prior to sintering. Tape-cast Bioglass® (no metallic lamination) was sintered in air at 1000°C.

Ductile metallic layers of copper or stainless steel 316L were incorporated into the Bioglass® structure by hot pressing under vacuum between 700°-1250°C and pressures up to 50 MPa. Samples were then cut to approximately 3 mm x 4 mm x 25 mm using an diamond saw and polished to 1 micron for flexural testing.

Bioactivity was tested in vitro by suspending sintered Bioglass® disks (10 mm dia x 2 mm) in a simulated body fluid (Table I) for times ranging from 2 hours to 8 weeks. Following immersion, the samples were dried with acetone and analyzed using FTIR spectroscopy (Nicolet 20SXB). Formation and growth of the surface hydroxyapatite was then easily monitored.

Table I. Ionic Concentrations (mM) of Human Blood Plasma and SBF

	Na^+	K^+	Mg^{2+}	Ca^{2+}	Cl^-	HCO^{3-}	HPO_4^{2-}	SO_4^{2-}
SBF	142	5	1.5	2.5	147.8	4.2	1	0.5
Plasma	142	5	1.5	2.5	103	27	1	0.5

Hardness was measured using a Vicker's diamond indent produced by a Micromet®3 microhardness tester (Buehler, Lake Bluff, IL). Samples were loaded to 100 g and held for 30 seconds. Elastic modulus was determined using ultrasonic methods.

Bend samples were then tested in four point flexure (inner span 6.67 mm and outer span 19.9 mm) using an Instron 1125 at a loading rate of 0.5 mm/min. The strength of tape-cast sintered Bioglass® (no metallic lamination) was determined using samples 1.5 x 2.0 x 23 mm.

RESULTS

Several properties of the tape-cast sintered Bioglass® (no metallic lamination) were measured. The standard deviation (S.D.) and number (n) of samples tested are given following each property value. The average strength of tape-cast sintered Bioglass® was 114.1 MPa (S.D. 23.0 MPa, n=18). For comparison, the flexural strength of cortical bone ranges from 100-200 MPa. The average hardness (Vicker's) was 5.73 GPa (S.D. 0.36 GPa, n=10). The elastic modulus of tape-cast sintered Bioglass® was 83 GPa (S.D. 5, n=3). This is higher than the elastic modulus of cortical bone (10-20 GPa).

The load-deflection diagrams for two Bioglass® laminates are shown in Figures 1a and 1b. In each figure, the cross-hatched area represents the area which would result if the tape-cast sintered Bioglass® was not reinforced with metallic layers; i.e., it would exhibit brittle behavior. Both Figures 1a and 1b clearly show that with incorporation of a ductile metal layer, the work of fracture (as measured by the area under the load deflection diagram) increased significantly compared to tape-cast sintered Bioglass® alone.

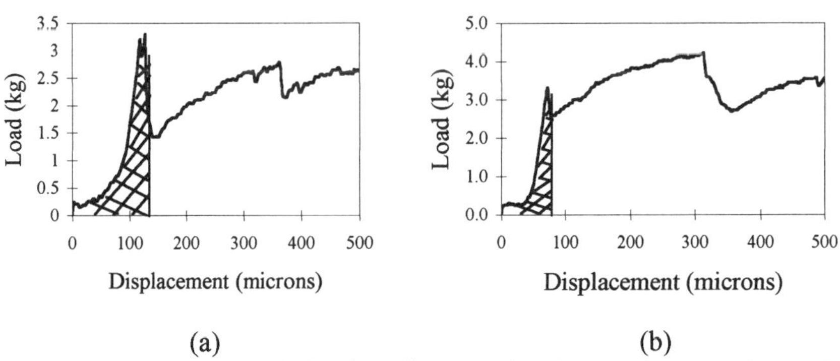

Figure 1. Laminate load-deflection diagram showing an increase in work of fracture as compared to Bioglass® alone.

In Figure 1b the peak load was also increased by incorporation of the metallic layer; i.e. the effective strength of the laminate was higher than that of tape-cast sintered Bioglass® alone. A laminate sample fractured in flexure is shown in Figure 2.

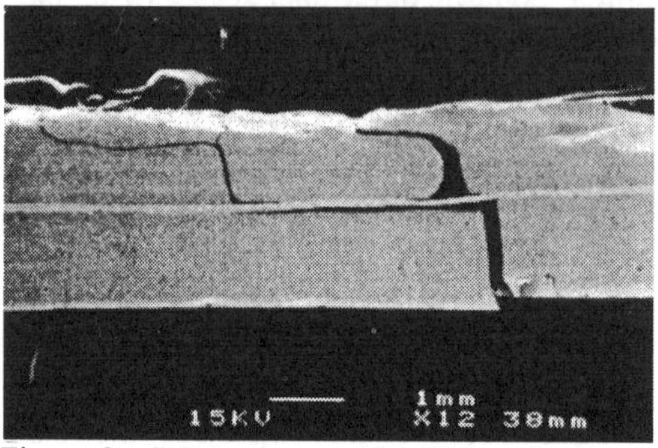

Figure 2. SEM micrograph of bioactive laminate composite fractured in flexure. The tensile side was located at the bottom of the micrograph. The outer layers (tape-cast sintered Bioglass®) are bioactive, whereas the middle metallic layer provides ductility.

The FTIR spectra of a tape-cast sintered Bioglass® sample is shown in Figures 3 and 4. The initial spectra of tape-cast sintered Bioglass® prior to reaction in SBF is shown in Figure 3. A small amount of calcium phosphate phase is present in initially (peaks near 575 and 610 cm^{-1}).

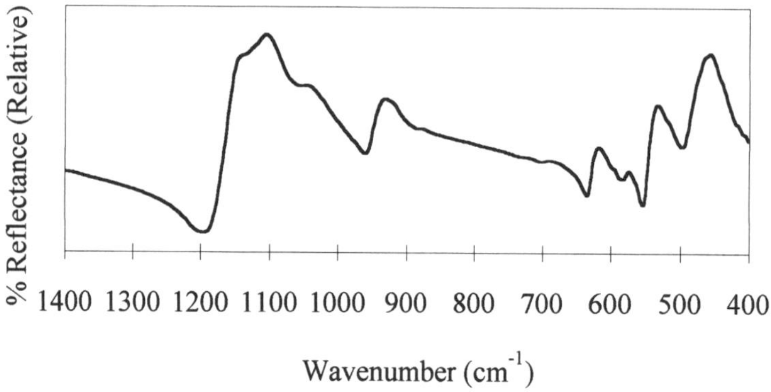

Figure 3. FTIR spectra of tape-cast sintered Bioglass® prior to immersion in SBF.

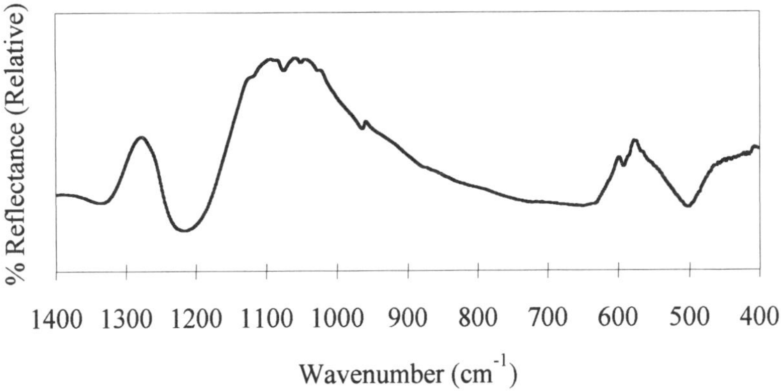

Figure 4. FTIR spectra of tape-cast sintered Bioglass® following SBF immersion.

The silica layer is represented by the larger peaks near 450 and 530 cm^{-1} in Figure 3. As the hydroxyapatite layer grows in thickness (with time) the relative intensity of the hydroxyapatite peaks (575 and 600 cm^{-1}) to the silica peak (450 cm^{-1}) increases.

CONCLUSIONS

Tape-cast sintered Bioglass® retains in vitro bioactivity as evident by the formation of crystalline hydroxyapatite following immersion in a simulated body fluid. The strength of tape-cast sintered Bioglass® (non-laminated) is within the flexural strength range of cortical bone. The results of this work indicate that the work of fracture of sintered tape-cast Bioglass® 45S5 can be increased by incorporation of one or more ductile metal layers.

REFERENCES

[1]W. Cao and L.L. Hench, "Bioactive Materials," *Ceram. Intl.*, **22** [6] 493-507 (1996).

[2]L.L. Hench, Ö.H. Anderson, and G.P. LaTorre, "The Kinetics of Bioactive Ceramics Part III: Surface Reactions for Bioactive Glasses compared with an Inactive Glass," Bioceramics, Vol. 4 (Ed. W. Bonfield, G.W. Hastings, and K.E. Tanner) Proceedings of the 4th International Symposium on Ceramics in Medicine, London, UK, September 1991.

[3]M.M. Pereira, A.E. Clark, and L.L. Hench, "Effect of Texture on the Rate of Hydroxyapatite Formation on Gel-Silica Surface," *J. Am. Ceram. Soc.*, **78** [9] 2463-2468 (1995).

[4]C.A. Shapoff, D.C. Alexander, and A.E. Clark, "Clinical Use of a Bioactive Glass Particulate in the Treatment of Human Osseous Defects," *Compendium*, **18** [4] 352-363 (1997).

[5]Z. Chen and J.J. Mecholsky, Jr., "Control of Strength and Toughness of Ceramic/Metal Laminates Using Interfacial Design," *J. Mater. Res.*, **8** 2362-2269 (1993).

ACKNOWLEDGMENTS

This research was funded in part by USBiomaterials, Alachua, FL. The authors wish to thank D.C. Greenspan, G.P. LaTorre, and J. Zhong for useful discussions concerning this work.

KEYWORD AND AUTHOR INDEX